W9-CSP-224

Mechanism-Based Enzyme Inactivation: Chemistry and Enzymology

Volume I

Author

Richard B. Silverman

Professor of Chemistry and of
Biochemistry, Molecular Biology, and Cell Biology
Northwestern University
Evanston, Illinois

CRC Press, Inc.
Boca Raton, Florida

02957498

CHEMISTRY

Library of Congress Cataloging-in-Publication Data

Silverman, Richard B.
  Mechanism-based enzyme inactivation.

  Includes bibliographies and index.
  1. Enzyme inhibitors. I. Title. [DNLM: 1. Enzyme
Inhibitors. QU 143 S587m]
QP601.5.S55 1988   591'.1925   87-13825

ISBN 0-8493-4543-X (v. 1)
ISBN 0-8493-4544-8 (v. 2)

Direct all inquiries to CRC Press, Inc., 2000 Corporate Blvd., N.W., Boca Raton, Florida, 33431.

© 1988 by CRC Press, Inc.

International Standard Book Number 0-8493-4543-X (v. 1)
International Standard Book Number 0-8493-4544-8 (v.2)

Library of Congress Card Number 87-13825
Printed in the United States

PREFACE

This book is intended to be used as a reference text by those in the field interested in applying mechanism-based inactivation approaches to studies of a particular enzyme or by those with an interest in this exciting area of enzyme inactivation in order to gain an overview of the field and to learn the fundamentals of the method. The text also can be used in a graduate-level course concerned with mechanism-based enzyme inactivation or as one of several texts in an advanced undergraduate or graduate course on enzyme inhibition in general. Since it is assumed in this text that the reader is familiar with enzyme mechanisms, it may be helpful to precede a course that utilizes this text by a course on enzyme mechanisms. At least, the students should be familiar with writing organic chemical mechanisms.

It may be useful to the reader to know how the information in this text was collected. It was observed that a computer search was not very useful, since most of the papers in the field do not have the words "mechanism-based," "enzyme-activated," or "suicide" in their titles or abstracts (and I did not have the inclination to peruse every paper that mentioned "enzyme inhibition or inactivation"). The approach taken was a more direct one, namely, that letters were sent requesting reprints (or publication lists) from 146 different principal authors of papers published in the field. Of the 146 requests, all but six were answered (I have restrained myself from revealing the names of those six individuals). These references were used as primary sources from which other relevant references were extracted. This process continued, i.e., obtaining references from within the references, until all possibilities were exhausted. I am indebted to the 140 individuals who responded to my request; much time in collecting references was saved by their assistance. When these sources were expended, the references in available reviews were checked to be certain that no obvious omissions were made. The literature through the first half of 1987 is covered here (the work cited from mid-1986 and references therein were added to the galley proofs). These endeavors should have been sufficient to uncover most of the relevant references. If I missed your contribution to this field, please accept my apology. Rather than sending me a nasty note regarding my oversight, kindly send me copies of your relevant publications; maybe, some day an addendum will be published.

Richard B. Silverman
Evanston, IL
August, 1986

THE AUTHOR

Richard B. Silverman, Ph.D., is Professor of Chemistry and Professor of Biochemistry, Molecular Biology, and Cell Biology at Northwestern University.

Dr. Silverman received his B.S. degree in chemistry at the Pennsylvania State University in 1968, then spent two years in the U.S. Army as a medicinal chemist at the Walter Reed Army Institute of Research in Washington, D.C. He obtained his M.A. degree in 1972 and Ph.D. degree in 1974 in chemistry from Harvard University. Prior to joining the faculty of Northwestern University in 1976, Dr. Silverman was a National Institutes of Health post-doctoral research fellow in the Graduate Department of Biochemistry at Brandeis University.

Dr. Silverman's current research interests are the design and synthesis of mechanism-based enzyme inactivators and their mechanisms of enzyme inactivation, the elucidation of enzyme mechanisms, and the design of potential new pharmaceutical agents.

ACKNOWLEDGMENTS

I am grateful to Dr. Stephen J. Hoffman and especially to Dr. Mark A. Levy for their critical comments on the first chapter of the original draft (and for not discouraging the writing of the remainder of the book), to Carol Lewis for typing the manuscript, to Karen Heneghan and Dianne Deplewski for drawing most of the schemes and figures, and to the National Institutes of Health (grants GM 32634, HL 27108, NS 15703, and GM 35844) for financial support of my research during the writing of this work.

To loving Barbara for her devotion, her understanding, and her radiance, to Matt and Marggie for their love, their wit and their exuberance, and to Philly for his giant smiles, his hugs, and his belly laughs.

MECHANISM-BASED ENZYME INACTIVATION: CHEMISTRY AND ENZYMOLOGY

Volume I

Introduction
Protonation and Deprotonation Reactions
Phosphorylation Reactions
Addition Reactions
Acylation Reactions
Elimination Reactions

Volume II

Isomerization Reactions
Decarboxylation Reactions
Oxidation Reactions
Oxygenation Reactions
Polymerization Reactions

TABLE OF CONTENTS

Chapter 5
Acylation Reactions 95

ABBREVIATIONS AND SHORTHAND NOTATIONS
USED IN THE BOOK

Because of the myriad of abbreviations commonly used, particularly in synthetic organic chemistry, this alphabetized compilation of abbreviations is included for easy referral while reading the book.

ACC	1-aminocyclopropane-1-carboxylate
Ad	adenine
Ado	adenosine
B:	active site base
BHT	butylated hydroxytoluene (2,6,-di-*tert*-butyl-4-methylphenol)
Bn	benzyl
Boc	*tert*-butoxycarbonyl
Bz	benzoyl
Cbz	carbobenzoxyl
CH$_2$H$_4$folate	5,10-methylenetetrahydrofolate
m-CPBA	*meta*-chloroperoxybenzoic acid
Cy	cyclohexyl
Dabco	1,4-diazabicyclo[2.2.2]octane
DBN	1,5-diazabicyclo[4.3.0]non-5-ene
DBU	1,8-diazabicyclo[5.4.0]undec-7-ene
DCC	dicyclohexylcarbodiimide
DDEP	3,5-dicarbethoxy-2,6-dimethyl-4-ethyl-1,4-dihydropyridine
DEAD	diethyl azodicarboxylate
Dibal	diisobutyl aluminum hydride
DMAP	4-dimethylaminopyridine
DMSO	dimethyl sulfoxide
DON	6-diazo-5-oxo-*L*-norleucine
DONV	5-diazo-4-oxo-*L*-norvaline
dR	deoxyribosyl
dRP	deoxyribose phosphate
DTNB	5,5'-dithiobis(2-nitrobenzoate)
DTT	dithiothreitol
EDTA	ethylenediaminetetraacetic acid
EPR	electron paramagnetic resonance
ESR	electron spin resonance
FdUMP	5-fluoro-2'-deoxyuridine monophosphate
Fl	oxidized flavin
FlH· or Fl$^{\div}$	flavin semiquinone
FlH$^-$ or FlH$_2$	reduced flavin
GABA	γ-aminobutyric acid
GTP	guanosine triphosphate
HPLC	high performance liquid chromatography
Im	imidazole
IR	infrared
LDA	lithium diisopropylamide
MAO	monoamine oxidase
MPDP$^+$	1-methyl-4-phenyl-2,3-dihydropyridinium ion
MPP$^+$	1-methyl-4-phenylpyridinium ion

MPTP	1-methyl-4-phenyl-1,2,3,6-tetrahydropyridine
Ms	mesyl (methanesulfonyl)
NAC	*N*-acetylcysteamine
NAD$^+$	nicotinamide adenine dinucleotide
NADH	reduce form of NAD$^+$
NaDodSO$_4$	sodium dodecyl sulfate
NBS	*N*-bromosuccinimide
NMR	nuclear magnetic resonance

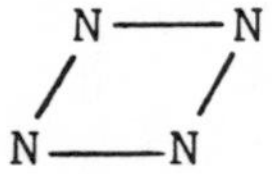

N———N
/   /
/   /
N———N
        porphyrin ring system (usually protoporphyrin IX)

Np	naphthyl
NXS	*N*-halosuccinimide
ORD	optical rotatory dispersion
PAGE	polyacrylamide gel electrophoresis
Pant	pantetheine
PhthN	phthalimido
P$_i$	phosphate ion
PLP	pyridoxal 5'-phosphate
PMP	pyridoxamine 5'-phosphate
PPA	polyphosphoric acid
PQQ	pyrroloquinoline quinone
Pyr	the substituted pyridine nucleus of PLP or PMP
RBL	rat basophilic leukemia
TFA	trifluoroacetic acid
THF	tetrahydrofuran
THP	tetrahydropyranyl
TMEDA	N,N,N',N'-tetramethylethylenediamine
TMS	trimethylsilyl
TPCK	L-1-chloro-3-tosylamido-4-phenyl-2-butanone
TPP	thiamin pyrophosphate
Ts	tosyl (*p*-toluenesulfonyl)
X:	active site nucleophile

Chapter 1

INTRODUCTION

I. THE BIRTH OF MECHANISM-BASED ENZYME INACTIVATION

The concept of mechanism-based enzyme inactivation has been known for many more years than the term has been defined. As early as the 1940s, it was known that L-serine was a time-dependent inactivator of threonine dehydratase.[1] In the 1950s, 6-diazo-5-oxo-L-nor-leucine (DON) was isolated from an unidentified *Streptomyces*[2] and shown to be an inactivator of various glutamine-dependent enzymes.[3] A variety of hydrazines[4] and cyclopropylamines[5] were tested in the 1950s and 1960s as inactivators of mitochondrial monoamine oxidase for future use in the treatment of depression.[6] It was not until 1970 that the pioneering paper by Konrad Bloch and co-workers[7] at Harvard appeared which described the mechanism for inactivation of β-hydroxydecanoylthioester dehydrase by 3-decynoylthioesters. This work resulted in the blossoming of a new type of enzyme inactivation about which this book is concerned. A mechanism-based enzyme inactivator is a relatively unreactive compound, having a structural similarity to the substrate or product for a particular enzyme that, via its normal catalytic mechanism of action, converts the inactivator molecule into a species which, without prior release from the active site, binds most often covalently, to that enzyme.

II. NOMENCLATURE

A. Is "Mechanism-Based Enzyme Inactivation" the Best Terminology?

While reading the references for this book, it became quite apparent that there was a major nomenclature problem in the field. As far as I can tell, the term "mechanism-based inactivator" was first used in print by Rando.[8] However, this is not the only terminology utilized. No less than 20 different names have been employed in reference to this type of enzyme inhibition. For the trivia buffs I have enumerated below the other terms that I have encountered. The person initially responsible for the anthropomorphic references to the enzyme committing "suicide" is Massey.[9] "Suicide substrate"[10] and the term "suicide enzyme inactivator"[11] were first used by Abeles. Another early nomenclature suggestion by Rando[12] for this type of inactivator is "k_{cat} inhibitor." A term introduced by the Merrell Dow Pharmaceuticals research group is "enzyme-activated irreversible inhibitor."[13] These are the most commonly used terms; alternative names published for either the process of inactivation or the inactivator include the following: Trojan Horse inactivator,[14] enzyme-induced inactivation,[15] enzyme-activated substrate inhibition,[16] latent irreversible enzyme inhibition,[17] dynamic affinity labeling,[18] suicide-inducing substrate,[19] autocatalytic inactivation,[20] self-catalyzed inactivation,[21] trap substrates,[22] inhibitory substrates,[23] enzymatically transformed irreversible inhibition,[24] substrate-induced irreversible inhibition,[25,26] catalytic irreversible inhibition,[26a] proinhibitors,[27] and thanatophore.[28] The term "suicide substrate" in a French paper[29] was translated in its English summary as "booby trap substrate." Enclosed with the letter I sent to 146 principal investigators in this field requesting their relevant reprints (see the Preface) was a nomenclature poll. Each was asked to select his or her preference for terminology from among the following choices (the number in parentheses is the number of those polled who selected a particular term): mechanism-based inactivation (44), enzyme-activated irreversible inhibition (43), suicide inactivation (9), k_{cat} inhibition (2), and other (10). Those who chose the "other" category suggested the following alternative names: latent inactivators (Hanzlik), trap inhibitors (Vilkas), enzyme-generated inactivators (Kalman, Covey, Orlowski), enzyme-activated recurrent inhibitor or inactivating substrate (de Groot, Noll), enzyme-activated self inactivation (Levy), and booby trap substrates (Belleau, Alworth). Although my slight preference is reflected in the title of this book, I must concur with the respondants of the poll regarding the two favorite names. However, I do not believe that either of these terms is fully accurate. As pointed out by many of those polled, a knowledge of the mechanism of enzyme action also is necessary in the design of transition state analogues, so "mechanism-based inactivators" need not only refer to the type of inactivator to which this book refers. Some "mechanism-based inactivators" are not **activated** by the enzyme prior to irreversible enzyme inhibition; but, instead, are converted into stable complexes; consequently, "enzyme-activated irreversible inhibition" also is not a completely accurate term. I,[30] and others,[14,31,31a] have objected to the terminology that comprises the word "suicide." The main objections are the use of an anthropomorphic term for a chemical phenomenon and the incorrect usage, on semantic grounds, of the word suicide. Suicide is the **voluntary** and **intentional** taking of one's own life. A target enzyme, for the case in point, is being deceived into transforming a compound that appears to be only a slightly modified substrate into an altered species which then causes enzyme inactivation. The more accurate anthropomorphic analogy is homocide and, maybe, "enzymocide substrate" is the best descriptor. Several of those polled who did not select the terminology "suicide inactivation," as well as those who did, felt that, although that term was not accurate, it was a catchy name that had been in use for a long time and "everyone" knew what it meant. Unfortunately, "everyone" only includes those who currently are aware of this type of inactivation; but this may be only a small percentage of those who, in the future, will become aware, and, in my opinion, is the reason not to perpetuate its usage. The "catchy name" argument often was suggested as useful to the biologists and pharmacologists who were not interested in the chemical and enzymological bases for the terminology, but rather, found "suicide substrate" an easy term to use. In the response to my poll, Klinman (Berkeley) relayed an anecdote of a postdoc in her lab who was asked at a job interview if suicide inhibitors were used to prevent people from jumping off of the Golden Gate Bridge.

By choosing the title for the book that I did, I joined the ranks of those who perpetuate a less-than-optimal term for this type of inactivation (the term "inactivation" refers to an **irreversible** inhibition process). The most accurate names, in my opinion, are Trojan Horse inactivation (if you need a "catchy" name), enzyme-induced inactivation, enzymatically transformed irreversible inhibition, and latent inactivation. All of the names starting with "enzyme" (e.g., enzyme-activated, enzyme-induced, etc.) could be ambiguous, however, since they do not make it clear that the enzyme that is doing the activation (or induction) is the enzyme that is being inactivated. Another class of important enzyme inactivators, for example, are metabolically activated inhibitors,[32] in which one enzyme transforms the inactivator to an activated form that, then, reacts with a different protein. These compounds are not classified as mechanism-based inactivators.

B. Rate Constant and Dissociation Constant Terminology

An appropriate name for mechanism-based inactivation is not the only source of confusion in the field regarding nomenclature. The name "k_{cat} inhibitor" has not received support because k_{cat} is not really the rate constant to which this type of enzyme inactivation refers. When an enzyme catalyzes a reaction, there is, generally, a rapid equilibrium binding between enzyme and substrate, followed by a rate-determining catalytic step, defined by the rate constant, k_{cat}, which is the step in which product is formed (Scheme 1).

$$E + S \underset{k_{-1}}{\overset{k_1}{\rightleftharpoons}} E \cdot S \underset{k_{-2}}{\overset{k_2}{\rightleftharpoons}} E \cdot P \underset{k_{-3}}{\overset{k_3}{\rightleftharpoons}} E + P$$

Scheme 1.

In the above equation, if the rate-determining step is k_2, then k_2 is k_{cat}. When an enzyme catalyzes the conversion of a mechanism-based inactivator to its reactive form, there are two processes that then can follow. The altered form can be released as a product or it can react with the enzyme (Scheme 2). In this case, k_2 is similar to k_{cat} in that it is the rate constant for conversion of the inactivator to its activated form (I′), which often is the rate determining step; I′ can either be released as a product (k_3) or react with the enzyme (k_4). In order to distinguish the rate constant for conversion of substrate to product in Scheme 1 (i.e., k_{cat}) from that for transformation of the mechanism-based inactivator into its activated form (Scheme 2), k_{inact} appears to be an appropriate and commonly used nomenclature. The rate constant, k_{inact}, however, is not simply k_2, but rather is a complex mixture of k_2, k_3, and k_4. The derivation and kinetic implications of k_{inact} are described in Section VI.A.

$$E + I \underset{k_{-1}}{\overset{k_1}{\rightleftharpoons}} E \cdot I \overset{k_2}{\longrightarrow} E \cdot I' \overset{k_4}{\longrightarrow} E\text{--}I''$$
$$k_3 \downarrow$$
$$E + P$$

Scheme 2.

Analogous to the separate identities of k_{cat} and k_{inact}, a corresponding differentiation must be made among the constants that define the dissociation of the enzyme·substrate complex (i.e., Michaelis-Menton constant, K_m), the dissociation of the enzyme·reversible inhibitor complex (the K_i), and a term that usually is thought to be the dissociation constant for the initial reversible enzyme·mechanism-based inactivator complex. Most investigators in this field use the term K_i to refer to the latter constant as well. However, since these last two constants are different (see Section VI.B. for a detailed discussion), the nomenclature should be different to avoid confusion. The symbol K_I has been employed by some for the mechanism-based inactivator dissociation constant, and this terminology is supported here.

C. Partition Ratio

An important concept related to mechanism-based inactivation is the partition ratio. This is a term introduced by Walsh[33,34] to describe the ratio of the number of latent inactivator molecules converted and released as product relative to each turnover leading to enzyme inactivation. According to Scheme 2, this would be the ratio k_3/k_4. The partition ratio, therefore, is a measure of the efficiency of the inactivator and will depend upon the reactivity of the activated intermediate, its rate of diffusion from the active site, and the proximity of an appropriate nucleophile, radical, or electrophile on the enzyme for covalent bond formation. The most efficient mechanism-based inactivators are those having partition ratios of 0 (every turnover produces inactivated enzyme).

III. TYPES OF ENZYME INACTIVATORS

All of the inactivators described in this section are "active-site directed", i.e., they bind at the active site and render the enzyme effectively inactive.

A. Transition-State Analogues and Slow, Tight-Binding Inhibitors

These first two categories are included as enzyme inactivators even though they usually involve noncovalent interactions with enzymes, because in some cases the binding is so tight as to be functionally covalent. Detailed discussions of transition-state inhibition can be found in the works of Lienhard,[35] Wolfenden,[36] and Lindquist.[37] A transition-state analogue is a stable compound whose structure resembles the conformation of a substrate at a postulated transition state of the reaction that the target enzyme catalyzes. It also has been generalized to include compounds whose structures are similar to known or postulated fleeting intermediates along the pathway of enzyme catalysis. Sometimes these inhibitors are composites of intermediates or transition-state structures for enzymes utilizing more than one substrate; in these cases they are referred to as multisubstrate analogues. The basis for the exceedingly tight binding of transition-state analogues to the target enzyme derives from the early hypotheses of Pauling[38] and Ogston[39] who suggested that an enzyme achieves its great rate enhancements by changing its conformation in such a way so that the strongest interactions occur between the substrate and enzyme active site at the transition state of the reaction. Therefore, if a compound is designed to have a structure similar to that of the substrate at a transition state of the reaction, it will bind much more tightly than will the substrate in the ground state. In a sense, then, these compounds also could be termed "mechanism-based enzyme inactivators" because the mechanism of the enzyme reaction must be understood so that a structure can be designed to mimic the conformation of the substrate at a transition state.

Slow, tight-binding inhibitors are compounds that bind slowly (i. e., produce time-dependent inhibition) and either noncovalently[40,40a] or covalently,[40b,40c] and are released very slowly because of the exceedingly tight interaction. It is believed that a conformational change in the enzyme is responsible for both the slowness and tightness of the binding; a change in the protonation state of the enzyme may be involved.[40] Other explanations include displacement of a water molecule[40a] and reversible, covalent bond formation.[40b,40c] Both these inhibitors and transition-state analogues bind so tightly that stoichiometric titrations with the enzyme are possible.

B. Affinity Labeling Agents

This class of enzyme inactivators is comprised exclusively of covalent inactivators. They are compounds that contain a reactive functional group, e.g., an α-haloketone or isocyanate group, and react with active site nucleophiles, generally via S_N2 alkylation or acylation mechanisms. Often more than one nucleophile and/or protein undergo reactions, but when working with a homogeneous enzyme, this is an effective inactivation approach, and specificity can be achieved. Increased specificity can be attained by attaching the reactive functional group to a compound whose structure resembles that of a substrate for the target enzyme. In these cases, under certain reaction conditions, it is possible to limit the reaction to a 1:1 stoichiometry of inactivator to enzyme, and therefore, these modified affinity-labeling agents can be used with partially purified or crude enzyme mixtures. A more detailed description of these inactivators has been presented.[41] A variation on affinity labeling in which enzyme modification occurs only in the presence of a substrate, which converts the enzyme into an active form for inactivation, has been termed syncatalytic enzyme modification.[41a]

C. Mechanism-Based Enzyme Inactivators

The essential, and unique, feature of a mechanism-based enzyme inactivator, which was defined earlier, is that the target enzyme must chemically transform the molecule into the actual inactivating species and that inactivation must occur by this species prior to its release from the active site. This is not to say that the activated species cannot be released from the enzyme; only that the molecule ultimately producing inactivation has to be one that was not earlier released and then returned to cause inactivation. Release of the activated species from the enzyme to produce a stable product only lowers the efficiency of the inactivator, i.e., raises its partition ratio. It should be noted that there are a few inactivators that are transformed by the target enzyme via mechanisms unrelated to the normal catalytic mechanisms. These inactivators are not, strictly speaking, mechanism-based enzyme inactivators. Nonetheless, they are included here for completeness.

D. Metabolically Activated and Multienzyme-Activated Inactivators

In the case described above, where the activated species is released from the enzyme, this species could return to the active site of the same enzyme or, when a mixture of enzymes is present, could lead to inactivation of a different enzyme. When one enzyme activates a compound, then this reactive species is released and proceeds to inactivate another enzyme, it is a metabolically activated inactivator.[32]

More recently, a variation of mechanism-based inactivation has been described in which one compound is metabolically converted by one or more enzymes into the mechanism-based inactivator for a different enzyme. These compounds, e.g., (*E*)-β-(fluoromethylene)-*m*-tyrosine,[42] were termed dual enzyme-activated inhibitors, because two enzymes were involved in the mechanism-based inactivation (in the above case, aromatic L-amino acid decarboxylase catalyzed the conversion of this compound to a mechanism-based inactivator of monoamine oxidase). Earlier examples of this sort of in vivo activation are α-fluoromethylputrescine[43] and 5-hexyne-1,4-diamine,[44] both of which are converted by mono-amine oxidase and aldehyde dehydrogenase to the corresponding γ-aminobutyric acid an-alogues which are mechanism-based inactivators of γ-aminobutyric acid aminotransferase. Since three enzymes are involved in this inactivation process, and other examples, such as hypoglycin[45] and 5-fluorouracil,[46] are known where at least three enzymes are involved, the name for this approach could be generalized to "multienzyme-activated inactivators". In analogy to "prodrugs" in medicinal chemistry,[47] these compounds are "promechanism-based inactivators," compounds that require metabolic conversion into the actual mechanism-based inactivator. These inactivators are discussed in the sections corresponding to the actual enzyme inactivation rather than in sections appropriate for the metabolic activation of the multienzyme-activated inactivator.

IV. USES AND ADVANTAGES OF THE MECHANISM-BASED ENZYME INACTIVATION APPROACH

Mechanism-based enzyme inactivators have been proven to be most useful in the study of enzyme mechanisms and in the design of highly specific, low-toxicity drugs.

A. In Enzyme Mechanism Studies

The value of these inactivators in the study of enzyme mechanisms is derived from the fact that they are substrates for the target enzyme. As substrates, they are converted by the catalytic mechanism of the enzyme into products. However, the products of these enzyme-catalyzed reactions are generally very reactive species that then become attached to the enzyme. The important feature, though, is that the conversion to the activated form is, at least, initiated by the same catalytic steps involved in the reaction with normal substrates; the products just happen to be more reactive than those generated from normal substrates. Therefore, whatever mechanistic information that can be gleaned from inactivation studies is generally directly related to the catalytic mechanism of the enzyme. It should be kept in mind, though, that a few inactivators proceed via mechanisms different than the normal catalytic mechanism. Studies with mechanism-based inactivators can be a two-way street. Either the inactivators can be designed on the basis of a known mechanism of action of the target enzyme, or the mechanism of action of the target enzyme can be elucidated by an understanding of the inactivation mechanism. Often, a combination of the two approaches is taken: a hypothesis regarding the enzyme mechanism is formulated in the design of the inactivator, and, then, the results of the inactivation studies can be used to support or modify the hypothesis. Many of the compounds described in this book were designed to test specific mechanistic postulates and this will be discussed where appropriate.

B. In Drug Design

Many enzyme-inhibitor drugs are specific reversible competitive inhibitors, where inhibition arises from the formation of an enzyme-inhibitor complex in preference to an enzyme-substrate complex. According to the Le Chatelier Principle, these equilibria depend upon the concentrations of enzyme, substrates, and inhibitor. Since the enzyme concentration is usually low and fixed, the equilibrium constant and, therefore, the enzyme·inhibitor concentration, will depend upon the inhibitor concentration. When the inhibitor concentration diminishes, the enzyme·inhibitor concentration diminishes, and the effect of the inhibitor can be overcome by substrate. If the inhibitor is a drug, the maximal pharmacological effect will occur when the drug concentration is maintained at a saturating level at the site of the enzyme. This may require administration of the drug repeatedly for an extended period of time, which may lead to patient noncompliance. In theory, an improved approach would be to use a specific irreversible enzyme-inactivator drug, i.e., one that forms a covalent bond to the target enzyme. In this case, once the enzyme reacts with the inactivator, a process that ideally could require only one inactivator molecule per enzyme active site, it would not be necessary to maintain the high drug concentration in the body. Of course, the gene encoding the inactivated enzyme will produce new enzyme, so additional drug will be needed, but this process could take several days, or, at least, several hours. As mentioned above, the two principal classes of irreversible inactivators are affinity-labeling agents and mechanism-based inactivators. Many cancer chemotherapeutic agents are affinity-labeling agents. The major disadvantage of an affinity-labeling agent is that not only can it react with the target enzyme in vivo, but because of its reactivity, it also can react with other enzymes and biomolecules, resulting in toxicity and side effects. The numerous side effects of cancer chemotherapeutic agents are well known. However, mechanism-based inactivators are unreactive compounds, and this is the key feature that makes them so amenable to drug design. Because of their unreactivity, nonspecific alkylations of other proteins are generally not a problem. Ideally, only the one target enzyme will be capable of catalyzing the appropriate reaction that converts the mechanism-based inactivator into its reactive form and also will have an appropriately juxtaposed nucleophile that can react with the incipient electrophilic center. If the inactivator has a partition ratio near zero, in which case the production of undesirable metabolites that may lead to toxic side effects is minimized, this compound will have the desired drug properties of high specificity and low toxicity. α-Difluoromethylornithine, a highly specific mechanism-based inactivator of ornithine decarboxylase, has been administered in clinical trials for the treatment of various protozoal infections in amounts of 30 g/day for several weeks with only minor side effects.[48]

The rational design of mechanism-based inactivators as drugs is a relatively recent approach in the pharmaceutical industry. Merrell Dow Pharmaceuticals (Strasbourg and Cincinnati) appears to have been a pioneer in this regard, having begun its program for the **design** of mechanism-based inactivators as drugs in 1973. The corresponding program at Merck & Co. began in 1970 with a patent on the antibacterial agent 3-fluoro-D-alanine.[49] However, although this compound was designed on the basis of the ''molecular physiology of bacteria'', it does not appear to have been designed as a mechanism-based inactivator of alanine racemase from which its antibacterial activity is derived. This compound was withdrawn from clinical development in 1984. (S)-α-Fluoromethylhistidine, on the other hand, is a mechanism-based inactivator of histidine decarboxylase that was designed by Merck chemists[50] in the mid-1970s as a potential antiulcer drug. It currently is in clinical trials. To date (mid 1987), no drugs, **specifically designed as mechanism-based enzyme inactivators** for target enzymes, are on the American drug market. Several drugs in current medical use, however, have been determined *ex post facto* to be mechanism-based inactivators of certain enzymes. These include the antidepressant agents, tranylcypromine and phenelzine, the antihypertensive agents, hydralazine and pargyline, and the antiparkinsonian drug, deprenyl (they all

Table 1
ENZYMES ALREADY TARGETED FOR MECHANISM-BASED INACTIVATION

Enzyme	Therapeutic goal
Monoamine oxidase	Antidepressant agent/antihypertensive agent/antiparkinsonian agent
γ-Aminobutyric acid aminotransferase	Anticonvulsant agent
Thymidylate synthetase	Anticancer agent
Ornithine decarboxylase	Anticancer-antiprotozoal agent
Xanthine oxidase	Uricosuric agent
Aromatic amino acid decarboxylase	Synergistic with antiparkinson drug
β-Lactamase	Synergistic with antibiotics
Histidine decarboxylase	Antihistamine; anti-ulcer drug
Testosterone 5α-reductase	Anticancer agent
Serine Proteases	Anticoagulant agent, antiviral agent, and treatment of emphysema, inflammation, arthritis, adult respiratory distress syndrome, pancreatitis, certain degenerative skin disorders, and digestive disorders
Vitamin K epoxide reductase	Anticoagulant agent
Dopamine β-hydroxylase	Pheochromocytoma agent/antihypertensive agent
Aromatase	Anticancer agent
DNA Polymerase I	Antiviral agent
Thyroid peroxidase	Antithyroid agent
D-Amino acid aminotransferase	Antibacterial agent
Arginine decarboxylase	Antibacterial agent
S-Adenosylhomocysteine hydrolase	Antiviral agent
Dihydroorotate dehydrogenase	Antiparasitic agent/anticancer agent
Dihydrofolate reductase	Anticancer agent/antibacterial agent/antiprotozoal agent

inactivate monoamine oxidase); the compound, clavulanic acid, used to protect penicillins and cephalosporins against bacterial degradation (β-lactamases); the antitumor drug 5-fluoro-2′-deoxyuridylate (5-fluorouracil is a prodrug), and the antiviral agent, 5-trifluoromethyl-2′-deoxyuridylate (thymidylate synthetase); the uricosuric agent, allopurinol (xanthine oxidase); the antithyroid drugs, methimazole, methylthiouracil, and propylthiouracil (thyroid peroxidase); and the antibiotic, chloramphenicol, the antifertility drug, norethindrone, the anesthetics, halothane and fluroxene, the sedative, ethclorvynol, the diuretic and antihypertensive agent, spironolactone, the pituitary suppressant, danazol, the pigmentation agent, methoxsalen, and the hypnotic, novonal (they all inactivate cytochrome P-450). Those drugs that inactivate cytochrome P-450 do not derive their pharmaceutical effect by that inactivation. Mother Nature is responsible for the design of chloramphenicol and clavulanic acid. Two compounds from Merrell Dow, namely, γ-vinyl GABA (vigabatrin; inactivates GABA transaminase) and α-difluoromethylornithine (eflornithine; inactivates ornithine decarboxylase), are to date in clinical trials for the treatment of seizures and protozoal infections, respectively.

The rational selection of appropriate target enzymes for design of mechanism-based inactivator drugs depends upon the same criteria as those for any type of enzyme inhibitor. The goal is to deplete the organism of a specific product or to accumulate a substrate of the target enzyme. In Table 1 are summarized some examples of enzymes targeted for mechanism-based inactivation and the therapeutic goal.

The importance of mechanism-based inactivators in drug design is only just surfacing; once these initial efforts prove to be valuable additions to the pharmaceutical armamentarium, it should be a much greater impetus for drug companies to employ this approach in the design of drugs.

V. CRITERIA FOR MECHANISM-BASED ENZYME INACTIVATION

Since the elucidation of mechanism-based enzyme inactivation, a set of criteria has evolved for the evaluation of this type of inactivation. More often than not, only a few of these criteria are utilized when a potential mechanism-based inactivator is studied. However, in order to truly characterize an inactivator as falling into this class, most or all of the criteria should be satisfied. The criteria are as follows:

1.   A time-dependent loss of enzyme activity (usually, but not necessarily, pseudo first-order with respect to enzyme) is observed.
2.   The rate of inactivation is proportional to low concentrations of inhibitor, but independent at high concentrations (saturation kinetics).
3.   The rate of inactivation is slower in the presence of substrate than in its absence.
4.   Enzyme activity does not return upon dialysis or gel filtration.
5.   A 1:1 stoichiometry of radioactively labeled inactivator to active site usually results (see discussion in Section VII.E. concerning stoichiometries of less than 1:1) after inactivation followed by dialysis or gel filtration.
6.   A catalytic step for conversion of the inactivator to a reactive intermediate can be demonstrated.
7.   There is no lag time for inactivation, the presence of exogenous nucleophiles has no effect on the inactivation rate, and, following inactivation, a second equal addition of enzyme results in the same rate of inactivation as the first addition.

The first five criteria would define any covalent or tightly bound, slow-binding, noncovalent competitive inhibitor. Criteria 6 and 7 characterize the inactivator as being mechanism-based. Although in all but a relatively few cases, mechanism-based inactivation results in covalent attachment of the inactivator to the enzyme, it should be noted that this type of inactivation need not be covalent and need not be irreversible in nature. There are several examples of mechanism-based inactivators that follow all of the above kinetic criteria, yet the enzyme activity returns spontaneously or after further treatment of the inactivated enzyme. This return of enzyme activity could indicate the cleavage of the new bond of the modified enzyme or that the enzyme-inactivator complex was a tight, noncovalent one. The reversibility of certain mechanism-based inactivators may be advantageous in drug design; it may be possible to inactivate a specific target enzyme at low doses, but then, after a particular therapeutic effect has been realized, to administer a reactivator that releases the inactivator from the enzyme prior to the onset of unwanted side effects. This would be analogous to the administration of compounds such as pyridine 2-aldoximine methiodide to reactivate acetylcholinesterase following inactivation by certain organophosphates.[51]

Each of the criteria for mechanism-based inactivation is discussed in more detail below. Experimental protocols to support each of these criteria can be found in Section VII.

A. Time Dependence

In Section II.B. of this chapter (Scheme 2), the reaction of an enzyme with a mechanism-based inactivator was given. Since the rate of k_1 is usually much faster than k_2, formation of the E·I complex occurs much faster than conversion of the inactivator to the reactive intermediate. Consequently, the time dependence, in the simplest case, is a measure of the rate of this activation step, although, as discussed in Section IV.A., k_2 does not have to be the rate-determining step.

B. Saturation

If formation of the E·I complex is fast (i.e., time independent), the equilibrium k_1/k_{-1} will be diffusion limited (i.e., on the time scale of the assay) and will depend upon the concentration of I. As the concentration of I increases, more and more enzyme molecules will be tied up in the reversible E·I complex form and, therefore, the fate of inactivation will increase (k_2 only occurs from the E·I complex). When a point is reached where essentially all of the enzyme is in the E·I complex form, addition of more inactivator will no longer affect the rate of inactivation. At this point the maximum rate of inactivation is achieved and this phenomenon is known as saturation — the enzyme is saturated with inactivator (the term saturation is usually used in reference to the normal substrate conversion to product).

C. Substrate Protection

Since a mechanism-based inactivator is a substrate for the enzyme, it must be competitive with the normal substrate for the active site. Consequently, when substrate is added concomitant with inactivator, the competition for binding at the active site prevents E·I formation and, thereby, decreases the rate of inactivation. A competitive, reversible inhibitor should have the same effect as a substrate in protecting the enzyme from inactivation.

D. Irreversibility

In almost all cases, mechanism-based inactivation results in covalent attachment of the inactivator to the enzyme. Therefore, removal of excess inactivator (inactivation experiments generally involve the use of a large excess of inactivator over enzyme) by dialysis or gel filtration will not affect the inactivated enzyme because it no longer is in equilibrium with the free enzyme and inactivator. In a few cases, an exceedingly tight, noncovalent complex is formed following enzyme activation that is stable to dialysis or gel filtration.

Since many enzymes catalyze reactions via covalent intermediates, an enzyme-inactivator complex which is not stable to dialysis or gel filtration would then be defined as an enzyme-product complex, and the inactivator really would be just a substrate. Since there is no set definition as to how long an enzyme-inactivator complex must persist before the reaction is classified as an irreversible inactivation, many philosophical discussions can arise. At some point with increasing time, a covalent enzyme-product complex must be redefined as an enzyme-inactivator complex. There will be a continuum, then, of reversible enzyme-inactivator complexes (those that persist beyond an arbitrary stability cutoff, but then decompose to products) to irreversible enzyme-inactivator complexes (those that are stable to denaturation and even acid hydrolysis of the inactivated enzyme). Some enzyme-inactivator complexes are stable to dialysis or gel filtration, but decompose upon addition of another compound, e.g., the cofactor or substrates for the enzyme.

E. Stoichiometry of Inactivation

Since mechanism-based inactivation requires that the enzyme catalyze a reaction on the inactivator, active-site attachment is most likely. This, in general, prevents further reactions from taking place since the active site is blocked. Therefore, if a radioactively labeled inactivator and a homogeneous enzyme were mixed, a 1:1 stoichiometry of inactivator:active site would be expected. Frequently, "half-sites" reactivity is observed, so that 100% inactivation occurs when only half of the active sites are labeled. This is believed to be the result of negative cooperativity upon binding to one active site.

F. Involvement of a Catalytic Step

By definition, a mechanism-based inactivator is converted by the enzyme into the actual inactivating species. Therefore, some catalytic step must be demonstrated as a requirement for inactivation.

G. Inactivation Occurs Prior to Release of the Activated Species

If the inactivator is converted into an activated form that escapes the active site, and then returns to inactivate the enzyme, the inactivator is considered to be metabolically activated,[32] not mechanism-based. In this case, inactivation could occur by attachment to a residue at a site other than the active site. Furthermore, the product may have rearranged to another form, so that enzyme-inactivator structure identification may be misleading. The most important problem with this situation is that these inactivators, then, would be no more specific than affinity-labeling agents, and in fact, these compounds are really proaffinity-labeling agents.

Ideally, all of these criteria should be satisfied before one can be confident that the inactivator is mechanism-based. Specific experimental procedures, which are short and simple, have been established to test each of these criteria and are discussed in Section VII. Once the inactivator and enzyme are in hand, often the synthesis of a radioactively labeled analogue is the most time-consuming aspect of the initial studies. A complete investigation usually requires radioactive compounds in order to establish covalent modification of the enzyme, to determine the stoichiometry between enzyme and inactivator, and to carry out experiments directed at the elucidation of the mechanism of action of the inactivator.

VI. KINETICS FOR MECHANISM-BASED ENZYME INACTIVATION

In Section II.B. the kinetic constants for mechanism-based inactivation, k_{inact} and K_I, were introduced. In this section these terms are defined and their kinetic implications are discussed. Since much of the discussion in this section would require reference to Scheme 2 (Section II.B.), which describes the general reaction of an enzyme with a mechanism-based inactivator, it has been rewritten here as Scheme 3.

$$E + I \underset{k_{-1}}{\overset{k_1}{\rightleftharpoons}} E \cdot I \overset{k_2}{\longrightarrow} E \cdot I' \overset{k_4}{\longrightarrow} E\text{-}I''$$
$$k_3 \downarrow$$
$$E + P$$

Scheme 3.

A. k_{inact}

A common misconception is that k_{inact} is k_2 in Scheme 3. The term k_{inact} is not just one rate constant, but rather is a complex mixture of k_2, k_3, and k_4. On the basis of the work of Kitz and Wilson,[52] Jung and Metcalf[53] derived the expression shown in Equation 1 (k_{cat} and K_i were used by Jung and Metcalf[53] instead of k_{inact} and K_I, respectively),

$$\frac{\partial \ln E}{\partial t} = \frac{k_{inact} I}{K_I + I} \tag{1}$$

where E is the concentration of active enzyme, I is the inactivator concentration, k_{inact} is the rate of inactivation, and K_I is defined later in this section. The half-life for inactivation ($t_{1/2}$), then, is described by Equation 2:

$$t_{1/2} = \frac{0.69}{k_{inact}} + \frac{0.69}{k_{inact}} \frac{K_I}{I} \tag{2}$$

Waley[54] applied the steady-state hypothesis to Scheme 3 and derived an expression related to Equation 1 for the loss of enzyme activity with time in terms of the microscopic rate constants, k_2, k_3, and k_4 (Equation 3).

$$\frac{-d \ln a}{dt} = \frac{As}{B + s} \qquad a = \text{enzyme activity}$$

s = inactivator concentration
at a given time

$$A = k_2 k_4/(k_2' + k_3 + k_4)$$

$$B = [(k_{-1} + k_2)/k_1][(k_3 + k_4)/(k_2 + k_3 + k_4)] \qquad (3)$$

If Equations 1 and 3 are compared, then it can be seen that K_I = B and k_{inact} = A = $k_2 k_4/(k_2 + k_3 + k_4)$. If k_2 is rate determining (i.e., $k_2 \ll k_4$), which often is the case, then k_{inact} reduces to $k_2 k_4/(k_3 + k_4)$. Therefore, k_{inact} = k_2 only if $k_2 \ll k_4$ and k_3 is very slow or zero. If k_4 is the rate determining step (i.e., $k_4 \ll k_2$), then k_{inact} is $k_2 k_4/(k_2 + k_3)$. In this case, if k_3 is slow or zero, then k_{inact} = k_4. It is assumed that the values given for k_{inact} represent the inactivation rate constants at infinite concentrations of inactivator (see Figure 3A in Section VII.B.).

Equation 3 shows that inactivation is not necessarily first order. The extent of deviation from first-order kinetics depends upon how much the inactivator concentration is altered during the course of the inactivation. If the initial inactivator concentration (s_o) is much greater than $k_3/k_4 e_o$, where e_o is the initial enzyme concentration, the alteration will be small and the k_{inact} will approximate $As_o/(B + s_o)$, a term that varies hyperbolically with s_o. If one of the steps is principally rate determining, the parameter expressions are simplified. From Equation 3, the half-life in the progress curve for inactivation and product formation can be derived[54] (Equation 4).

$$t_{1/2} = \frac{1}{A}\left[\frac{\left(\dfrac{B}{s_o}\right)\ln\left(2 - \dfrac{re_o}{s_o}\right)}{1 - \dfrac{re_o}{s_o}} + \ln 2 \right] \qquad (4)$$

A, which is effectively the rate constant for the overall flux, may be much less than the k_{cat} for a good substrate without inactivation being too slow. This suggests that structure-activity relationships of substrates may not be very important in designing mechanism-based inactivators. For example, benzylamine is an excellent substrate for monoamine oxidase and α-methylbenzylamine is hardly a substrate at all,[55] yet N-cyclopropyl-α-methylbenzylamine is an excellent mechanism-based inactivator of monoamine oxidase.[55] The parameter B/s_o is a measure of the fraction of uncombined enzyme. When high initial inactivator concentrations are used ($s_o \gg B$), then $t_{1/2}$ approaches $(1/A)\ln 2$, the same expression to which Equation 2 reduces at infinite inactivator concentration. As discussed in Section VII.B., inactivator concentrations can be extrapolated to infinity, so $t_{1/2} = \ln 2/k_{inact}$ is the expression used to determine the k_{inact} for a particular inactivator. The parameter most characteristic of mechanism-based inactivators is re_o/s_o, where r = k_3/k_4 (the partition ratio). This term is the magnitude of the partition ratio compared with the ratio of the initial inactivator concentration to that of enzyme. If $s_o \gg re_o$, then Equation 3 reduces to $t_{1/2} = [\ln 2 (1 + B/s_o)]/A$, which is equivalent to Equation 2, where A is k_{inact} and B is K_I. Therefore, an unfavorable (i.e., high) partition ratio can be compensated for by using a high inactivator to enzyme (s_o/e_o) ratio. When the enzyme is completely inactivated, the partition ratio is given by P_∞/e_o, where P_∞ is the total amount of product formed. This ratio is independent of the inactivator concentration. The initial work of Waley[54] was extended[56] and a linear plot for the steady-state kinetics of mechanism-based inactivation was described. The $t_{1/2}$ for enzyme inactivation, which also is the $t_{1/2}$ for product formation, was measured in a series of experiments in which the inactivator concentration (I_o) was varied, keeping the ratio of enzyme concentration to inactivator concentration fixed. A plot of $[I_o]t_{1/2}$ vs. (I_o) is linear and the kinetic parameters, k_{inact} and K_I, can be obtained from the slope and intercept, respectively. Simplifications of progress curve equations were described that are valid when the inactivator concentration is high or low or when the extent of reaction is low.

According to Tatsunami et al.,[57] a problem with the earlier work of Waley[54] is that the sum of the concentrations of unreacted inactivator, of that bound to the inactivated enzyme, and of that converted to products, measured at different times after the initiation of the inactivation, is different than the initial total concentration. On the basis of Equation 5,

$$\frac{dI}{dt} = -C\left(1 + \frac{1}{r}\right)\frac{aI}{B + I} \quad I = \text{concentration of inactivator}$$

$$a = E_o - (E - I''), \ E_o = e \quad \text{at} \quad t = 0$$

$$B = \left(\frac{k_{-1} + k_2}{k_1}\right)\left(\frac{k_3 + k_4}{k_2 + k_3 + k_4}\right)$$

$$C = \frac{k_2 k_3}{k_2 + k_3 + k_4}$$

$$r = \frac{k_3}{k_4} \tag{5}$$

progress curves of inactivator consumption, product formation, and enzyme inactivation were calculated for a hypothetical model system and were compared with the exact solution. The results were much more consistent with the exact solutions than those obtained with Waley's equations.[54] The most important factor derived is the term $(1 + r)e_o/I_o$ (Waley proposed re_o/I_o). When $(1 + r)e_o/I_o$ is greater than 1, all of the inactivator molecules are consumed, leaving some enzyme molecules active; when this term is less than 1, all of the enzyme molecules are inactivated, leaving excess inactivator; when it equals 1, then all of the enzyme is inactivated and all of the inactivator is consumed. A plot of P_f/I_o (the final product to the initial inactivator concentration) vs. e_o/I_o gives a straight line with slope r.

Equations were derived by Galvez et al.[57a] for the kinetics of an enzyme reaction that leads to inactivation. The overlap between the transient phase of inactivation and the steady-state phase of catalysis, where the initial concentration of the inactivation ≫ the initial concentration of the enzyme (and vice versa), either in the presence or absence of an auxiliary substrate, was studied by Tudela et al.[57b] Equations ere obtained showing the dependence of the product concentration on time.

The transient phase approach of Tudela et al.[57b] was applied by García Cánovas et al.[57c] to the study of the inactivation of mushroom tyrosinase by o-diphenols. Equations of product vs. time were developed for the multisubstrate mechanism, and the kinetic parameters for inactivation were determined.

B. K_I

Now consider the term K_I. As long as k_1 and k_{-1} in Scheme 1 (Section II.B.) are much greater than k_2, then the K_m value and reversible inhibitor K_i value **do** represent the dissociation constants for breakdown of enzyme · substrate or enzyme · reversible inhibitor complexes, respectively. However, the use of the same nomenclature (K_i) to define the dissociation constants for both the enzyme · reversible inhibitor complex and the enzyme · mechanism-based inactivator complex, which is a common practice, is misleading. The K_i in the former case is obtained experimentally by determining the effect on the rate of conversion of subsaturating concentrations of substrate to product upon addition of a constant amount of inhibitor. Therefore, this is measuring the effect on the E·S complex. The K_I for inactivation of an enzyme by a mechanism-based inactivator is obtained experimentally, as described in Section VII.B., by determining the effect on the rate of inactivation of a change in the inactivator concentration. As long as k_2 in Scheme 3 is rate-determining, then the values of K_i and K_I will be the same. However, as discussed above, the rate of inactivation by a mechanism-based inactivator is a function of both k_2 and k_4, so if k_4 becomes rate-determining or partially rate-determining, then the value of K_I increases and these two constants have different values, i.e., $K_I \geqslant K_i$. Just as the K_m represents the concentration of substrate that gives half the maximal velocity, the K_I is the concentration of inactivator that produces half the maximal rate of inactivation.

Meloche[58] has made a similar analysis for a comparison of reversible inhibitors and affinity-labeling agents (see Section III.B. for the definition of an affinity-labeling agent). Scheme 4 describes the reaction of a typical affinity-labeling agent with an enzyme.

$$E + I \underset{k_{-1}}{\overset{k_1}{\rightleftharpoons}} E \cdot I \overset{k_2}{\longrightarrow} E\text{–}I$$

Scheme 4.

In this case there is an initial reversible reaction between the enzyme and the inactivator (k_1/k_{-1}), followed by an irreversible reaction (k_2). The dissociation constant for the initial reversible E·I complex, which is the same as for any reversible enzyme · inhibitor complex, is $K_i = k_{-1}/k_1$. The corresponding inactivation constant was assigned[58] the symbol K_{inact} in order to differentiate it from K_i, because it actually represents the concentration of inactivator giving the half-maximum inactivation rate and, presumably, half-saturation of the enzyme. This term is equal to $(k_{-1} + k_2)/k_1$. Therefore, K_{inact} can be greater than K_i if k_2 is partially rate limiting. From Equations 1 and 3 for a mechanism-based inactivator, the term $K_I = B$, which equals $K_{inact}[(k_3 + k_4)/(k_2 + k_3 + k_4)]$. It is obvious from this expression that K_I also is different than K_i.

Another theoretical treatment of the kinetics of affinity labeling was reported,[59] which may be applicable to the kinetics of mechanism-based inactivation. In this treatment of time-dependent inactivation, a system of linear differential equations are obtained when the inactivator concentration is much greater than the enzyme concentration. It is shown that log plots are not straight, which accounts for why pseudo first-order kinetics often are not observed. It is suggested that a wider range of inactivator concentrations and extended time intervals be used. Also, a mathematical approach is suggested for the solution of the equations when the inactivator concentration is not much greater than the enzyme concentration.

C. Kinetics for Two-Site Inactivators of Allosteric Enzymes

A kinetic expression for two-site irreversible inhibitors of allosteric enzymes was derived by Kuo and Jordan[60] on the basis of Scheme 5, where $K_{i,r}$

$$E + I \underset{K_{i,r}}{\rightleftharpoons} E \cdot I \underset{K_{i,c}}{\overset{I}{\rightleftharpoons}} E \cdot I_2 \overset{k_c}{\longrightarrow} E' \cdot I_2$$

Scheme 5.

is the dissociation constant for I from the regulatory site, $K_{i,c}$ is that from the catalytic site, and k_c is the first-order rate constant for irreversible inactivation. These terms can be inserted into a modified form of the equations for irreversible inactivation of Kitz and Wilson[52] and Jung and Metcalf[53] (see Equation 1) with the assumption that $I \gg E$ (Equation 6),

$$\ln(\epsilon/E_o) = k_c t/(1 + K_{i,c}/[I] + K_{i,c}K_{i,r}/[I]^2) \tag{6}$$

where E_o is the stoichiometric amount of enzyme and ϵ is the total active enzyme ($E + E \cdot I + E \cdot I_2$) remaining at time t. The observed half-life for each concentration of I is shown in Equation 7:

$$t_{1/2} = t_{1/2,\infty}(1 + K_{i,c}/[I] + K_{i,c} K_{i,r}/[I]^2) \tag{7}$$

where $t_{1/2,\infty}$ is the half-life at saturation ($0.693/k_c$). At early times, inactivation follows first-order kinetics, but at later times, the kinetics of inactivation become biphasic. In the presence of a substrate or allosteric effector, the inactivation kinetics become much more complicated (Equation 8),

$$k'_{i,c} = \cfrac{k_{i,c}}{\cfrac{K_{i,r}K_{i,c}}{[I]^2} + \left(1 + \cfrac{K_{i,r}\,[S]}{[I]\,K_{s,r}}\right)\left(1 + \cfrac{K_{i,c}}{[I]}\left(1 + \cfrac{[S]}{K_{s,c}}\right)\right)}$$

$$+ \cfrac{k_{i,c}}{\cfrac{K_{i,c}K_{s,r}}{[I]\,[S]} + \left(1 + \cfrac{K_{s,r}\,[I]}{[S]\,K_{i,r}}\right)\left(1 + \cfrac{K_{i,c}}{[I]}\left(1 + \cfrac{[S]}{K_{s,c}}\right)\right)} \tag{8}$$

where $K_{s,r}$ is the dissociation constant for the substrate in the regulatory site and $K_{s,c}$ is that for the substrate in the catalytic site.

D. Relationship of Inactivation Kinetics to Steady-State Kinetics

Villafranca and co-workers[61,62] have pointed out that inactivation kinetic data and steady-state kinetic data are related by the partition ratio. However, this is valid only under certain circumstances. In one case,[61] it was suggested that the inactivation rate data could be converted to steady-state rate data simply by multiplying the measured constants k_{inact}/K_m by V_{max}/k_{inact}, which was stated to be equal to the partition ratio. However, V_{max} is a rate for a given enzyme concentration and k_{cat} is a rate constant; therefore, this is valid only if the rate depends exclusively on k_{cat} or if the enzyme concentration is 1 (see Equation 9).

$$\frac{k_{inact}}{K_m}\left(\frac{k_{cat}}{k_{inact}}\right) = \frac{k_{cat}}{K_m} = \frac{k_{cat}[\text{enzyme}]}{K_m[\text{enzyme}]} = \frac{V_{max}}{K_m[\text{enzyme}]} \tag{9}$$

In another example,[62] inactivation rate data were converted into steady-state rate by multiplying the partition ratio by k_{inact}/K_I. When the k_{cat} can be equated to the V_{max} (as above) and the K_m has the same value as the K_I (see Section VI.B.), then this relationship is valid.

VII. EXPERIMENTAL PROTOCOLS FOR MECHANISM-BASED ENZYME INACTIVATION

The following discussion describes the experiments that are carried out in order to test each of the criteria for mechanism-based enzyme inactivation.

A. Time-Dependent Loss of Enzyme Activity

These studies become much simpler if a continuous assay, e.g., a spectrophotometric assay, for the enzyme is available. The experiment for time dependence involves incubating

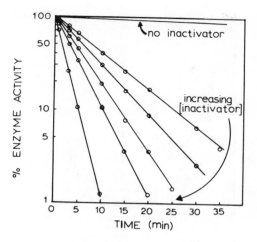

FIGURE 1.   Time-dependent inactivation of enzymes by mechanism-based enzyme inactivators.

the enzyme in buffer solution with all components required for the substrate reaction except for the substrate. The inactivator is then added and this point is taken as time 0. In some cases it is easier to add one of the required components other than the inactivator to initiate the reaction. For example, if the enzyme is unstable, it can be the last component added. As soon as everything is mixed (preferably on the order of 15 sec), an aliquot is removed, diluted (preferably greater than 50-fold) into the enzyme assay mixture, and the enzyme activity is assayed. The purpose for diluting the preincubation mixture into a high concentration of substrate is to quench the inactivation reaction immediately by allowing substrate to occupy the remaining active sites. All free enzyme molecules or those involved in reversible E·I complex formation will be protected instantly from further inactivation, because the inactivator concentration is diluted and the substrate concentration is saturating. In order to get enough dilution and still have reasonable enzyme activity, a fairly concentrated enzyme solution is necessary during the preincubation. Too concentrated an enzyme solution, however, in the author's experience, leads to diminished rates of inactivation, presumably because of increased viscosity and protein interactions. Identical aliquots then are removed periodically (from periods of seconds to hours, depending upon the inactivation rate), and the same dilution and assay is run. With a continuous assay, the progress of the inactivation can be followed immediately, so the appropriate time intervals between aliquots can be assessed. When a timed assay is required, a large number of points need to be taken over an extended time period, since it generally is not known what the results are until after the complete experiment is concluded. It is best to obtain at least five data points and to observe inactivation to completion. In some cases, the rate of inactivation changes after the enzyme has been greater than 90% inactivated; therefore, loss of enzyme activity for at least one concentration of inactivator should be followed to completion. When a crude enzyme mixture is used, it is best to make duplicate or triplicate runs and average the results. A control, which includes everything in the preincubation mixture except the inactivator, must be run at each time point, since all inactivation rates are relative to this control.

It is generally convenient to determine the concentration range of inactivator to be used prior to setting up the entire experiment. This can be done quickly when a continuous assay is available, since, if the rate is too slow or too fast, the experiment can be terminated and a different concentration of inactivator used. A plot of the **log** percent of enzyme activity remaining (as measured by a slope of a spectrophotometric trace, or the amount of product formed in a given time interval, or by whatever means the enzyme activity is determined relative to the noninactivated control) vs. time can be constructed (Figure 1) as evidence

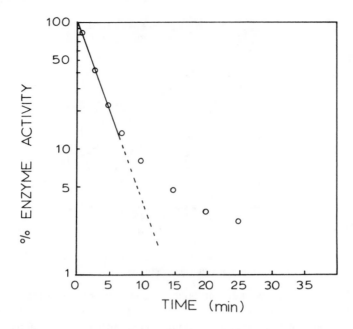

FIGURE 2.   Nonpseudo first-order loss of enzyme activity.

for time-dependent inactivation. Often these plots show pseudo first-order kinetics, but this is not universal.[54]

Several problems can arise in obtaining the kinetic constants. When an inactivator has a very low K_I and a large k_{inact}, it may not be possible to measure the inactivation rate fast enough, unless the inactivator concentration is lowered to a point where pseudo first-order kinetics are no longer observed, i.e., the inactivator concentration is no longer much greater than the enzyme concentration. Often this can be remedied by lowering the temperature or by using a less-than-optimum pH value in order to slow down the rate of catalysis so that a higher inactivator concentration can be used and the rate of inactivation still be measurable. However, since these inactivators act as substrates, the values obtained for K_I and k_{inact} at a higher temperature or different pH cannot be compared to constants obtained at maximal catalytic efficiency because of the effects of temperature[63] and pH[64] on kinetic constants. Therefore, any comparison of inactivators must be made with constants obtained under identical experimental conditions, namely, pH, temperature, ionic strength, etc. Another problem may arise from an inactivator with a high partition ratio. In this case, nonpseudo first-order kinetics may result because the inactivator concentration may be decreased below its K_I. This problem can be detected by measuring the concentration of inactivator at different time points during the inactivation by some analytical procedure, and calculating whether the change in inactivator concentration is significant. If so, a much larger inactivator concentration should be used. In these cases though, the kinetic constants are not very useful, unless initial rates are used, since the inactivator concentration is not constant throughout the experiment.

When an inactivator generates a metabolite that binds much more tightly to the enzyme than does the inactivator, nonpseudo first-order kinetics also can be observed. This is because as inactivation progresses, a higher concentration of the metabolite is formed and this competes with the inactivator for the active site. In this case the values obtained at earlier time points and higher inactivator concentrations are most reliable and kinetic constant values can be estimated from these points. A similar problem can arise when an inactivator has a high partition ratio, even if the product does not bind very effectively to the enzyme. In both of these situations, plots similar to that shown in Figure 2 are obtained.

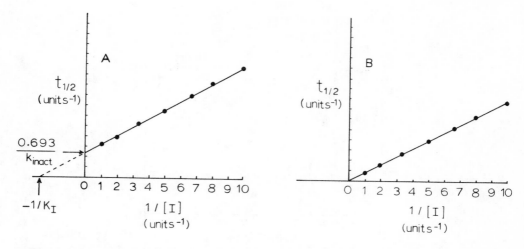

FIGURE 3. Kitz and Wilson[52] plot for an inactivator (A) exhibiting saturation or (B) not exhibiting saturation.

Another cause for nonpseudo first-order kinetics is observed with pyridoxal phosphate (PLP)-dependent aminotransferases, when there is a slow conversion of the PLP to pyridoxamine phosphate (PMP). In the absence of an α-ketoacid, the enzyme is protected from inactivation, since it is in its PMP form. If the inactivation is carried out in the presence of an α-ketoacid, this problem can be alleviated.

Biphasic kinetics can arise from several different reasons, including that (1) two or more different inactivation processes are occurring simultaneously, (2) the inactivated enzyme is not stable, and breakdown is rate determining, (3) there is negative cooperativity between two subunits in a multi-subunit enzyme, whereby attachment to one subunit renders an adjoining subunit less active, and (4) there is heterogeneity of subunit composition that results in nonequivalent binding to the subunits.

B. Saturation

In order to show saturation, the time-dependent experiment described above is repeated at each of several concentrations of inactivator with a fixed enzyme concentration. The inactivator concentrations selected, of course, should be considerably below saturation so that a change in inactivator concentration produces a large enough difference in the rate of inactivation for results to be unambiguous. If the half-life ($t_{1/2}$) for inactivation at each inactivator concentration is plotted against 1/[inactivator], referred to as a Kitz and Wilson plot,[52] two possibilities can result (Figure 3). Figure 3A represents a plot as observed for enzyme saturation; at infinite inactivator concentration there is a finite half-life for inactivation (intercept of ordinate is greater than 0). The point of intersection with the ordinate is the half-life at saturation, from which the rate constant at saturation (k_{inact}) can be calculated ($k_{inact} = 0.693/t_{1/2}$). This relationship is a special case (i.e., for infinite inactivator concentration) of the general expressions given by Equations 2 and 4 (see Section VI.A.). In general, the values of k_{inact} are quite low (on the order of 10^{-2} to 1 min^{-1}) relative to values of k_{cat} for normal substrates (10^2 to 10^4 min^{-1}, when pseudo first-order). The intercept with the X-axis is equal to $-1/K_I$, from which the K_I value at saturation is easily obtained. One common pitfall in the design of these saturation experiments often is seen in the Kitz and Wilson[52] plots (Figure 3A), namely, that most of the data points are bunched up at one end of line. For example, if the inactivator concentrations are conveniently chosen to be 1.0, 0.8, 0.6, 0.4, 0.2, 0.1, and 0.05 mM, then 1/[I] = 1, 1.25, 1.67, 2.5, 5, 10, and 20. The slope of this line would be difficult to draw accurately because of the weighting factors at the low end (most of the data points are between 1/[I] = 1 to 5). Inactivator concentrations should be selected so as to give an even distribution of data points along the X-axis.

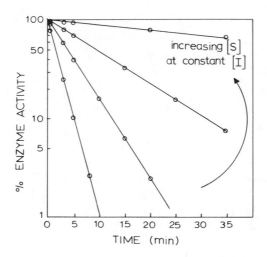

FIGURE 4.   Substrate protection during mechanism-
based inactivation.

When the data points are connected by a line passing through the origin (Figure 3B), saturation is not observed, but rather a bimolecular reaction may be evident. This implies that k_{inact} is fast relative to formation of the initial E·I complex, which occurs prior to any catalytic event. Often, the apparent absence of saturation can be remedied by carrying out the inactivation experiments at a lower temperature,[53] thereby lowering k_{inact}. When an inactivator has a low K_I and a large k_{inact}, all of the concentrations of inactivator used that give measureable rates may lead to the same rate of inactivation. This indicates that saturation has already been reached. If the concentration is lowered, pseudo first-order kinetics may be lost, because the inactivator and enzyme concentrations may approach stoichiometry. Repeating the experiment at a lower temperature or different pH may allow subsaturating concentrations of inactivator to be used.

When saturation occurs at low concentrations of inactivator, even at low temperature, then a lower limit for k_{inact} can be estimated. Owens and Barden[65] showed that for affinity labeling of an enzyme, if it is assumed that $[I] \ll K_i$, then the relationship expressed in Equation 10, which is based on the theory of Kitz and Wilson,[52] is valid.

$$\log E^a = \frac{k_2 t}{3.4\, K_i}\,(I) + \log E^\circ \tag{10}$$

In this equation, E^a is the enzyme activity at time t, E° is the enzyme activity at t = 0, K_i = (I)(E)/E·I, and k_2 is the rate constant for covalent bond formation from the E·I complex (see Scheme 4 in Section VI.B.). The enzyme activity remaining after a short fixed incubation time period is determined. If a plot of the log of the percent enzyme activity remaining vs. [I] is linear, then the slope = $k_{inact} t/2.3\, K_I$. The value of k_{inact} can be estimated for an assumed value of K_I.

C. Substrate or Competitive Inhibitor Protection

The experiment carried out to show substrate protection is the same as the inactivation experiment except that the resulting rate is compared with the rate from identical preincubation mixtures containing varying concentrations of substrate. If there is protection from inactivation, the rate of inactivation will be diminished relative to the corresponding rate of inactivation in the absence of substrate (Figure 4). In lieu of substrate, a known reversible competitive inhibitor can be used to demonstrate an active site reaction.

D. Irreversibility

Following inactivation, the preincubation mixture either can be dialyzed exhaustively (several buffer changes) or chromatographed on a gel filtration column (size exclusion chromatography) in order to remove excess and reversibly bound ligands. A noninactivated control must be carried through the same operation for comparison of enzyme activity, which is set equal to 100%. Sometimes the enzyme-inactivator adduct is marginally stable. It has been observed in several cases that when dialysis or gel filtration is carried out at room temperature, partial or complete enzyme activity may return, but when carried out at 4°C, inactivation may persist. If activity returns at room temperature, a low temperature experiment should be run to determine stability. The rates of reactivation (if observed) at various temperatures can be used to calculate the Arrhenius activation energy and change in free energy for the decomposition of the inactivated enzyme complex. Also, changing the pH after inactivation can have an important effect on the stability of the enzyme-inactivator complex. Once again, making these changes in temperature or pH may result in a semantic argument regarding whether an enzyme-inactivator complex or an enzyme-product complex is obtained.

E. Stoichiometry of Inactivation

The difficult (or, at least, most time-consuming) part of this experiment is often the synthesis of the radiolabeled-inactivator molecule with known specific activity. Once this compound is in hand, the enzyme is incubated with a concentration of the inactivator shown to produce complete inactivation. Gel filtration or dialysis of the inactivated enzyme to constant specific activity is carried out, and from a protein assay and radioactivity determination, the stoichiometry can be ascertained. If some enzyme activity remains after these procedures, the amount of protein must be multiplied by the fraction of enzyme inactivated in order to calculate the radioactivity bound per inactive enzyme. A 1:1 stoichiometry of radioactivity to active site is usually predicted. When greater than a 1:1 stoichiometry results, it suggests that a nonspecific covalent reaction may have taken place in addition to active-site labeling. This can arise from release of the incipient reactive species from the active site a fraction of the turnovers. As discussed in Section VII.G., this often can be prevented by having an electrophilic trapping agent, e.g., a thiol, present during inactivation. When the stoichiometry is less than 1:1, it may mean that some inactive enzyme was initially present or, in the case of multimeric enzymes, some form of negative cooperativity is occurring (see Section V.E.).

F. Involvement of a Catalytic Step

Depending upon the reaction catalyzed by the target enzyme, experimental procedures to demonsrate the involvement of a catalytic step will vary. For example, if a C–H bond is broken during inactivation, the inactivator can be synthesized specifically deuterated at that position. If that bond cleavage is partially rate determining, there should be a deuterium isotope effect on the rate of inactivation. If no deuterium isotope effect is observed, it may indicate that k_4 in Scheme 3 (Section VI) is rate determining. Also, see Section VII.H. concerning deuterium isotope effects on the partition ratio. If the reaction involves a reduction or oxidation of a cofactor, the corresponding change in the optical spectrum of the cofactor during inactivation can be monitored. For whatever chemical steps are being catalyzed, some evidence for their involvement during inactivation must be presented.

G. Inactivation Occurs Prior to Release of the Activated Species

Three tests can be used to determine if the activated species was released from the active site prior to inactivation. Should this species be released, a time-dependent increase in the rate of inactivation would be observed as the concentration of the reactive species builds up

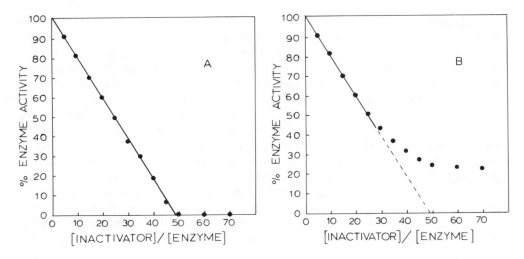

FIGURE 5.   Determination of the partition ratio by the titration method with (A) linear and (B) nonlinear loss of enzyme activity.

in solution. This will only occur if the activated species is not so reactive that it is rapidly quenched by some component in the incubation buffer. Likewise, if a reactive species builds up in solution, the rate of inactivation of a fresh aliquot of enzyme added to the initial inactivated enzyme solution should be faster than that of the initial aliquot. With a mechanism-based inactivator, the rates of the two inactivations will be identical (unless the inactivator concentration has been depleted). A third test is to have a trapping agent in the buffer to react with any reactive intermediates as they are released into solution. Because of the superior nucleophilicity of thiols, β-mercaptoethanol or dithiothreitol are generally selected to trap reactive electrophiles. However, it is possible that the thiol may enter the active site and compete with active site nucleophiles for the activated species. In these cases, the reactive species is being trapped prior to release and, therefore, this defeats the purpose of the experiment; furthermore, the wrong conclusion may be made based on the result. If one of these small thiols does protect the enzyme, it is advisable to try a bulkier thiol, such as reduced glutathione. The larger thiol would, most likely, be incapable of entering the active site. If protection still is observed, the activated species is, most likely, being released into solution. However, if protection is not observed, it does not indicate definitively that release of the reactive species does not occur. The reactive species may return to and react more rapidly with an active site nucleophile than with the added thiol.[65a]

H. Determination of the Partition Ratio

There are several ways to determine the partition ratio for an inactivator. The most common method is by titration. Increasing amounts of inactivator are added to a known amount of enzyme, and the reaction is allowed to go to completion. After dialysis or gel filtration, a plot of the percent of enzyme activity remaining vs. the ratio of the moles of inactivator per mole of enzyme active sites (all samples must be normalized to a constant protein concentration) can be constructed (Figure 5A). Often, the higher ratio points deviate from a straight line because of product inhibition or product protection of the enzyme towards further inactivation (Figure 5B); extrapolation of the lower ratio linear part will give the turnover number, i.e., the number of inactivator molecules required for complete inactivation. This number includes the one molecule (assuming a 1:1 stoichiometry) of inactivator that inactivates the enzyme. The partition ratio, therefore, is the turnover number minus one, the number of molecules leading to product per each inactivation event. If the enzyme aliquots are not dialyzed or gel filtered prior to measurement of the enzyme activity remaining, and a product formed is a potent reversible inhibitor of the enzyme, then each assay will show an artificially low enzyme activity. This then will result in a falsely low partition ratio. The titration method is limited in use if the inactivation rate is so slow that other events, such as enzyme or inactivator stability, become factors. A second method is to determine the molar amount of products generated and divide by the molar amount of enzyme used, assuming a 1:1 stoichiometry of inactivator to enzyme and 100% inactivation. Since a large excess of inactivator is generally used, the products must be distinguished from the inactivator molecules; this is accomplished most often by chromatographic or ion exchange separation of the compounds. The radioactive small molecules can be separated from the labeled enzyme either by gel filtration, ultrafiltration, protein precipitation, or microdialysis. The latter technique involves suspending a narrow dialysis bag containing the protein solution into 3 to 5 mℓ of a stirred buffer solution (flea stirring bar) in a test tube; the buffer is changed once. If 300 $\mu\ell$ of enzyme solution are dialyzed twice against 3 mℓ of buffer each time, greater than 99% of the small molecules will be removed into the 6 mℓ of dialysate. Gel filtration requires more of the worker's time to set up, run, and analyze. The microdialyzed enzyme solution can then be dialyzed against a large volume of buffer to remove the residual unbound radioactivity so that stoichiometry (see Section VII.E.) with the enzyme can be measured. A dilution problem is not a factor with ultrafiltration. A third measure of the partition ratio is to determine experimentally the values for k_{cat} and k_{inact} for the reaction; this gives the partition ratio, the ratio of k_{cat}/k_{inact}, directly.

When the inactivation mechanism involves carbon-hydrogen bond cleavage, a study of deuterium isotope effects on the partition ratio, the rates of product formation, and the rates of inactivation can be quite useful for the establishment of the point in the inactivation mechanism where product formation branches from inactivation.[61,62] If there is no isotope effect on the partition ratio or on V_{max}, but the same isotope effect on k_{inact}/K_I and V/K_I, then inactivation is occurring from a species in which C–H bond cleavage has already occurred, and both product formation and inactivation occur from a common intermediate. If there is no isotope effect on the partition ratio, but a different isotope effect on V_{max} than on k_{inact}, the product formation and inactivation occur from different species and both pathways involve C–H bond cleavage. If there is a deuterium-dependent decrease in the partition ratio, an inverse isotope effect on k_{inact}/K_I and a normal isotope effect on V_{max}, then partitioning is occurring at the point of C–H bond cleavage and this bond breakage only is involved in product formation, not in inactivation.

VIII. CONTENT AND ORGANIZATION OF THE BOOK

The title of the book is ''Mechanism-Based Enzyme **Inactivation**'' rather than ''Mechanism-Based Enzyme **Inactivators**,'' because the book is devoted to the chemistry and mechanism of the inactivation process rather than to the inactivators themselves and what uses have been found for them. An attempt has been made to be as complete as possible in covering the literature on mechanism-based enzyme inactivation. However, the word ''complete'' must be qualified. The intent in writing this book was to offer a summary of the **chemistry** and **enzymology** of mechanism-based inactivation. Because of the usefulness of this type of enzyme inactivaton to the study of metabolic pathways and to the design of drugs, much of the literature in this field is concerned with the physiological, pharmacological, and clinical aspects of these inactivators. This literature is too vast to be included in one book and because my personal interest lies in the chemical and enzymological regime, I have limited this book to those aspects. Consequently, no in vivo work is described. Tissue culture or crude homogenate studies where the effect of the inactivator is on the organism or on a metabolic pathway rather than on a specific enzyme, or studies involving whole animals, are **not** included.

The purpose of this book is not to reiterate published experiments in detail, but rather to summarize results and present important chemical and biochemical conclusions for the reader. If specific experiments and interpretations of results are desired, the reader is referred to the original publication. Frequently, only a small section of a paper is devoted to experiments involving mechanism-based inactivation. In these cases, just the relevant aspects are cited, even if the results comprise a minor portion of the entire paper.

Only research journals are used as sources of information for inclusion in this book. Access to monographs, reviews, and patents is often difficult, and the patent literature is often, shall we say, imaginary. Data found in meeting abstracts and proceedings is not included because often it is preliminary, and is later modified. It is my bias that if the work is important enough to be mentioned at a meeting, it will be submitted for publication in a research journal when it is completed.

There are numerous ways in which the information regarding mechanism-based inactivation could be organized, e.g., according to the type of functional group in the molecule, the class of enzyme, or the chemical reaction involved. Because this book is concerned with mechanism-based **inactivation** rather than **inactivators**, and because my interest is in the organic chemistry of these inactivators, I have organized the chapters on the basis of the type of chemical reactions involved. Since the design of new inactivators must be derived from the type of reactions that the enzyme is capable of catalyzing, it seems to me that this organization would be most useful to those interested in finding out what mechanistic approaches have been utilized in the past so that new approaches can be contemplated and vice versa. For each inactivator discussed, a synthesis of the compound is given; if a synthesis is not shown, it generally indicates that the compound is commercially available or it was obtained as a gift. The source of the enzyme in each case is mentioned, because often there are specificities of a particular inactivator for any enzyme from different sources. Since it often is difficult to ascertain what purity of enzyme has been used in the experiments, this information is not provided here. The minimal criteria accepted for mechanism-based inactivation are time-dependent inactivation that is competitive with a substrate or competitive inhibitor. When more data are present in a paper, that information is stated. If a test for reversibility, such as dialysis or gel filtration, is performed, the stability of the inactivated enzyme is mentioned.

The mechanisms for inactivation depicted in this book are generally those that were suggested by the authors of the papers from which the information was extracted. Since mechanism-based inactivations are generally not simple one-step mechanisms, there may be confusion as to how to determine in which reaction type the mechanism is to be categorized. If the only "mechanism" that is required for inhibitor activation prior to covalent bond formation is one involving nonchemical phenomena, such as binding to the active site or to a coenzyme or an enzyme conformational change, then the compound is characterized as an affinity labeling agent and is not included in this book. The **initial** catalytic step in the **proposed** mechanism is what determines the category (i.e., chapter) into which the inactivation reaction fits. Therefore, all inactivations beginning with, for example, an elimination of H–X from the molecule are described in the chapter on elimination reactions (Chapter 6). Within a chapter the initial reaction category is subdivided according to the subsequent reactions in the mechanism. For example, if, after the elements of H–X are eliminated, an active-site nucleophile adds across the double bond in a Michael reaction, this inactivator would be listed in Chapter 6 under the heading, Elimination/Addition. When there is more than one inactivator in a subcategory, the information is discussed compound by compound. If there is more than one paper regarding a particular inactivator, they are arranged according to the enzyme source. Certain chemical transformations are disregarded in the determination of the reaction category. In the above example, the proton removal, even if stepwise, is considered to be part of the elimination process; therefore, it is not a deprotonation reaction, but rather an elimination reaction. In the case of PLP-dependent enzymes, the initial Schiff base formation of the inactivator with the coenzyme is considered to be part of the active-site binding and is therefore ignored for purposes of categorization. Since this step is the first step in all PLP-dependent enzyme reactions, it would not be useful to include it when trying to differentiate the important steps that follow. Because this Schiff base formation is recognized as simple active-site binding, compounds whose mechanism of inactivation of PLP-dependent enzymes is the formation of a stable Schiff base to the coenzyme are not classified as mechanism-based inactivators.

Some mechanism-based inactivators initially acts as affinity labeling agents, but then, after this first covalent reaction is completed, a mechanism-based reaction ensues (e.g., many of the inactivators described in Chapter 5 on acylation reactions). In these cases, the affinity labeling reaction, e.g., acylation, defines the category (i.e., chapter) for the inactivator. Since flavin, nicotinamide, pyridoxal, tetrahydrofolate, and metal ion enzyme cofactors are electrophilic in nature, it is possible for initial covalent reactions to take place by a nucleophilic inactivator. This situation can be difficult to demonstrate, however, since a two-electron transfer from the inactivator to the cofactor could occur prior to addition, thereby interchanging the polarities of the inactivator and cofactor. Because of this possibility, several inactivators are cross-referenced in more than one chapter.

There are many examples of compounds that were designed as mechanism-based inactivators for specific enzymes and that are quite effective, yet inactivation is protected by thiols. Although these compounds are not true mechanism-based inactivators, the inactivation chemistry proposed is still valid and often quite clever. Therefore, these inactivators are included, but are placed in a special section entitled "Not Mechanism-Based Inactivation" toward or at the end of all appropriate chapters. Also included in these sections are compounds designed as mechanism-based inactivators that do not inactivate the enzyme, and compounds that follow all of the criteria for mechanism-based inactivation, but are activated by mechanisms other than the normal catalytic mechanisms.

Since it is somewhat unusual for mechanism-based inactivators to be transformed into noncovalent inactivators, these examples are presented separately at the end of each chapter for which these compounds are most suitable in a section entitled "Noncovalent Inactivation." The chapters are arranged in order of increasing complexity of the reactions involved, and this order also defines the order in which subcategories are discussed within a chapter.

Before showing an example, it must be pointed out that just because an enzyme converts a potential mechanism-based inactivator into a reactive intermediate, it does not guarantee that a covalent reaction to the enzyme will take place. Pig brain γ-aminobutyric acid aminotransferase converts β-halo-γ-aminobutyric acids[66] and 4-amino-2-(X methyl)-2-butenoic acids[67] into reactive pyridoxal phosphate-bound Michael acceptors, which are released from the enzyme and hydrolyzed. Hog liver olefinic thiolester isomerase converts 3-decynoyl-*N*-acetylcysteamine to (+)-2,3-decadienyl-*N*-acetylcysteamine, an allenic thioester, which is released from the unharmed enzyme.[14] Haloperoxidase and lactoperoxidase produce α-bromoketones, compounds that are potent affinity labeling agents, from acetylenes and hydrogen peroxide,[68] and these enzymes appear to be unscathed.

As an example of how the inactivation mechanisms are categorized, consider the inactivation of γ-cystathionase by propargylglycine[69] (Structural Formula 1.1, Scheme 6).

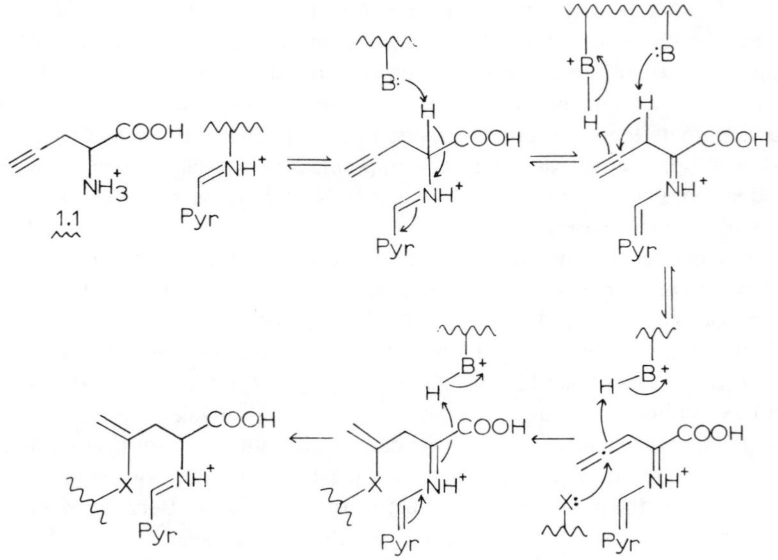

Scheme 6.   Containing Structural Formula 1.1.

The first step in the inactivation mechanism is an azallylic isomerization and, therefore, this inactivation scheme would be in Chapter 7 on isomerization reactions. The next step is an acetylene-allene isomerization and, therefore, the subcategory would be Isomerization/ Isomerization Reactions. The third step is a Michael addition of an active site nucleophile to the activated allene and, therefore, this inacitvation scheme would be found under the subheading Isomerization/Isomerization/Addition in the Isomerization/Isomerization sub-category of Chapter 7. Note that the Schiff base formation with the PLP is not relevant to the categorization of the reaction. Also consider the inactivation of histidine decarboxylase by α-trifluoromethylhistamine[70] (Scheme 7). Although the enzyme catalyzes the decarbox-ylation of histidine to histamine, this inactivator is discussed in Chapter 6 on elimination reactions, because the first step is elimination of HF. Thus, it is the first step in the inactivation mechanism, not the enzyme mechanism, that dictates the category into which the inactivator falls. Also, note that a more accurate description of the mechanism would be deprotonation to a stabilized anion, followed by elimination (i.e., an ElcB mechanism) rather than the concerted (i.e., E2) mechanism shown. However, since the deprotonation is not involved in categorization, the anion stabilization is not shown. Another typically omitted intermediate throughout the book, in order to conserve space, is the tetrahedral intermediate formed during the reaction of nucleophiles with carboxylic acid derivatives and related compounds.

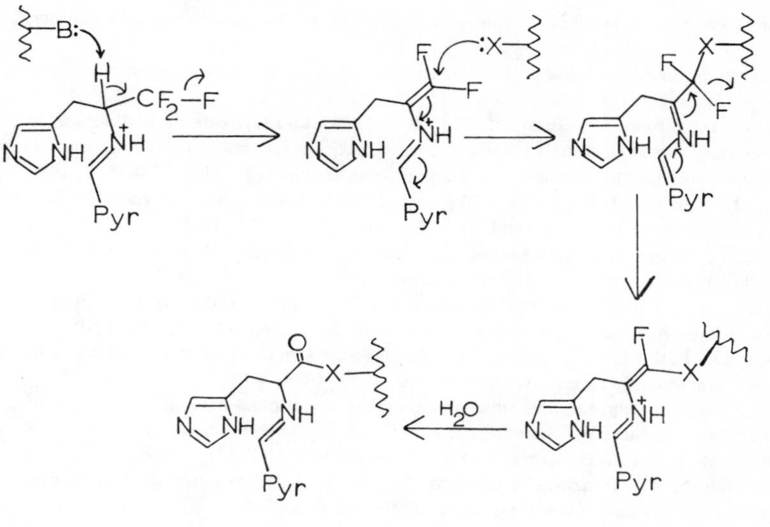

Scheme 7.

REFERENCES

1. **Wood, W. A. and Gunsalus, I. C.,** Serine and threonine deaminases of *Escherichia coli:* activators for a cell-free enzyme, *J. Biol. Chem.,* 181, 171, 1949.
2. **Dion, H. W., Fusari, S. A., Jakubowski, Z. L., Zora, J. G., and Bartz, Q. R.,** 6-Diazo-5-oxo-L-norleucine, a new tumor-inhibitory substance. II. Isolation and characterization, *J. Am. Chem. Soc.,* 78, 3075, 1956.
3. **Buchanan, J. M.,** The amidotransferases, *Adv. Enzymol.,* 39, 91, 1973.
4. **Barsky, J., Pacha, W. L., Sarkar, S., and Zeller, E. A.,** Amino oxidases. XVII. Mode of action of 1-isonicotinyl-2-isopropylhydrazine on monoamine oxidase, *J. Biol. Chem.,* 34, 389, 1959.
5. **Zeller, E. A., Sarkar, S., and Reinen, R. M.,** Amine oxidases. XIX. Inhibition of monoamine oxidase by phenylcylopropylamines and iproniazid, *J. Biol. Chem.,* 237, 2333, 1962.
6. **Kaiser, C. and Setler, P. E.,** Antidepressant agents, in *Burger's Medicinal Chemistry,* Part 3, 4th ed., Wolff, M. E., Ed., John Wiley & Sons, New York, 1981, 997.
7. **Endo, K., Helmkamp, G. M., Jr., and Bloch, K.,** Mode of inhibition of β-hydroxydecanoyl thioester dehydrase by 3-decynoyl-N-acetylcysteamine, *J. Biol. Chem.,* 245, 4293, 1970.
8. **Rando, R. R.,** Mechanism-based irreversible enzyme inhibitors, *Methods Enzymol.,* 46, 28, 1977.
9. **Massey, V., Komai, H., Palmer, G., and Elion, G. B.,** On the mechanism of inactivation of xanthine oxidase by allopurinol and other pyrazolo(3,4-*d*)-pyrimidines, *J. Biol. Chem.,* 245, 2837, 1970.
10. **Walsh, C. T., Abeles, R. H., and Kaback, H. R.,** Mechanisms of active transport in isolated bacterial membrane vesicles. X. Inactivaton of D-lactate dehydrogenase and D-lactate dehydrogenase-coupled transport in *Escherichia coli* membrane vesicles by an acetylenic substrate, *J. Biol. Chem.,* 247, 7858, 1972.
11. **Abeles, R. H. and Maycock, A. L.,** Suicide enzyme inactivators, *Acc. Chem. Res.,* 9, 313, 1976.
12. **Rando, R. R.,** Chemistry and enzymology of k$_{cat}$ inhibitors, *Science,* 185, 320, 1974.
13. **Seiler, N., Jung, M. J., and Koch-Weser, J., Eds.,** *Enzyme-Activated Irreversible Inhibitors,* Elsevier/North-Holland, Amsterdam, 1978.
14. **Miesowicz, F. M. and Bloch, K.,** Purification of hog liver isomerase. Mechanism of isomerization of 3-alkenyl and 3-alkynyl thioesters, *J. Biol. Chem.,* 254, 5868, 1979.
15. **John, R. A., Jones, E. D., and Fowler, L. J.,** Enzyme-induced inactivation of transaminases by acetylenic and vinyl analogues of 4-aminobutyrate, *Biochem. J.,* 177, 721, 1979.
16. **White, E. H., Jelinski, L. W., Politzer, I. R., Branchini, B. R., and Roswell, D. F.,** Active-site-directed inhibition of α-chymotrypsin by deaminatively produced carbonium ions: an example of suicide or enzyme-activated-substrate inhibition, *J. Am. Chem. Soc.,* 103, 4231, 1981.
17. **Schorstein, D. E., Suckling, C. J., and Wrigglesworth, R.,** 1-Chloro-3-nitropropane: an unusual latent irreversible enzyme inhibitor, *J. Chem. Res. (S),* 264, 1978.

18. **Morino, Y.**, Affinity labeling of aspartate aminotransferase and structure of its active site, *Seikagaku*, 45, 993, 1973.

19. **Rando, R. R. and de Mairena, J.**, Propargyl amine-induced irreversible inhibition of non-flavin-linked amine oxidases, *Biochem. Pharmacol.*, 23, 463, 1974.

20. **Reichart, D., Simon, A., Durst, F., Mathews, J. M., and Ortiz de Montellano, P. R.**, Autocatalytic inactivation of plant cytochrome P-450 enzymes: Selective inactivation of cinnamic acid 4-hydroxylase from *Helianthus tuberosus* by 1-aminobenzotriazole, *Arch. Biochem. Biophys.*, 216, 522, 1982.

21. **Chase, J. F. A. and Tubbs, P. K.**, Conditions for the self-catalysed inactivation of carnitine acetyltransferase, a novel form of enzyme inhibition, *Biochem. J.*, 111, 225, 1969.

22. **Béchet, J.-J., Dupaix, A., and Blagoeva, I.**, Inactivation of α-chymotrypsin by new bifunctional reagents: halomethylated derivatives of dihydrocoumarins, *Biochimie*, 59, 231, 1977.

23. **Yaouanc, J. J., Dugenet, P., and Kraus, J. L.**, Observations on the "in vitro" inhibition of dimethylglycine oxidase by "suicide-substrates," *Pharmacol. Res. Commun.*, 11, 115, 1979.

24. **King, S. and Phillips, A. T.**, Aromatic aminotransferase activity of rat brain cytoplasmic and mitochondrial aspartate aminotransferases, *J. Neurochem.*, 30, 1399, 1978.

25. **Bey, P.**, Substrate-induced irreversible inhibition of α-aminoacid decarboxylases. Application to glutamate, aromatic-L-α-aminoacid and ornithine decarboxylases, in *Enzyme-Activated Irreversible Inhibitors*, Seiler, N., Jung, M. J., and Koch-Weser, J., Eds., Elsevier/North-Holland, Amsterdam, 1978, 27.

26. **Palfreyman, M. G., Danzin, C., Jung, M. J., Fozard, J. R., Wagner, J., Woodward, J. K., Aubry, M., Dage, R. C., and Koch-Weser, J.**, Substrate-induced irreversible inhibition of aromatic, L-amino acid decarboxylase by α-difluoromethyl-DOPA, in *Enzyme-Activated Irreversible Inhibitors*, Seiler, N., Jung, M. J., and Koch-Weser, J., Eds., Elsevier/North-Holland, Amsterdam, 1978, 221.

26a. **Bitonti, A. J., Casara, P. J., McCann, P. P., and Bey, P.**, Catalytic irreversible inhibition of bacterial and plant arginine decarboxylase activities by novel substrate and product analogues, *Biochem. J.*, 242, 69, 1987.

27. **Thelander, L., Larsson, B., Hobbs, J., and Eckstein, F.**, Active site of ribonucleoside diphosphate reductase from *Escherichia coli*. Inactivation of the enzyme by 2'-substituted ribonucleoside diphosphates, *J. Biol. Chem.*, 251, 1398, 1976.

28. **Sinnott, M. L.**, Problems with terminology, *Chem. Eng. News*, 62, 2, 1984.

29. **Kraus, J.-L., Yaouanc, J.-J., and Sturtz, G.**, Rôle des substituants *N*-allyl et *N*-propargyl dans le mécanisme d'inhibition d'enzymes flavoprotéiques: *N*-déméthylase monoamineoxydase, *Eur. J. Med. Chem. Chim. Ther.*, 10, 507, 1975.

30. **Silverman, R. B. and Hoffman, S. J.**, The organic chemistry of mechanism-based enzyme inhibition: a chemical approach to drug design, *Med. Res. Rev.*, 4, 415, 1984.

31. **John, R. A.**, Enzyme-induced inactivation of pyridoxal phosphate-dependent enzymes: approaches to the design of specific inhibitors, in *Enzyme Inhibitors as Drugs*, Sandler, M., Ed., Macmillan, London, 1979, 73.

31a. **Albert, A.**, *Selective Toxicity*, 7th ed., Chapman and Hall, London, 1985, 373.

32. **Nelson, S. D.**, Metabolic activation and drug toxicity, *J. Med. Chem.*, 25, 753, 1982.

33. **Walsh, C., Cromartie, T., Marcotte, P., and Spencer, R.**, Suicide substrates for flavoprotein enzymes, *Methods Enzymol.*, 53D, 437, 1978.

34. **Wang, E. and Walsh, C.**, Suicide substrates for the alanine racemase of *Escherichia coli* B, *Biochemistry*, 17, 1313, 1978.

35. **Lienhard, G. E.**, Enzymatic catalysis and transition-state theory, *Science*, 180, 149, 1973.

36. **Wolfenden, R.**, Transition state analog inhibitors and enzyme catalysis, *Ann. Rev. Biophys. Bioeng.*, 5, 271, 1976.

37. **Lindquist, R. N.**, Design of enzyme inhibitors. Transition state analogs, in *Drug Design*, Vol. 5, Ariëns, E. J., Ed., Academic Press, New York, 1975, 24.

38. **Pauling, L.**, Molecular architecture and biological reactions, *Chem. Eng. News*, 24, 1375, 1946.

39. **Ogston, A. G.**, Activation and inhibition of enzymes, *Discuss. Faraday Soc.*, 20, 161, 1955.

40. **Morrison, J. F.**, The slow-binding and slow, tight-binding inhibition of enzyme-catalyzed reactions, *Trends Biochem. Sci.*, 7, 102, 1982.

40a. **Rich, D. H.**, Pepstatin-derived inhibitors of aspartic proteinases. A close look at an apparent transition-state analogue inhibitor, *J. Med. Chem.*, 28, 263, 1985.

40b. **Imperiali, B. B. and Abeles, R. H.**, Inhibition of serine proteases by peptidyl fluoromethyl ketones, *Biochemistry*, 25, 3760, 1986.

40c. **Stein, R. L., Strimpler, A. M., Edwards, P. D., Lewis, J. J., Mauger, R. C., Schwartz, J. A., Stein, M. M., Trainor, D. A., Wildonger, R. A., and Zottola, M. A.**, Mechanism of slow-binding inhibition of human leukocyte elastase by trifluoromethyl ketones, *Biochemistry*, 26, 2682, 1987.

41. **Baker, B. R.**, *Design of Active-Site Directed Irreversible Enzyme Inhibitors*, John Wiley & Sons, New York, 1967.

41a. **Birchmeier, W. and Christen, P.,** Syncatalytic enzyme modification: characteristic features and differentiation from affinity labeling, *Methods Enzymol.,* 46, 41, 1977.

42. **McDonald, I. A., Lacoste, J. M., Bey, P., Wagner, J., Zreika, M., and Palfreyman, M. G.,** (E)-β-(Fluoromethylene)-*m*-tyrosine: a substrate for aromatic L-amino acid decarboxylase liberating an enzyme-activated irreversible inhibitor of monoamine oxidase, *J. Am. Chem. Soc.,* 106, 3354, 1984.

43. **Danzin, C., Bey, P., Schirlin, D., and Claverie, N.,** α-Monofluoromethyl and α-difluoromethyl putrescine as ornithine decarboxylase inhibitors: *in vitro* and *in vivo* biochemical properties, *Biochem. Pharmacol.,* 31, 3871, 1982.

44. **Danzin, C., Casara, P., Claverie, N., and Grove, J.,** Effects of enantiomers of 5-hexyne-1,4-diamine on ODC, GAD, and GABA-T activities in the rat, *Biochem. Pharmacol.,* 32, 941, 1983.

45. **Wenz, A., Thorpe, C., and Ghisla, S.,** Inactivation of general acyl-CoA dehydrogenase from pig kidney by a metabolite of hypoglycin A, *J. Biol. Chem.,* 256, 9809, 1981.

46. **Washtien, W. L. and Santi, D. V.,** Assay of intracellular free and macromolecular-bound metabolites of 5-fluorodeoxyuridine and 5-fluorouracil, *Cancer Res.,* 39, 3397, 1979.

47. **Stella, V. J. and Himmelstein, K. J.,** Prodrugs and site-specific drug delivery, *J. Med. Chem.,* 23, 1275, 1980.

48. **Sjoerdsma, A., Golden, J. A., Schechter, P. J., Barlow, J. L. R., and Santi, D. V.,** Successful treatment of lethal protozoal infections with the ornithine decarboxylase inhibitor, α-difluoromethylornithine, *Trans. Assoc. Am. Phys.,* 97, 70, 1984.

49. **Kollonitsch, J. and Kahan, F. M.,** Antibacterial 3-fluoro-L- and -D-alanine, *U.S. Patent Appl.,* 60, 645, 1970; *Chem. Abstr.,* 76, 100053g, 1972.

50. **Kollonitsch, J., Patchett, A. A., Marburg, S., Maycock, A. L., Perkins, L. M., Doldouras, G. A., Duggan, D. E., and Aster, S. D.,** Selective inhibitors of biosynthesis of aminergic neurotransmitters, *Nature (London),* 274, 906, 1978.

51. **Frode, H. C. and Wilson, I. B.,** Acetylcholinesterase, in *The Enzymes,* Vol. 5, 3rd ed., Boyer, P. D., Ed., Academic Press, New York, 1971, 87.

52. **Kitz, R. and Wilson, I. B.,** Esters of methanesulfonic acid as irreversible inhibitors of acetylcholinesterase, *J. Biol. Chem.,* 237, 3245, 1962.

53. **Jung, M. J. and Metcalf, B. W.,** Catalytic inhibition of γ-aminobutyric acid-α-ketoglutarate transaminase of bacterial origin by 4-aminohex-5-ynoic acid, a substrate analog, *Biochem. Biophys. Res. Commun.,* 67, 301, 1975.

54. **Waley, S. G.,** Kinetics of suicide substrates, *Biochem. J.,* 185, 771, 1980.

55. **Silverman, R. B.,** Effect of α-methylation on inactivation of monoamine oxidase by *N*-cyclopropylbenzylamine, *Biochemistry,* 23, 5206, 1984.

56. **Waley, S. G.,** Kinetics of suicide substrates. Practical procedures for determining parameters, *Biochem. J.,* 227, 843, 1985.

57. **Tatsunami, S., Yago, N., and Hosoe, M.,** Kinetics of suicide substrates. Steady-state treatments and computer-aided exact solutions, *Biochim. Biophys. Acta,* 662, 226, 1981.

57a. **Galvez, J., Varon, R., and Garcia Carmona, F.,** III. Kinetics of enzyme reactions with inactivation steps, *J. Theor. Biol.,* 89, 37, 1981.

57b. **Tudela, J., García Cánovas, F., Varón, R., García Carmona, F., Gálvez, J., and Lozano, J. A.,** Transient-phase kinetics of enzyme inactivation induced by suicide substrates, *Biochim. Biophys. Acta,* 912, 408, 1987.

57c. **García Cánovas, F., Tudela, J., Martinez Madrid, C., Varón, R., García Carmona, F., and Lozano, J. A.,** Kinetic study on the suicide inactivation of tyrosinase induced by catechol, *Biochim. Biophys. Acta,* 912, 417, 1987.

58. **Meloche, H. P.,** Bromopyruvate inactivation of 2-keto-3-deoxy-6-phosphogluconic aldolase. I. Kinetic evidence for active site specificity, *Biochemistry,* 6, 2273, 1967.

59. **Childs, R. E. and Bardsley, W. G.,** Time-dependent inhibition of enzymes by active-site-directed reagents. A theoretical treatment of the kinetics of affinity labeling, *J. Theor. Biol.,* 53, 381, 1975.

60. **Kuo, D. J. and Jordan, F.,** Active-site directed irreversible inactivation of brewers' yeast pyruvate decarboxylase by the conjugated substrate analogue (E)-4-(4-chlorophenyl)-2-oxo-3-butenoic acid: development of a suicide substrate, *Biochemistry,* 22, 3735, 1983.

61. **Fitzpatrick, P. F., Flory, D. R., Jr., and Villafranca, J. J.,** 3-Phenylpropenes as mechanism-based inhibitors of dopamine β-hydroxylase: evidence for a radical mechanism, *Biochemistry,* 24, 2108, 1985.

62. **Fitzpatrick, P. F. and Villafranca, J. J.,** The mechanism of inactivation of dopamine β-hydroxylase by hydrazines, *J. Biol. Chem.,* 261, 4510, 1986.

63. **Laidler, K. J. and Peterman, B. F.,** Temperature effects in enzyme kinetics, *Methods Enzymol.,* 63A, 234, 1979.

64. **Tipton, K. F. and Dixon, H. B. F.,** Effects of pH on enzymes, *Methods Enzymol.,* 63A, 183, 1979.

65. **Owens, M. S. and Barden, R. E.,** S-(4-Bromo-2,3-dioxobutyl)-CoA: an affinity label for certain enzymes that bind acetyl-CoA, *Arch. Biochem. Biophys.,* 187, 299, 1978.

65a. **Covey, D. F., McMullan, P. C., Weaver, A. J., and Chien, W. W.,** Inactivation of *Streptomyces* hydrogenans 20β-hydroxysteroid dehydrogenase by an enzyme-generated ethoxyacetylene ketone in the presence of a thiol scavenger, *Biochemistry,* 25, 7288, 1986.

66. **Silverman, R. B. and Levy, M. A.,** Substituted 4-aminobutanoic acids. Substrates for γ-aminobutyric acid α-ketoglutaric acid aminotransferase, *J. Biol. Chem.,* 256, 11565, 1981.

67. **Silverman, R. B., Durkee, S. C., and Invergo, B. J.,** 4-Amino-2-(*X* methyl)-2-butenoic acids: substrates and potent inhibitors of γ-aminobutyric acid aminotransferase, *J. Med. Chem.,* 29, 764, 1986.

68. **Geigert, J., Neidleman, S. L., and Dalietos, D. J.,** Novel haloperoxidase substrates. Alkynes and cyclopropanes, *J. Biol. Chem.,* 258, 2273, 1983.

69. **Washtien, W. and Abeles, R. H.,** Mechanism of inactivation of γ-cystathionase by the acetylenic substrate analogue propargylglycine, *Biochemistry,* 16, 2485, 1977.

70. **Metcalf, B. W., Holbert, G. W., and Lippert, B. J.,** α-Trifluoromethylhistamine: a mechanism-based inhibitor of mammalian histidine decarboxylase, *Bioorg. Chem.,* 12, 91, 1984.

Chapter 2

PROTONATION AND DEPROTONATION REACTIONS*

I. INTRODUCTION

Mechanism-based enzyme inactivation which is initiated by proton donation involves, almost exclusively, reactions of diazoketones. A distinction is made in this chapter between diazo compounds which require the presence of Cu(II) or Ag(I) for inactivation and those which do not. Because it is known that copper catalyzes the decomposition of diazoketones to carbenoid species,[1] those inactivations that require Cu(II) are considered to be affinity-labeling agents and are mentioned only briefly. A more detailed discussion of proton-initiated inactivation is provided for those compounds that function in the absence of Cu(II).

II. PROTONATION/SUBSTITUTION

A. Diazoketones that Require Cu(II) for Activity

The compound that most exemplifies this category is diazoacetyl-D,L-norleucine methyl ester. It has been used to inactivate pepsin from various sources,[2-4] pig thyroid acid proteinase,[5] peptidase A, the acid protease from *Penicillium janthinellum*,[6,7] other acid proteases from numerous sources,[8-13] and bovine rennin.[14,15] D,L-1-Diazo-3-tosylamido-2-heptanone and L-1-diazo-3-tosylamido-4-phenyl-2-butanone also inactivate Type A and B acid proteases from *Aspergillus niger* var. *macrosporus*;[13] the latter compound inactivates swine pepsin.[16] *N*-Diazoacetyl-*N'*-2,4-dinitrophenylethylenediamine,[17] α-diazo-*p*-bromoacetophenone,[18] *N*-diazoacetyl-L-phenylalanine methyl ester[19,20] and various other diazoacetyl compounds[21,22] also are inactivators of pepsin.

B. Diazoketones that Do Not Require Cu(II) for Activity

The general mechanism of inactivation in this category involves enzyme-catalyzed protonation of the diazoketone (Structural Formula 2.1) followed by active-site nucleophilic attack (Scheme 1).

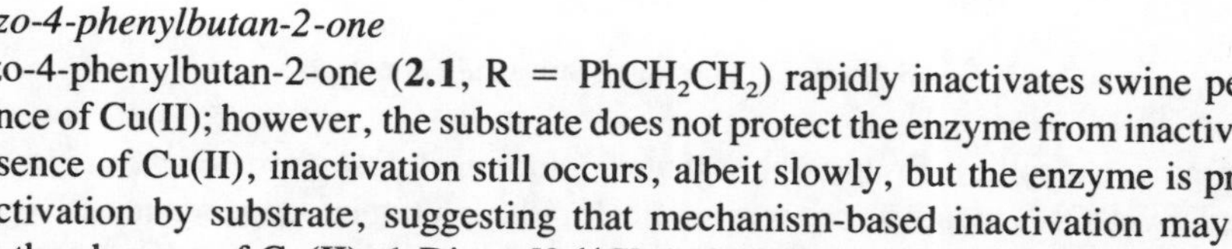

Scheme 1. Containing Structural Formula 2.1.

1. 1-Diazo-4-phenylbutan-2-one

1-Diazo-4-phenylbutan-2-one (**2.1**, R = PhCH$_2$CH$_2$) rapidly inactivates swine pepsin in the presence of Cu(II); however, the substrate does not protect the enzyme from inactivation.[23] In the absence of Cu(II), inactivation still occurs, albeit slowly, but the enzyme is protected from inactivation by substrate, suggesting that mechanism-based inactivation may be operative in the absence of Cu(II). 1-Diazo-[2-^{14}C]-4-phenylbutan-2-one inactivation results in the incorporation of 1 mol of [^{14}C] per mole of enzyme; the radioactivity is not released by hydroxylamine or thioethanol at pH 5.5.[23] At slightly basic pH, however, the radioactivity

* A list of abbreviations and shorthand notations can be found prior to Chapter 1.

is released; the recovered radioactive compound is 1-hydroxy-4-phenylbutan-2-one.[24] Peptic digestion of the [^{14}C]-labeled pepsin yields a radioactive tetrapeptide identified by Fry et al.[24] as Ile-Val-Asp-Thr, in which radioactivity is attached to the β-carbonyl group of the aspartate residue (Scheme 1, X = CO_2^-). Base-catalyzed hydrolysis of this keto ester would produce the observed 1-hydroxy-4-phenylbutan-2-one.

2. 6-Diazo-5-oxo-L-norleucine (DON)

The general synthetic routes to diazoketones are exemplified by the syntheses of one of the most well-studied diazoketone mechanism-based inactivators, namely, 6-diazo-5-oxo-L-norleucine[25] of DON (Structural Formula 2.2, Scheme 2), a natural product originally isolated from an unidentified *Streptomyces*.[26] Because of the structural similarity of DON to that of glutamine, many glutamine-dependent enzymes are inactivated by this compound. 5-Phosphoribosyl pyrophosphate amidotransferase from chicken liver undergoes time-dependent inactivation by DON.[27] D-DON, *N*-trifluoroacetyl-L-DON methyl ester, and 5-diazo-4-oxo-D,L-norvaline does not inactivate the enzyme. The K_I for DON decreases by two orders of magnitude in the presence of the complementary substrates, phosphoribosyl pyrophosphate and Mg(II). When [6-^{14}C]DON was used by Hartman[27,28] as the inactivator, the rate of inactivation corresponded to the rate of incorporation of radioactivity. One mole of radioactivity is incorporated per active site; prolonged incubation does not result in further incorporation of radioactivity. If the enzyme is first treated with the sulfhydryl reagent, *p*-hydroxymercuribenzoate, no incorporation of radioactivity from [6-^{14}C]DON occurs, suggesting that an essential active site cysteine may be labeled during inactivation by DON. Reactions catalyzed by this enzyme that do not utilize glutamine are not inhibited by DON (actually, they are accelerated), supporting the specificity of the DON interaction with the glutamine binding site.

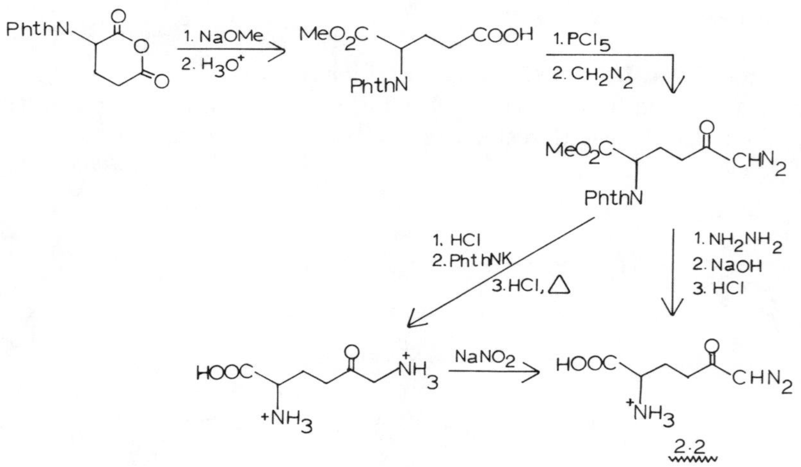

Scheme 2.   Containing Structural Formula 2.2.

Glutamine phosphoribosylpyrophosphate amidotransferase from *Escherichia coli* is inactivated by DON; inactivation is specific for the glutamine-dependent activity.[29] Using limiting amounts of DON, 0.93 mol of DON per mole of subunit results in complete enzyme inactivation. Because of the dependence of inactivation on the presence of Mg(II) and 5-phosphoribosyl-1-pyrophosphate, Messenger and Zalkin[29] suggested ordered binding. This same enzyme from cloned *E. coli pur F* also is inactivated by [6-^{14}C]DON with incorporation of 1 mol of radioactivity per subunit.[30] A cyanogen bromide peptide labeled with [6-^{14}C]DON was sequenced, and the amino-terminal cysteine residue in the enzyme was shown to be alkylated. This supports the suggestion by Tso et al.[30] that the N-terminal cysteine residue is involved in glutamine amide transfer. Similar results as described above were observed by Vollmer et al.[31] for [6-^{14}C]DON inactivation of glutamine phosphoribosylpyrophosphate amindotransferase from *Bacillus subtilis*. When the labeled enzyme is submitted to automated Edman degradation, all the radioactivity is released in the first cycle, thus supporting N-terminus alkylation. Although glutamine-dependent phosphoribosylamine formation and glutaminase activity are abolished by DON, the ammonia-dependent reaction and 5-phosphoribosyl-1-pyrophosphate hydrolysis reaction is not inhibited by DON.

Preincubation of γ-glutamyl transpeptidase from rat kidney with DON leads to time-dependent enzyme inactivation;[32-34] the affinity of the enzyme for L-DON is only four to five times greater than that for D-DON.[34] γ-Glutamyl substrates, but not acceptor substrates, protect the enzyme.[32] The rate of inactivation is accelerated by the presence of maleate,[33] which is known to enhance the glutaminase activity of this enzyme. Inactivation by [6-^{14}C]DON[32,33] or by [1,2,3,4,5-^{14}C]DON[33] results in stoichiometric incorporation of radioactivity after dialysis and denaturation. Only the light subunit of the enzyme (the enzyme is comprised of two nonidentical glycopeptide subunits) is labeled,[32] indicating that the γ-glutamyl binding site is on this subunit. Since the enzyme is not inactivated by various sulfhydryl reagents, and performic acid does not release the radioactivity incorporated by [6-^{14}C]DON inactivation, it was suggested by Inoue et al.[34] that a sulfhydryl group is not involved. An ester linkage (Asp or Glu) was ruled out because alkali does not release the radioactivity either; however, phenylmethanesulfonyl fluoride, a specific reagent for serine residues does inactivate the enzyme.[34] Similar results were obtained with γ-glutamyl transpeptidase from human kidney.[32,35] [6-^{14}C]DON also inactivates γ-glutamyl transpeptidase from rat renal brush border membranes with the incorporation of 1 mol of radioactivity per 102,000 g of detergent-solubilized enzyme or 64,000 g of papain-solubilized enzyme.[36] Both enzyme forms are composed of two nonidentical subunits. DON also inactivates the γ-glutamyl transpeptidase of rat ascites tumor cell AH-130[37] and LY-5.[38] Inactivation of the rat ascites tumor cell LY-5 has no effect on the rate of transport of amino acids into the cells, indicating that this enzyme is not involved with amino acid transport.[38]

Cytidine triphosphate synthetase from *E. coli* B is inactivated by DON with incorporation of only 1 mol of inactivator per mole of dimeric enzyme,[39] even in the presence of GTP, which activates the glutamine reaction and enhances the rate of inactivation by DON. This half-of-the-sites reactivity was postulated by Levitski et al.[40] to be the result of negative cooperativity in which covalent labeling by one DON molecule induces a conformational change which is transmitted to the adjoining identical subunit where no reaction with DON occurs. Only the glutamine activity, not the ammonia activity, is abolished, indicating that the reaction occurs in the glutamine binding site. Of the ten sulfhydryls in the native enzyme, only one is lost after DON inactivation,[39,40] presumably a cysteine in the glutamine binding site.

Guanosine 5'-monophosphate synthetase from *E. coli* B-96 also is inactivated by DON, but irreversible binding depends upon the presence of the substrates XMP, ATP, and Mg(II).[41] NaDodSO$_4$-polyacrylamide gel electrophoresis of enzyme labeled with [6-^{14}C]DON gives one radioactive band corresponding to a molecular weight of 60,000 ± 6,000. Nearly parallel rates of inactivation of the glutamine and ammonia-dependent activities are obtained.[42] These results contrast those of Patel et al.[43,44] who have studied the DON inactivation of an enzyme from *E. coli* B-96 that they call xanthosine 5'-phosphate amidotransferase. This enzyme shows greater sensitivity of the glutamine-dependent activity over the ammonia-dependent activity towards DON, although DON is competitive with both glutamine and ammonia.[44] XMP, ATP, and Mg(II) also are required for DON inactivation.[43,44] It is believed that the substrates expose active site residues which DON alkylates. After DON inactivation, there are six less titratable sulfhydryl groups out of a total of 22. It is not clear if all six are alkylated by DON or only some are, the rest being sterically hindered from titration with DTNB. It was suggested by Lee and Hartman[41] that GMP synthetase and XMP amidotransferase are the same protein.

DON also inactivates the bifunctional, regulatory enzyme aggregate, anthranilate synthetase — anthranilate 5-phosphoribosylpyrophosphate phosphoribosyltransferase from *Salmonella typhimurium*.[45] The glutamine-dependent anthranilate synthetase activity is affected more so than the ammonia-dependent activity; anthranilate synthetase Component I is not inactivated at all, suggesting that it does not have a glutamine binding site. Since inactivation is dependent upon the presence of the substrate, chorismate, an ordered binding mechanism has been proposed by Nagano et al.[45] Inactivation by [6-14C]DON followed by gel electrophoresis indicates that most of the radioactivity is attached to the anthranilate 5-phosphoribosylpyrophosphate phosphoribosyltransferase polypeptide chain. Two moles of radioactivity bind per mole of enzyme concomitant with the alkylation of two cysteine residues. Similar results as above also are obtained when DON inactivates anthranilate synthetase from *Serratia marcescens*.[46] The anthranilate synthetase from *Pseudomonas putida* undergoes time-dependent inactivation by DON; only the aminotransferase reaction, not ammonia-dependent reaction, is affected.[47] The rate of inactivation is accelerated by Mg(II)[48] and chorismate.[47,48] Activity does not return upon dialysis, but addition of the 2-mercaptoethanol results in slow reactivation.[47] This may suggest that a methionine residue is involved.[49] As in the case of the *S. typhimurium* enzyme,[45] the primary site of action is Component II (called AS II), not AS I.[47] However, both AS I and II must be incubated with DON to achieve full inactivation. This suggests that either chorismate must bind prior to glutamine, or chorismate binding to AS I induces a conformational change allowing AS II to react with glutamine or DON.[47] The latter hypothesis appears to be more compatible with the data.

Glutaminase A from *E. coli* is another enzyme inactivated by DON;[50,51] both active sites are labeled by [6-14C]DON, and [14C] incorporation parallels enzyme inactivation.[50] In addition to enzyme inactivation, DON acts as a substrate; 70 ± 10 mol of DON are hydrolyzed for each mole that reacts with the enzyme[51] (i.e., the partition ratio is 70 ± 10). The products of hydrolysis are glutamate and diazomethane, identified by Hartman and McGrath[51] as methanol and methyl benzoate in the presence of benzoic acid. In order to determine if inacitvation results from reaction with DON or with the diazomethane generated, the enzyme was inactivated with [6-14C, 4-3H]DON, tryptic digested, and the peptides separated. Since the ratio of 3H/14C remains constant throughout purification, it indicates that DON is responsible. A mechanism to account for the partitioning of DON is shown in Scheme 3. A tetrahedral common intermediate has been suggested by Sinnott[49] in an alternate mechanism (Scheme 4).

DON also inactivates L-glutamine:D-fructose-6-phosphate aminotransferase from rat liver in a time-dependent manner.[52] Gel filtration does not restore enzyme activity.

Yeast nicotinamide adenine dinucleotide synthetase,[53] *Achromobacteraceae* glutaminase-asparaginase,[54] and 2-formamido-N-ribosylacetamide 5′-phosphate:L-glutamine amido ligase (adenosine diphosphate)[55] are irreversibly inactivated by DON as well. DON is a time-dependent inactivator of glucosamine-6-phosphate synthetase from *Salmonella typhimurium* LT2.[55a]

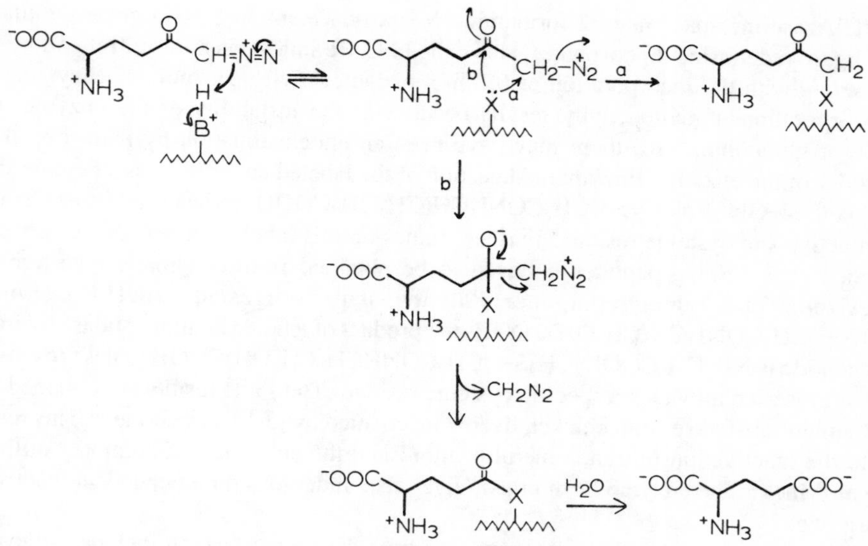

Scheme 3.

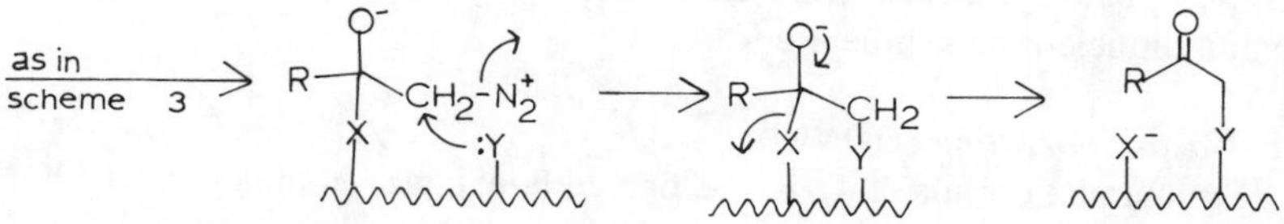

Scheme 4.

3. *O-Diazoacetyl-L-serine (Azaserine)*

The isosteric interchange of the 4-methylene group of DON for an oxygen atom gives another glutamine analogue, *O*-diazoacetyl-L-serine (Structural Formula 2.3, azaserine), which also was originally isolated from an unidentified *Stretpomyces*.[56] Its structure was identified,[57] and its synthesis was carried out[58] as shown in Scheme 5.

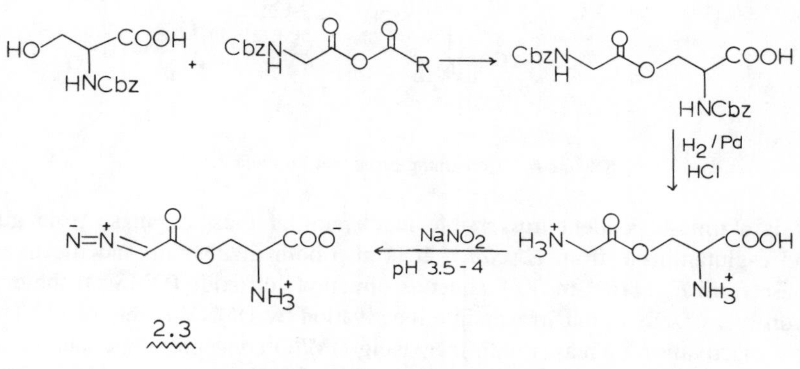

2.3

Scheme 5. Containing Structural Formula 2.3.

[2-^{14}C]Azaserine inactivates 2-formamido-*N*-ribosylacetamide 5′-phosphate:L-glutamine amido-ligase (adenosine diphosphate), also called FGAR amidotransferase, from *Salmonella typhimurium* with the incorporation of 0.3 mol of radioactivity per mole of enzyme.[59] The low incorporation of radioactivity results because of the instability of the enzyme in the absence of glutamine; also, there may have been an underestimation by Dawid et al.[59] of the activity of the enzyme. Enzymatic digestion of the labeled enzyme gives a peptide shown to be Ala–Leu–Gly–Val–Cys–[14]CH$_2$CONHCH(CH$_2$OH)COOH, indicating that azaserine labels an active site cysteine residue. Further studies on this labeled enzyme were carried out by French et al.[60] Two peptides believed to be obtained from enzymatic digestion were synthesized. The major component of the papain-digested labeled enzyme is Val–Cys–[14]CH$_2$CONHCH(CH$_2$OH)COOH; the product of leucine aminopeptidase hydrolysis of this peptide is NH$_2$CH(COOH)CH$_2$S–[14]CH$_2$CONHCH(CH$_2$OH)COOH. These results confirm that azaserine alkylates a specific cysteine residue. The same results are obtained when FGAR amidotransferase from chicken liver is inactivated by [2-^{14}C]azaserine.[61] This reaction leads to the inactivation of glutamine utilization, and the activation of ammonia utilization by the enzyme.[62] The enzyme from pigeon liver also undergoes time-dependent inactivation by azaserine.[55]

Other enzymes inactivated by azaserine in a time-dependent fashion include 5-phosphoribosylpyrophosphate amidotransferase from chicken liver,[27] γ-glutamyl transpeptidase from rat kidney[32] and human kidney,[35] anthranilate synthetase from *Pseudomonas putida*,[47] glucosamine-6-phosphate synthetase from *Salmonella typhimurium* LT2,[55a] and yeast nicotinamide adenine dinucleotide synthetase.[53]

4. 5-Diazo-4-oxo-L-norvaline (DONV)

DONV (Structural Formula 2.4) is the one-carbon lower homologue of DON; consequently, it is an analogue of aspartic acid. Unlike DON and azaserine, DONV is not a natural product; its synthesis[63,64] is shown in Scheme 6.

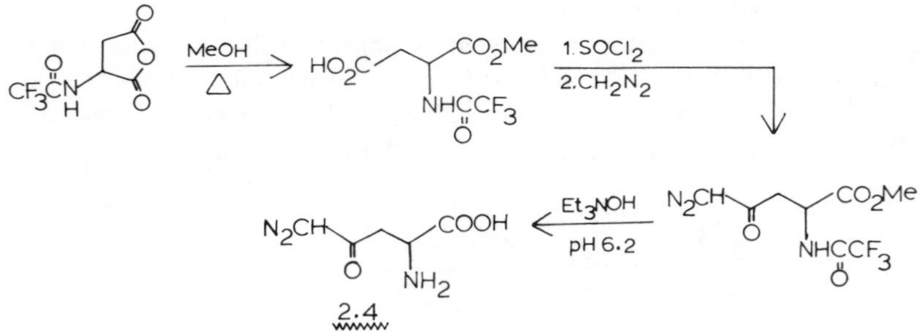

Scheme 6.   Containing Structural Formula 2.4.

DONV is a time-dependent irreversible inactivator of L-asparaginase from guinea pig serum and L-glutaminase from *E. coli*.[64] It is also both a substrate and inactivator of L-asparaginase from *E. coli*.[65] In 50% aqueous dimethyl sulfoxide (DMSO), the enzyme no longer hydrolyzes DONV, but irreversible inactivation by DONV is retained.[65] The rate of irreversible inactivation increases with increasing DMSO concentration concomitant with a decrease in the rate of asparagine and DONV hydrolysis.[66] Lachman and Handschumacher[66] suggested that the lower hydrolysis rate is the result of less water in the active site with increasing DMSO concentrations. Since water and DMSO form a 2:1 association complex, in 50% DMSO there is only 25% free water. Inactivation with [5-^{14}C]DONV results in the incorporation of 4 mol of radioactivity per mole of enzyme after acid precipitation.[65] This is consistent with polyacrylamide gel electrophoresis results which indicate that the enzyme is a tetramer of identical subunits. Chymotryptic digestion of the labeled enzyme gives two radioactive peptides, one was found by Peterson et al.[67] to be a degradation fragment of the other. The labeled decapeptide isolated (Val–Gly–Ala–Met–Arg–Pro–Ser–Thr–Ser–Met) corresponds to residues 111 to 120 of the enzyme. On the basis of the failure of carboxypeptidase C to degrade the peptide past the C-terminus methionine, the DONV is most likely attached to serine-119.[67] Model reactions by Chang et al.[68] of DONV with protected serine and threonine in the presence of boron trifluoride indicate that it is reasonable for DONV to alkylate the hydroxyl group of either serine or threonine in L-asparaginase.

5. Benzyloxycarbonyl-L-phenylalanyldiazomethylketone and Related Peptides

In an effort to prepare protease-specific inactivators, benzyloxycarbonyl-protected L-phenylalanyldiazomethylketone (Cbz–PheCHN$_2$) and related peptides were synthesized by Shaw[69] as shown in Scheme 7.

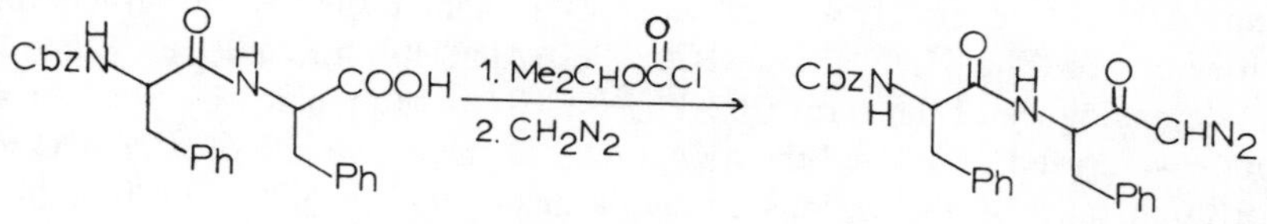

Scheme 7.

Porcine pepsin is inactivated by Cbz–PheCHN$_2$.[70] Cu(II) is not required, but inactivation occurs much more rapidly in its presence. In the absence of Cu(II) a 1:1 stoichiometry of inactivator to enzyme was observed. Changes in the ORD spectrum of inactivated pepsin indicate that attachment is to an active site amino acid residue involved in producing a conformational change in the enzyme.

Cbz–PheCHN$_2$ and Cbz–Phe–PheCHN$_2$ are pseudo first-order, time-dependent inactivators of papain, even in the presence of β-mercaptoethanol.[71] The rate of inactivation by Cbz–PheCHN$_2$ is pH-independent between pH 4.0 and 5.5, increases to a maximum at pH 6.5, then decreases sharply at higher pH values. Inactivation is accompanied by the loss of one cysteine residue. The pH-rate profile is similar for the dipeptide analogue, but it is 200 times more effective as an inactivator than is Cbz–PheCHN$_2$. With the use of Cbz–Phe–Phe[^{14}C]CHN$_2$, 1.02 mol of [^{14}C] is incorporated per mole of enzyme. The mechanism proposed by Leary et al.[71] is that shown in Scheme 1 (X = Cys-SH). A modified mechanism has been proposed by Brocklehurst and Malthouse[72] to account for the effect of pH change on the rate of inactivation (Scheme 8).

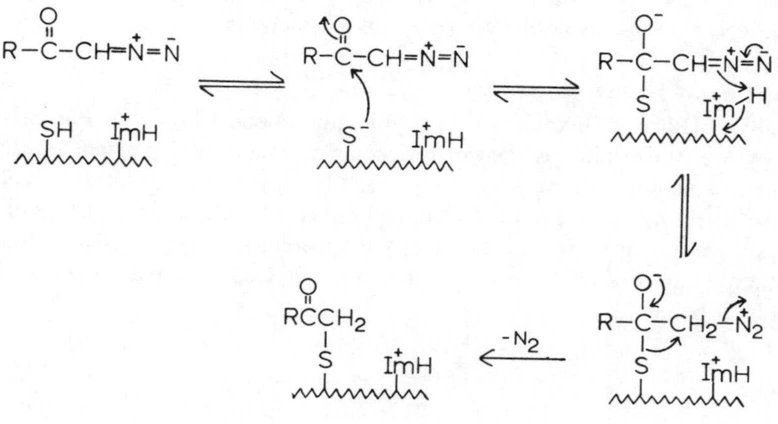

Scheme 8.

The thiol protease cathepsin B_1 from bovine spleen also is inactivated by Cbz–PheCHN$_2$ and the corresponding dipeptide, Cbz–Phe–PheCHN$_2$; the dipeptide is much more potent.[73] The presence of β-mercaptoethanol does not protect the enzyme. Several Cbz-protected di- and tripeptide diazomethyl ketones were prepared and tested as inactivators of cathepsin B from bovine spleen and chymotrypsin.[74] Chymotrypsin is not inactivated by any of these compounds, although the diazo ketone is decomposed. Cathepsin B, however, is inactivated by all seven of the compounds tested; the most potent compound is Cbz–Phe–AlaCHN$_2$. It appears that it is desirable to have a phenylalanyl residue in the pentultimate position. When Cbz–Phe–AlaCHN$_2$ is used in tissue culture, intracellular cathepsin B is almost completely inactivated prior to the inhibition of protein degradation.[75]

Cathepsin L from rat liver lyosomes is rapidly inactivated by Cbz–Phe–PheCHN$_2$ and Cbz–Phe–AlaCHN$_2$ ($t_{1/2}$ is 210 and 224 sec, respectively, at 5 nM).[76] The affinity of Cbz–Phe–AlaCHN$_2$ for cathepsin L is 2000-fold higher than for cathepsin B from rat liver and human liver. Cbz–Phe–PheCHN$_2$ is a selective inactivator of cathepsin L at low concentrations and reacts reversibly with cathepsin B from three species. Because of the differing specificities of Cbz–Phe–PheCHN$_2$ and Cbz–Phe–AlaCHN$_2$ for cathepsin B and L, a series of dipeptide analogues of the form Cbz–Phe–XCHN$_2$ was prepared by Shaw et al.[77] These compounds were used to defined the topography of the active site of beef spleen cathepsin B. Inactivation rates by the various dipeptide analogues correlate with binding affinities. Although alkylation of cathepsin B is not demonstrated, by analogy with papain,[71] it is assumed that inactivation is the result of active site alkylation. The best inactivator tested is the compound with X = *O*-benzylthreoninyl.[77]

A series of peptidyl diazomethyl ketones of the form Cbz–peptideCHN$_2$ was prepared and tested by Green and Shaw[78] as inactivators of clostripain, streptococcal proteinase, bovine spleen cathepsin B and C, β-trypsin, chymotrypsin, elastase, cathepsin D, thermolysin, and glyceraldehyde 3-phosphate dehydrogenase. Cbz–LysCHN$_2$ is the best inactivator of clostripain, Cbz–Ala–Phe–AlaCHN$_2$ is best for steptococcal proteinase, Cbz–Phe–Ala–CHN$_2$ for cathepsin B, and Gly–PheCHN$_2$ for cathepsin C. Cbz–Phe–AlaCHN$_2$ is inert to clostripain. In general, compounds which do not satisfy amino acid specificity requirements of an enzyme do not inactivate it, or do so very slowly. None of the compounds inactivate the serine proteases (chymotrypsin, trypsin, and elastase), a carboxyl protease (cathepsin D), the metalloprotease, thermolysin, or a thiolcontaining nonprotease, glyceraldehyde 3-phosphate dehydrogenase. Furthermore, all of the compounds are stable to β-mercaptoethanol and glutathione. Therefore, this approach to the design of specific thiol protease inactivation appears to be promising.

A prolyl endopeptidase in the soluble fraction of murine peritoneal macrophages is rapidly inactivated by Cbz–Ala–Ala–ProCHN$_2$.[79] Cbz–Phe–AlaCHN$_2$, Cbz–Ala–AlaCHN$_2$, and Cbz–LysCHN$_2$ also inactivate the enzyme, but at much slower rates.

6. 3-Keto-4-diazo-5α-dihydrosteroids

(5α, 20R)-4-Diazo-21-hydroxy-20-methylpregnan-3-one (Structural Formula 2.5), synthesized by Metcalf et al.[80] as shown in Scheme 9, is a time-dependent inactivator of rat prostate microsomal testosterone 5α-reductase.[81] However, because of the instability of the enzyme, reversibility could not be determined either by dialysis or gel filtration. The compound has a great affinity for the enzyme; the apparent K_I is 35 nM. No inhibition of 3α-hydroxysteroid oxidoreductase of rat prostate was detected. The mechanism of inactivation is shown in Scheme 1.

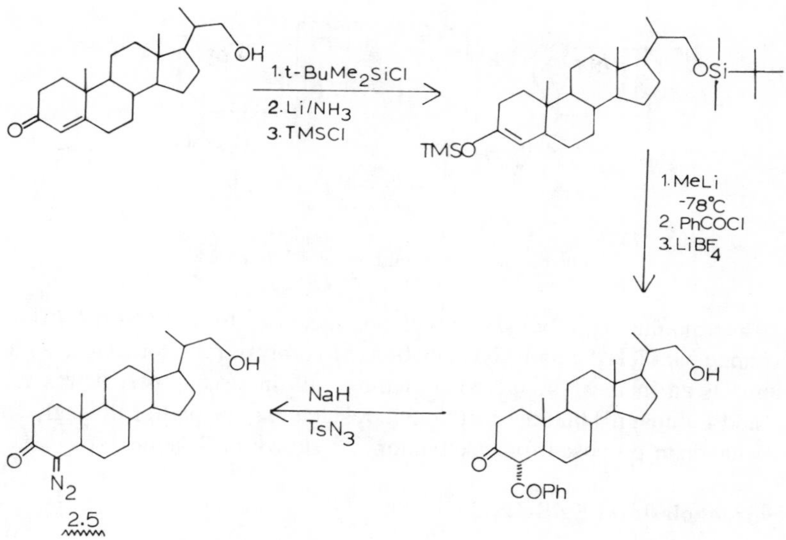

Scheme 9.   Containing Structural Formula 2.5.

C. Other Diazo Inactivators
1. 2,6-Anhydro-1-diazo-1-deoxy-D-glycero-L-manno-heptitol

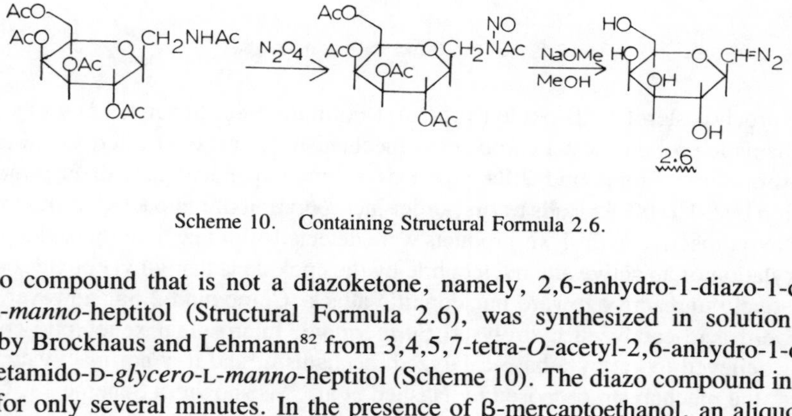

Scheme 10.   Containing Structural Formula 2.6.

A diazo compound that is not a diazoketone, namely, 2,6-anhydro-1-diazo-1-deoxy-D-*glycero*-L-*manno*-heptitol (Structural Formula 2.6), was synthesized in solution without isolation by Brockhaus and Lehmann[82] from 3,4,5,7-tetra-*O*-acetyl-2,6-anhydro-1-deoxy-1-nitrosoacetamido-D-*glycero*-L-*manno*-heptitol (Scheme 10). The diazo compound in solution is stable for only several minutes. In the presence of β-mercaptoethanol, an aliquot of this solution inactivates β-D-galactosidase from *E. coli*; dialysis does not regenerate enzyme activity. The specificity of the inactivator is supported by the observation that no inactivation of β-glucosidase from sweet almond occurs. When [7-³H]-inactivator is used, tritium is incorporated into the enzyme.[83] About 50% of the label is removed by treatment with hydroxylamine, suggesting an ester linkage.[83] The radioactive product released by this treatment was identified as 2,6-anhydro-D-*glycero*-L-*manno*-heptitol (Structural Formula 2.7, Scheme 11). All but 10% of the remaining radioactivity is detached from the enzyme by heating in buffer; the product released is 2,6-anhydro-1-deoxy-1-*S*-methyl-1-thio-D-*glycero*-L-*manno*- heptitol (Structural Formula 2.8, Scheme 11), suggesting that a methionine residue also is alkylated.

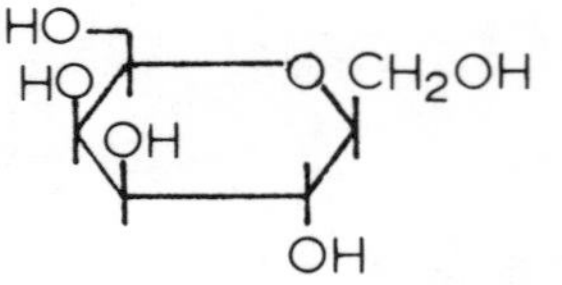

Scheme 11.   Structural Formulas 2.7 and 2.8.

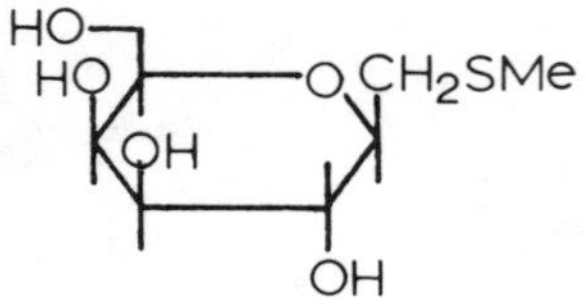

The ratio of methionine to carboxylate alkylation increases from 4:5 at pH 7.0 to 7:4 at pH 8.0. This change parallels the transglycosylation to hydrolysis ratio for lactose with increasing pH and suggests an increase in aglycone binding with increasing pH. It was suggested by Brockhaus and Lehmann[83] that the methionine residue may be part of this aglycone binding site. The mechanism proposed for inactivation[82] is shown in Scheme 1.

D. 7-Dehydrocholesterol 5,6β-Oxide

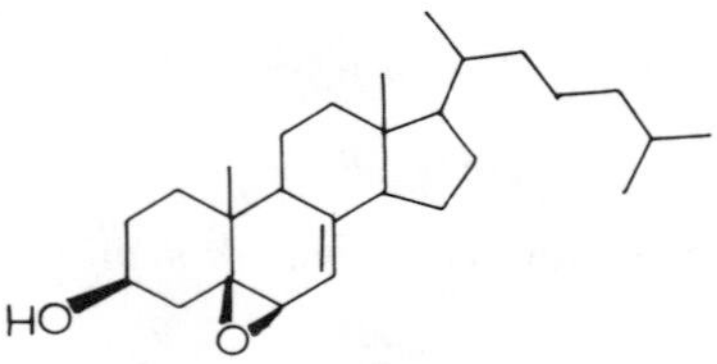

Scheme 12.   Structural Formula 2.8a.

7-Dehydrocholesterol 5,6β-oxide (Structural Formula 2.8a, Scheme 12) was synthesized as a mechanistic probe to test a carbocation mechanism for rat liver microsomal cholesterol oxide hydrolase.[83a] Compound **2.8a** produces a time-dependent loss of enzyme activity. Inactivation by 3-[^{3}H]-**2.8a** leads to first-order incorporation of radioactivity into microsomal protein. No radioactive hydrolysis products were detected after inactivation. Evidence against direct alkylation of an active site nucleophile by the epoxide is that vinyl epoxides are known to be relatively unreactive toward nucleophilic attack. Compound **2.8a,** however, is highly labile toward acid-catalyzed hydrolysis. Furthermore, microsomal xenobiotic epoxide hydrolase is believed to act by a base-catalyzed mechanism, and it is not inactivated by **2.8a.** Therefore, the mechanism proposed by Nashed et al.[83a] is shown in Scheme 13. Compound **2.8a** was synthesized by Michaud et al.[83b] by the route shown in Scheme 14.

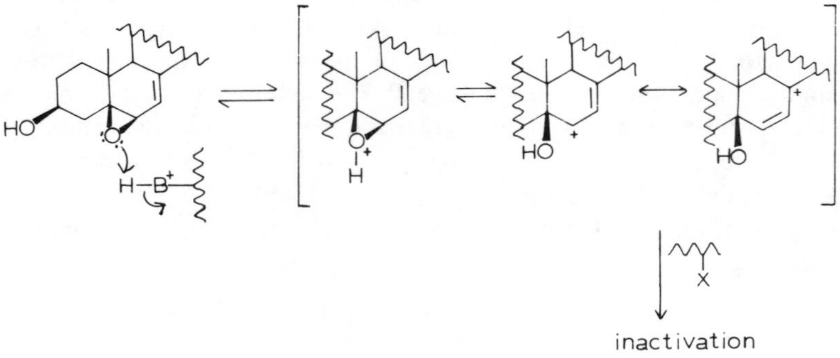

inactivation

Scheme 13.

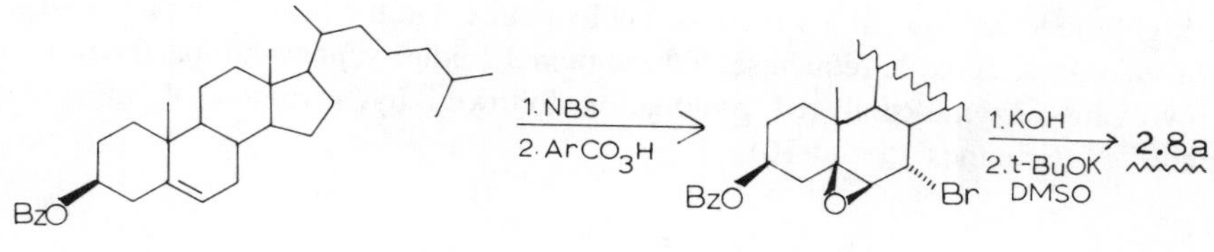

Scheme 14.

E. 2-Amino-7,8-dihydro-6-hydroxymethyl-7-spirocyclopropylpteridin-4(3H)-one

The first mechanism-based inactivator of dihydrofolate reductase, 2-amino-7,8-dihydro-6-hydroxymethyl-7-spirocyclopropylpteridin-4(3H)-one (**2.8b**) was reported by Haddow et al.[83c] to be activated by protonation (Scheme 14A), similar to the catalytic mechanism of the enzyme. This also is the first 2-amino-4-oxo pteridine to be an inhibitor of the enzyme. Compound **2.8b** was synthesized as shown in Scheme 14B.

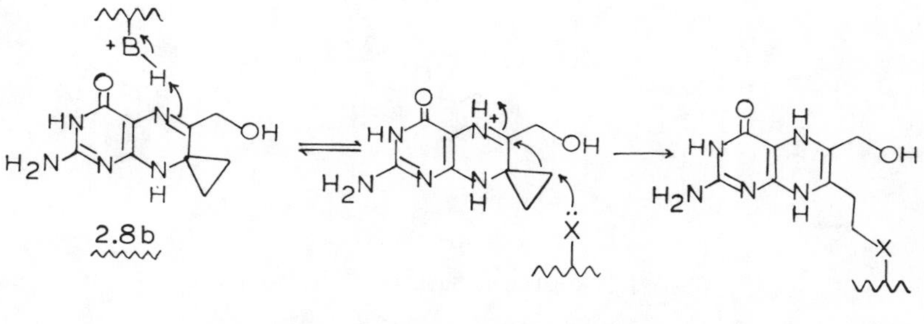

Scheme 14A.   Containing Structural Formula 2.8b.

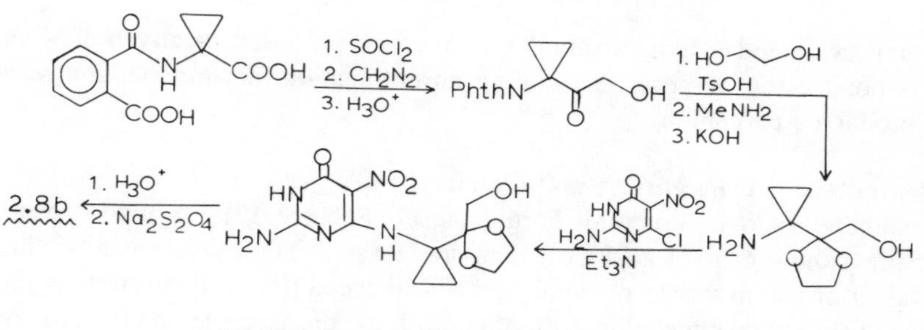

Scheme 14B.   Synthesis of Compound **2.8b**.

III. PROTONATION/ACYLATION

A. 3-Substituted-4-hydroxycoumarins and 2-Substituted-1,3-indanediones

The oral anticoagulant, warfarin, (Structural Formula 2.9, Scheme 15; R = $CH(Ph)CH_2COCH_3$) is known to inactivate the enzyme, vitamin K epoxide reductase. Chemical model studies were carried out by Silverman[84] to support a hypothesis that warfarin is a mechanism-based inactivator of this enzyme. This hypothesis was later extended by Silverman[85] to include 3-substituted-4-hydroxycoumarins (**2.9**) and 2-substituted-1,3-indanediones (Structural Formula 2.10, Scheme 16) in general. Both of these classes of compounds are stable to bases and nucleophiles (except for reversible deprotonation), suggesting that these compounds are not acting as affinity labeling agents. However, after electrophilic substitution at the 3-position of the coumarin analogues or the 2-position of the indanedione analogues (model reactions for enzyme-catalyzed protonation), these unreactive compounds become highly reactive toward a variety of nucleophiles. On the basis of a proposed mechanism for vitamin K epoxide reductase,[86] these model studies support an inactivation mechanism involving enzyme-catalyzed protonation followed by acylation of an active-site nucleophile[84,85] (Schemes 15 and 16).

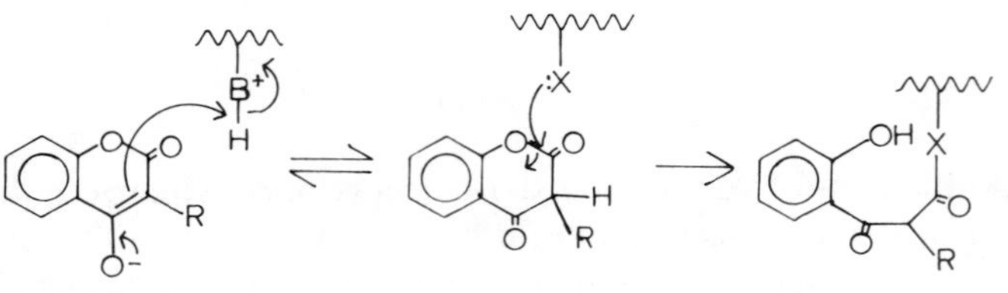

Scheme 15.   Containing Structural Formula 2.9.

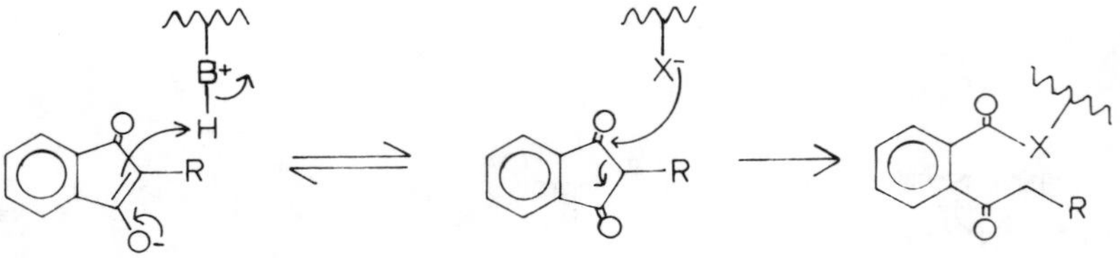

Scheme 16.   Containing Structural Formula 2.10.

IV. DEPROTONATION/ADDITION

This category includes those examples of mechanism-based inactivation where proton removal is not for the purpose of initiating an elimination or isomerization reaction, but rather to produce a carbanion.

A. 3,4-Pentadienoyl CoA and *trans*-3-Octenoyl CoA

3,4-Pentadienoyl CoA (Structural Formula 2.11, Scheme 17) is a time-dependent inactivator of pig kidney general acyl CoA dehydrogenase.[87] The enzyme can be titrated with one equivalent of inactivator to give enzyme with reduced flavin. If substrate is then added, the enzyme is slowly reactivated to 80% of control; the product released is 2,4-pentadienoyl CoA. The rate of reactivation is dependent upon the redox potential of the flavin. When the enzyme is reconstituted with various flavin analogues, the rate of reactivation decreases with increasing redox potential. This is consistent with a mechanism involving an equilibrium between inactivated enzyme and the inactivator anion plus flavin; the back reaction would be favored with decreasing redox potential. The 20% of enzyme activity that does not return upon reactivation could be the result of active site amino acids attacking the released $\alpha,\beta,\gamma,\delta$-unsaturated thioester; however, 2,4-pentadienoyl CoA does not inactivate the enzyme. Therefore, irreversible inactivation must occur at a point prior to release of product. The mechanism proposed by Wenz et al.[87] is shown in Scheme 17. N-5 substitution is suggested because of the ease of reoxidation of the flavin following denaturation of the inactivated enzyme. Also, the optical spectrum resembles that of N-(5)-acetyl-1,5-dihydroflavins.

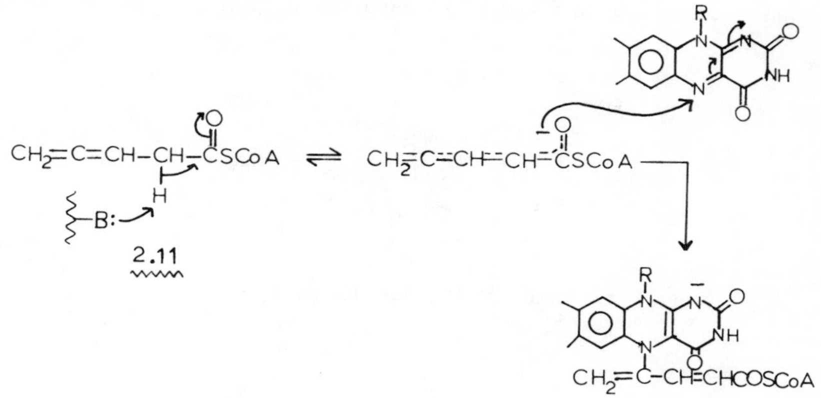

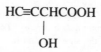

Scheme 17.   Containing Structural Formula 2.11.

Incubation of *trans*-3-octenoyl CoA with pig kidney medium chain acyl CoA dehydrogenase gives an immediate intense long-wavelength absorbance (λ = 820 nm) in the optical spectrum followed by a time-dependent decline of this absorbance with concomitant reduction of the bound flavin.[87a] The 820-nm species may be a charge-transfer complex between the inactivator and oxidized flavin. In order to test for the proposed stabilized carbanion intermediate related to the one shown in Scheme 17, the enzyme reaction was carried out by Powell et al.[87a] in D_2O and it was found that one α-proton exchanged without isomerization to the α,β-unsaturated compound.

B. Nitromethane

A reaction related to that shown in Scheme 17 is the inactivation of hog kidney D-amino acid oxidase by nitromethane; reduced FAD is formed.[88] Although no mechanism was given, it was suggested by Porter et al.[88] that the ionized form is responsible for alkylation of the flavin cofactor.

C. 2-Hydroxy-3-butynoic Acid

2-Hydroxy-3-butynoic acid (Structural Formula 2.12, Scheme 18) is a potent mechanism-based inactivator of the flavoenzymes, L-lactate oxidase,[89] D-lactate dehydrogenase,[90] L-amino acid oxidase,[91] L-α-hydroxy acid oxidase,[92] glycollate oxidase,[93] and flavocytochrome b_2.[94] Although a reasonable inactivation mechanism involves initial deprotonation of the α-carbon of the inactivator followed by carbanionic attack on the oxidized flavin, the oxidation mechanisms of these enzymes are not clearly understood; consequently, the studies with 2-hydroxy-3-butynoic acid are discussed in Chapter 9 on oxidation reactions (see Section II.C.1.a.(1)).

$$HC\equiv CCHCOOH$$
$$|$$
$$OH$$

Scheme 18.   Structural Formula 2.12.

V. NOT MECHANISM-BASED INACTIVATION

A. 2,4-Diamino-5-(4-chlorophenyl)-6-cyclopropylpyrimidine

2,4-Diamino-5-(4-chlorophenyl)-6-cyclopropylpyrimidine (Structural Formula 2.12a, Scheme 19) was prepared by Haddow et al.[83c] as an inactivator of dihydrofolate reductase via the same mechanism as in Scheme 14A; however, it was only a competitive reversible inhibitor.

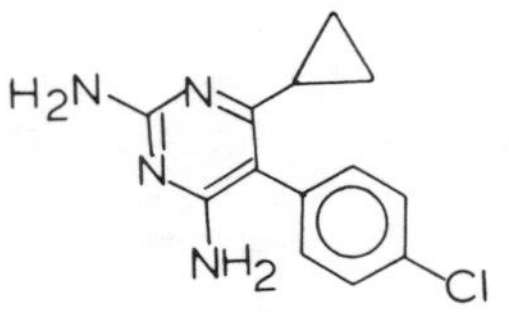

Scheme 19.   Structural Formula 2.12a.

REFERENCES

1. **Burke, S. D. and Grieco, P. A.,** Intramolecular reactions of diazocarbonyl compounds, *Org. React.,* 26, 361, 1979.
2. **Rajagopalan, T. G., Stein, W. H., and Moore, S.,** The inactivation of pepsin by diazoacetylnorleucine methyl ester, *J. Biol. Chem.,* 241, 4295, 1966.
3. **Hunkapiller, M., Heinz, J. E., and Mills, J. N.,** Comparative studies on the effect of specific inactivators of human gastricsin and pepsin, *Biochemistry,* 9, 2897, 1970.
4. **Kay, J. and Ryle, A. P.,** An active site peptide from pepsin C. *Biochem. J.,* 123, 75, 1971.
5. **Smith, G. D., Murray, M. A., Nichol, L. W., and Trikojus, V. M.,** Thyroid acid proteinase. Properties and inactivation by diazoacetylnorleucine methyl ester, *Biochim. Biophys. Acta,* 171, 288, 1968.
6. **Södek, J. and Hofmann, T.,** A pepsin-like enzyme from *Penicillium janthinellum, J. Biol. Chem.,* 243, 450, 1968.
7. **Södek, J. and Hoffman, T.,** Amino acid sequence around the active site aspartic acid and in penicillopepsin, *Can. J. Biochem.,* 48, 1014, 1970.
8. **Mizobe, F., Takahashi, K., and Ando, T.,** The structure and function of acid proteases. I. Specific inactivation of an acid protease from *Rhizopus chinensis* by diazoacetyl-DL-norleucine methyl ester, *J. Biochem. (Tokyo),* 73, 61, 1973.
9. **Takahashi, K., Mizobe, F., and Chang, W. J.,** Inactivation of acid proteases from *Rhizopus chinensis, Aspergillus saitoi* and *Mucor pusillus,* and calf rennin by diazoacetylnorleucine methyl ester, *J. Biochem. (Tokyo),* 71, 161, 1972.
10. **Takahashi, K. and Chang, W. J.,** Specific chemical modifications of acid proteases in the presence and absence of pepstatin, *J. Biochem. (Tokyo),* 73, 675, 1973.
11. **Takahashi, K., Chang, W.-J., and Arima, K.,** The structure and function of acid proteases. IV. Inactivation of the acid protease from *Mucor pusillus* by acid protease-specific inhibitors, *J. Biochem. (Tokyo),* 80, 61, 1976.
12. **Takahashi, K. and Chang, W.-J.,** The structure and function of acid proteases. V. Comparative studies on the specific inhibition of acid proteases by diazoacetyl-DL-norleucine methyl ester, 1,2-epoxy-3-(p-nitrophenoxy)propane and pepstatin, *J. Biochem. (Tokyo),* 80, 497, 1976.
13. **Chang, W.-J., Horiuchi, S., Takahashi, K., Yamasaki, M., and Yamada, Y.,** The structure and function of acid proteases. VI. Effects of acid protease-specific inhibitors on the acid proteases from *Aspergillus niger* var. *macrosporus, J. Biochem. (Tokyo),* 80, 975, 1976.
14. **Chang, W.-J. and Takahashi, K.,** The structure and function of acid proteases. II. Inactivation of bovine rennin by acid protease-specific inhibitors, *J. Biochem. (Tokyo),* 74, 231, 1973.
15. **Chang, W.-J. and Takahashi, K.,** the structure and function of acid proteases. III. Isolation and characterization of the active-site peptides from bovine rennin, *J. Biochem. (Tokyo),* 76, 467, 1974.
16. **Delpierre, G. R. and Fruton, J. S.,** Specific inactivation of pepsin by a diazoketone, *Proc. Natl. Acad. Sci. U.S.A.,* 56, 1817, 1966.
17. **Stepanov, V. M. and Vaganova, T. I.,** Identification of the carboxyl group of pepsin reacting with diazoacetamide derivatives, *Biochem. Biophys. Res. Commun.,* 31, 825, 1968.
18. **Erlanger, B. F., Vratsanos, S. M., Wassermann, N., and Cooper, A. G.,** Stereochemical investigation of the active center of pepsin using a new inactivator, *Biochem. Biophys. Res. Commun.,* 28, 203, 1967.
19. **Bayliss, R. S. and Knowles, J. R.,** An active site peptide from pepsin, *J. Chem. Soc. Chem. Commun.,* p. 196, 1968.

20. **Bayliss, R. S., Knowles, J. R., and Wybrandt, G. B.,** An aspartic acid residue at the active site of pepsin. The isolation and sequence of the heptapeptide, *Biochem. J.,* 113, 377, 1969.
21. **Lundblad, R. C. and Stein, W. H.,** On the reaction of diazoacetyl compounds with pepsin, *J. Biol. Chem.,* 244, 154, 1969.
22. **Kozlov, L. V., Ginodman, L. M., and Orekhovich, V. N.,** Inactivation of pepsin with aliphatic diazocarbonyl compounds, *Biokhimiya,* 32, 1011, 1967; *Biochemistry (USSR),* 32, 839, 1967.
23. **Hamilton, G. A., Spona, J., and Crowell, L. D.,** The inactivation of pepsin by an equimolar amount of 1-diazo-4-phenylbutanone-2, *Biochem. Biophys. Res. Commun.,* 26, 193, 1967.
24. **Fry, K. T., Kim, O. K., Spona, J., and Hamilton, G. A.,** A reactive aspartyl residue of pepsin, *Biochem. Biophys. Res. Commun.,* 30, 489, 1968.
25. **DeWald, H. A. and Moore, A. M.,** 6-Diazo-5-oxo-L-norleucine, a new tumor-inhibitory substance. Preparation of L-, D-, and DL-forms, *J. Am. Chem. Soc.,* 80, 3941, 1958.
26. **Dion, H. W., Fusari, S. A., Jakubowski, Z. L., Zora, J. G., and Bartz, Q. R.,** 6-Diazo-5-oxo-L-norleucine, a new tumor-inhibitory substance. II. Isolation and characterization, *J. Am. Chem. Soc.,* 78, 3075, 1956.
27. **Hartman, S. C.,** The interaction of 6-diazo-5-oxo-L-norleucine with phosphoribosyl pyrophosphate amidotransferase, *J. Biol. Chem.,* 238, 3036, 1963.
28. **Hartman, S. C.,** Phosphoribosyl pyrophosphate amidotransferase, purification and general catalytic properties, *J. Biol. Chem.,* 238, 3024, 1963.
29. **Messenger, L. J. and Zalkin, H.,** Glutamine phosphoribosylpyrophosphate amidotransferase from *Escherichia coli.* Purification and properties, *J. Biol. Chem.,* 254, 3382, 1979.
30. **Tso, J. Y., Hermodson, M. A., and Zalkin, H.,** Glutamine phosphoribosylpyrophosphate amidotransferase from cloned *Escherichia coli purF.* NH$_2$-terminal amino acid sequence, identification of the glutamine site, and trace metal analysis, *J. Biol. Chem.,* 257, 3532, 1982.
31. **Vollmer, S. J., Switzer, R. L., Hermodson, M. A., Bower, S. G., and Zalkin, H.,** The glutamine-utilizing site of *Bacillus subtilis* glutamine phosphoribosylpyrophosphate amidotransferase, *J. Biol. Chem.,* 258, 10582, 1983.
32. **Tate, S. S. and Meister, A.,** Affinity labeling of γ-glutamyl transpeptidase and location of the γ-glutamyl binding site on the light subunit, *Proc. Natl. Acad. Sci. U.S.A.,* 74, 931, 1977.
33. **Inoue, M., Horiuchi, S., and Morino, Y.,** Affinity labeling of rat kidney γ-glutamyl transpeptidase, *Eur. J. Biochem.,* 73, 335, 1977.
34. **Inoue, M., Horiuchi, S., and Morino, Y.,** Affinity labeling of rat kidney γ-glutamyl transpeptidase by 6-diazo-5-oxo-D-norleucine, *Eur. J. Biochem.,* 99, 169, 1979.
35. **Tate, S. S. and Ross, M. E.,** Human kidney γ-glutamyl transpeptidase. Catalytic properties, subunit structure, and localization of the γ-glutamyl binding site on the light subunit, *J. Biol. Chem.,* 252, 6042, 1977.
36. **Horiuchi, S., Inoue, M., and Morino, Y.,** γ-Glutamyl transpeptidase: Sidedness of its active site on renal brush-border membrane, *Eur. J. Biochem.,* 87, 429, 1978.
37. **Inoue, M., Horiuchi, S., and Morino, Y.,** Affinity labeling of γ-glutamyl transpeptidase of tumor cell AH-130 and transport activity of glutathione and amino acids, *Biochem. Biophys. Res. Commun.,* 79, 1104, 1977.
38. **Inoue, M., Horiuchi, S., and Morino, Y.,** γ-Glutamyl transpeptidase in rat ascites tumor cell LY-5. Lack of functional correlation of its catalytic activity with the amino acid transport, *Eur. J. Biochem.,* 78, 609, 1977.
39. **Long, C. W., Levitzki, A., and Koshland, D. E., Jr.,** The subunit structure and subunit interactions of cytidine triphosphate synthetase, *J. Biol. Chem.,* 245, 80, 1970.
40. **Levitzki, A., Stallcup, W. B., and Koshland, D. E., Jr.,** Half-of-the-sites reactivity and the conformational states of cytidine triphosphate synthetase, *Biochemistry,* 10, 3371, 1971.
41. **Lee, B. H. and Hartman, S. C.,** Preferential utilization of glutamine for amination of xanthosine 5'-phosphate to guanosine 5'-phosphate by purified enzymes from *Escherichia coli, Biochem. Biophys. Res. Commun.,* 60, 918, 1974.
42. **Zalkin, H. and Truitt, C. D.,** Characterizaton of the glutamine site of *Escherichia coli* guanosine 5'-monophosphate synthetase, *J. Biol. Chem.,* 252, 5431, 1977.
43. **Patel, N., Moyed, H. S., and Kane, J. F.,** Xanthosine-5'-phosphate amidotransferase from *Escherichia coli, J. Biol. Chem.,* 250, 2609, 1975.
44. **Patel, N., Moyed, H. S., and Kane, J. F.,** Properties of xanthosine 5'-monophosphate-amidotransferase from *Escherichia coli, Arch. Biochem. Biophys.,* 178, 652, 1977.
45. **Nagano, H., Zalkin, H., and Henderson, E. J.,** The anthranilate synthetase-anthranilate-5-phosphoribosylpyrophosphate phosphoribosyltransferase aggregate. On the reaction mechanism of anthranilate synthetase from *Salmonella typhimurium, J. Biol. Chem.,* 245, 3810, 1970.
46. **Zalkin, H. and Hwang, L. H.,** Anthranilate synthetase from *Serratia marcescens.* On the properties and relationship to the enzyme from *Salmonella typhimurium, J. Biol. Chem.,* 246, 6899, 1971.

47. **Queener, S. W., Queener, S. F., Meeks, J. R., and Gunsalus, I. C.,** Anthranilate synthase from *Pseudomonas putida*. Purification and properties of a two-component enzyme, *J. Biol. Chem.,* 248, 151, 1973.

48. **Goto, Y., Zalkin, H., Keim, P. S., and Heinrikson, R. L.,** Properties of anthranilate synthetase component II from *Pseudomonas putida, J. Biol. Chem.,* 251, 941, 1976.

49. **Sinnott, M. L.,** Affinity labeling via deamination reactions, *CRC Crit. Rev. Biochem.,* 12, 327, 1982.

50. **Hartman, S. C.,** Glutaminase of *Escherichia coli*. I. Purification and general catalytic properties, *J. Biol. Chem.,* 243, 853, 1968.

51. **Hartman, S. C. and McGrath, T. F.,** Glutaminase A of *Escherichia coli*. Reactions with the substrate analogue, 6-diazo-5-oxonorleucine, *J. Biol. Chem.,* 248, 8506, 1973.

52. **Bates, C. J. and Handschumacher, R. E.,** Inactivation and resynthesis of glucosamine-6-phosphate synthetase after treatment with glutamine analogs, *Adv. Enz. Regul.,* 7, 183, 1969.

53. **Yu, C. K. and Dietrich, L. S.,** Purification and properties of yeast nicotinamide adenine dinucleotide synthetase, *J. Biol. Chem.,* 247, 4794, 1972.

54. **Roberts, J., Holcenberg, J. S., and Dolowy, W. C.,** Isolation, crystallization, and properties of *Achromobacteraceae* glutaminase-asparaginase with antitumor activity, *J. Biol. Chem.,* 247, 84, 1972.

55. **Levenberg, B., Melnick, I., and Buchanan, J. M.,** Biosynthesis of purines. XV. The effect of aza-L-serine and 6-diazo-5-oxo-L-norleucine on inosinic acid biosynthesis de novo, *J. Biol. Chem.,* 225, 163, 1957.

55a. **Chamara, H., Andruszkiewicz, R., and Borowski, E.,** Inactivation of glucosamine-6-phosphate synthetase form *Salmonella typhimurium* LT2 by fumaroyl diaminopropanoic acid derivatives, a novel group of glutamine analogs, *Biochim. Biophys. Acta,* 870, 357, 1986.

56. **Fusari, S. A., Frohardt, R. P., Ryder, A., Haskell, T. H., Johannessen, D. W., Elder, C. C., and Bartz, Q. R.,** Azaserine, a new tumor-inhibitory substance. Isolation and characterization, *J. Am. Chem. Soc.,* 76, 2878, 1954.

57. **Fusari, S. A., Haskell, T. H., Frohardt, R. P., and Bartz, Q. R.,** Azaserine, a new tumor-inhibitory substance. Structural studies, *J. Am. Chem. Soc.,* 76, 2881, 1954.

58. **Nicolaides, E. D., Westland, R. D., and Wittle, E. L.,** Azaserine, synthetic studies. II, *J. Am. Chem. Soc.,* 76, 2887, 1954.

59. **Dawid, I. B., French, T. C., and Buchanan, J. M.,** Azaserine-reactive sulfhydryl group of 2-formamido-N-ribosylacetamide 5'-phosphate: L-glutamine amido-ligase (adenosine diphosphate). II. Degradation of azaserine-C^{14}-labeled enzyme, *J. Biol. Chem.,* 238, 2178, 1963.

60. **French, T. C., Dawid, I. B., and Buchanan, J. M.,** Azaserine-reactive sulfhydryl group of 2-formamido-N-ribosylacetamide 5'-phosphate: L-glutamine amido-ligase (adenosine diphosphate). III. Comparison of degradation products with synthetic compounds, *J. Biol. Chem.,* 238, 2186, 1963.

61. **Mizobuchi, K. and Buchanan, J. M.,** Biosynthesis of the purines. XXX. Isolation and characterization of formylglycinamide ribonucleotide amidotransferase-glutamyl complex, *J. Biol. Chem.,* 243, 4853, 1968.

62. **Mizobuchi, K., Kenyon, G. L., and Buchanan, J. M.,** Biosynthesis of purines, XXXI. Binding of formylglycinamide ribonucleotide and adenosine triphosphate to formylglycinamide ribonucleotide amidotransferase, *J. Biol. Chem.,* 243, 4863, 1968.

63. **Liwschitz, Y., Irsay, R. D., and Vincze, A. I.,** Synthesis of 5-diazo-4-oxo-L-norvaline, *J. Chem. Soc.,* 1308, 1959.

64. **Handschumacher, R. E., Bates, C. J., Chang, P. K., Andrews, A. T., and Fischer, G. A.,** 5-Diazo-4-oxo-L-norvaline: reactive asparagine analog with biological specificity, *Science (Washington, D.C.),* 161, 62, 1968.

65. **Jackson, R. C. and Handschumacher, R. E.,** *Escherichia coli* L-asparaginase. Catalytic activity and subunit nature, *Biochemistry,* 9, 3585, 1970.

66. **Lachman, L. B. and Handschumacher, R. E.,** The active site of L-asparaginase: dimethyl sulfoxide effect of 5-diazo-4-oxo-L-norvaline interactions, *Biochem. Biophys. Res. Commun.,* 73, 1094, 1976.

67. **Peterson, R. G., Richards, F. F., and Handschumacher, R. E.,** Structure of peptide from active site region of *Escherichia coli* L-asparaginase, *J. Biol. Chem.,* 252, 2072, 1977.

68. **Chang, P. K., Lachman, L. B., and Handschumacher, R. E.,** Synthesis of model compounds relevant to the active-site-directed inactivation of L-asparaginase by 5-diazo-4-oxo-L-norvaline, *Int. J. Pept. Protein Res.,* 14, 27, 1979.

69. **Shaw, E.,** Site-specific reagents for chymotrypsin and trypsin, *Methods Enzymol.,* 11, 677, 1967.

70. **Ong, E. B. and Perlmann, G. E.,** Specific inactivation of pepsin by benzyloxycarbonyl-L-phenylalanyl-diazomethane, *Nature (London),* 215, 1492, 1967.

71. **Leary, R., Larsen, D., Watanabe, H., and Shaw, E.,** Diazomethyl ketone substrate derivatives as active-site directed inhibitors of thiol proteases. Papain, *Biochemistry,* 16, 5857, 1977.

72. **Brocklehurst, K. and Malthouse, J. P. G.,** Mechanism of the reaction of papain with substrate-derived diazomethyl ketones. Implications for the difference in site specificity of halomethyl ketones for serine proteinases and cysteine proteinases and for stereoelectronic requirements in the papain catalytic mechanism, *Biochem. J.,* 175, 761, 1978.

73. **Leary, R. and Shaw, E.,** Inactivation of cathepsin B_1 by diazomethyl ketones, *Biochem. Biophys. Res. Commun.,* 79, 926, 1977.

74. **Watanabe, H., Green, G. D. J., and Shaw, E.,** A comparison of the behavior of chymotrypsin and cathepsin B towards peptidyl diazoketones, *Biochem. Biophys. Res. Commun.,* 89, 1354, 1979.

75. **Shaw, E. and Dean, R. T.,** The inhibition of macrophage protein turnover by a selective inhibitor of thiol proteinases, *Biochem. J.,* 186, 385, 1980.

76. **Kirschke, H. and Shaw, E.,** Rapid inactivation of cathepsin L by Z-Phe-PheCHN$_2$ and Z-Phe-AlaCHN$_2$, *Biochem. Biophys. Res. Commun.,* 101, 454, 1981.

77. **Shaw, E., Wikstrom, P., and Ruscica, J.,** An exploration of the primary specificity site of cathepsin B, *Arch. Biochem. Biophys.,* 222, 424, 1983.

78. **Green, G. D. J. and Shaw, E.,** Peptidyl diazomethyl ketones are specific inactivators of thiol proteinases, *J. Biol. Chem.,* 256, 1923, 1981.

79. **Green, G. D. J. and Shaw, E.,** A prolyl endopeptidase from murine macrophages, its assay and specific inactivation, *Arch. Biochem. Biophys.,* 225, 331, 1983.

80. **Metcalf, B. W., Jund, K., and Burkhart, J. P.,** Synthesis of 3-keto-4-diazo-5-α-dihydrosteroids as potential irreversible inhibitors of steroid 5-α-reductase, *Tetrahedron Lett.,* 21, 15, 1980.

81. **Blohm, T. R., Metcalf, B. W., Laughlin, M. E., Sjoerdsma, A., and Schatzman, G. L.,** Inhibition of testosterone 5α-reductase by a proposed enzyme-activated, active site-directed inhibitor, *Biochem. Biophys. Res. Commun.,* 95, 273, 1980.

82. **Brockhaus, M. and Lehmann, J.,** 2,6-Anhydro-1-diazo-1-deoxy-D-*glycero*-L-*manno*-heptitol: a specific blocking agent for the active site of β-galactosidase, *FEBS Lett.,* 62, 154, 1976.

83. **Brockhaus, M. and Lehmann, J.,** Ester and sulfonium salt formation in the active-site labeling of β-D-galactosidase from *Escherichia coli* by 2,6-anhydro-1-deoxy-1-diazo-D-*glycero*-L-*manno*-heptitol, *Carbohydr. Res.,* 63, 301, 1978.

83a. **Nashed, N. T., Michaud, D. P., Levin, W., and Jerina, D. M.,** 7-Dehydrocholesterol 5,6β-oxide as a mechanism-based inhibitor of microsomal cholesterol oxide hydrolase, *J. Biol. Chem.,* 261, 2510, 1986.

83b. **Michaud, D. P., Nashed, N. T., and Jerina, D. M.,** Stereoselective synthesis and solvolytic behavior of the isomeric 7-dehydrocholesterol 5,6-oxide, *J. Org. Chem.,* 50, 1835, 1985.

83c. **Haddow, J., Suckling, C. J., and Wood, H. C. S.,** Latent inhibition of dihydrofolate reductase by a spirocyclopropylpteridine, *J. Chem. Soc. Chem. Commun.,* p. 478, 1987.

84. **Silverman, R. B.,** A model for a molecular mechanism of anticoagulant activity of 3-substituted 4-hydroxycoumarins, *J. Am. Chem. Soc.,* 102, 5421, 1980.

85. **Silverman, R. B.,** Model studies for a molecular mechanism of action of oral anticoagulants, *J. Am. Chem. Soc.,* 103, 3910, 1981.

86. **Silverman, R. B.,** Chemical model studies for the mechanism of vitamin K epoxide reductase, *J. Am. Chem. Soc.,* 103, 5939, 1981.

87. **Wenz, A., Ghisla, S., and Thorpe, C.,** Studies with general acyl-CoA dehydrogenase from pig kidney. Inactivation by a novel type of "suicide" inhibitor, 3,4-pentadienoyl-CoA, *Eur. J. Biochem.,* 147, 553, 1985.

87a. **Powell, P. J., Lau, S.-M., Killian, D., and Thorpe, C.,** Interaction of acyl coenzyme A substrates and analogues with pig kidney medium-chain acyl-CoA dehydrogenase, *Biochemistry,* 26, 3704, 1987.

88. **Porter, D. J. T., Voet, J. G., and Bright, H. J.,** Nitromethane, a novel substrate for D-amino acid oxidase, *J. Biol. Chem.,* 247, 1951, 1972.

89. **Walsh, C. T., Schonbrunn, A., Lockridge, O., Massey, V., and Abeles, R. H.,** Inactivation of a flavoprotein, lactate oxidase, by an acetylenic substrate, *J. Biol. Chem.,* 247, 6004, 1972.

90. **Walsh, C. T., Abeles, R. H., and Kaback, H. R.,** Mechanisms of active transport in isolated membrane vesicles. X. Inactivation of D-lactate dehydrogenase and D-lactate dehydrogenase-coupled transport in *Escherichia coli* membrane vesicles by an acetylenic substrate, *J. Biol. Chem.,* 247, 7858, 1972.

91. **Cromartie, T., Fisher, J., Kaczorowski, G., Laura, R., Marcotte, P., and Walsh, C.,** Synthesis of α-hydroxy-β-acetylenic acids and their oxidation by and inactivation of flavoprotein oxidases, *J. Chem. Soc. Chem. Commun.,* 597, 1974.

92. **Cromartie, T. H., and Walsh, C. T.,** Rat kidney L-α-hydroxy acid oxidase: isolation of enzyme with one flavine coenzyme per two subunits, *Biochemistry,* 14, 2588, 1975.

93. **Jewess, P. J., Kerr, M. W., and Whitaker, D. P.,** Inhibition of glycollate oxidase from pea leaves, *FEBS Lett.,* 53, 292, 1975.

94. **Pompon, D. and Lederer, F.,** On the mechanism of flavin modification during inactivation of flavocytochrome b_2 from baker's yeast by acetylenic substrates, *Eur. J. Biochem.,* 148, 145, 1985.

Chapter 3

PHOSPHORYLATION REACTIONS*

I. PHOSPHORYLATION/PHOSPHORYLATION

A. Phosphate Esters

In general, phosphate esters are affinity-labeling agents. However, a process known as "aging" has been described in which, following enzyme active site phosphorylation, a second reaction takes place within the active site that usually results in the formation of a more stable adduct. This "aging" process, then, redefines the phosphorylating agents as mechanism-based enzyme inactivators.

α-Chymotrypsin from bovine pancreas is inactivated by triaryl phosphate esters (Structural Formula 3.1, R = Ar; Scheme 1) which involves initial phosphorylation of the active site serine, followed by "aging" with histidine-57 participation.[1]

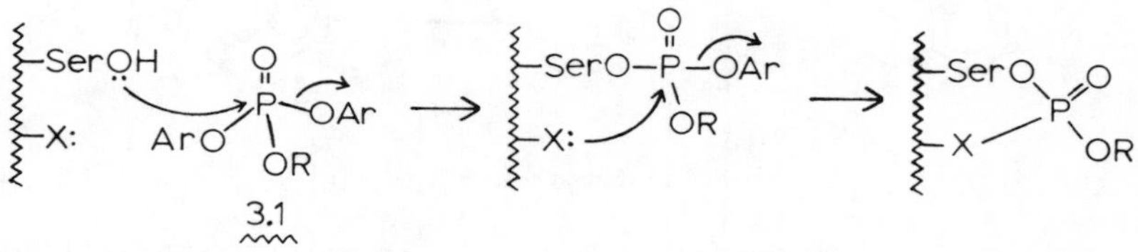

Scheme 1.   Containing Structural Formula 3.1.

Bis(p-nitrophenyl)methyl phosphate (**3.1**, R = Me) inactivates bovine pancreatic α-chymotrypsin and liver carboxylesterases from chicken, pig, sheep, and horse; 2 mol of p-nitrophenol are released per active site phosphorylated.[2] With the use of bis(p-nitrophenyl) [³H]methyl [³²P]phosphate, inactivation of the chicken liver enzyme results in incorporation of one methyl group and one phosphate. Once the "aging" process has occurred, nucleophiles, e.g., hydroxylamine, do not reactivate the enzyme. The mechanism proposed by Hamilton et al.[2] is shown in Scheme 1. The results also could be explained by demethylation during the "aging" process (Scheme 2).

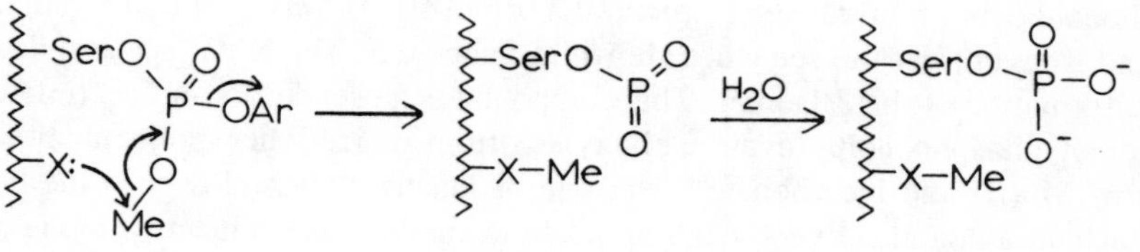

Scheme 2.

Saligenin cyclic phosphorus esters (Structural Formula 3.2, Scheme 3) are time-dependent inactivators of bovine pancreatic α-chymotrypsin, producing a phosphoenzyme which undergoes a rapid and quantitative "aging" process leading to a stable adduct (i.e., hydroxylamine does not regenerate enzyme activity).[3] The "aging" process results in attachment of the organic part of the molecule as an o-hydroxybenzyl group equivalent to ~20% of the amount

---

* A list of abbreviations and shorthand notations can be found prior to Chapter 1.

of phosphoenzyme produced. The mechanism produced by Toia and Casida[2] is shown in Scheme 3. Saligenin and its analogues were synthesized by Eto and Oshima[4] by phosphorylation of the corresponding *o*-hydroxybenzyl alcohols (Scheme 4).

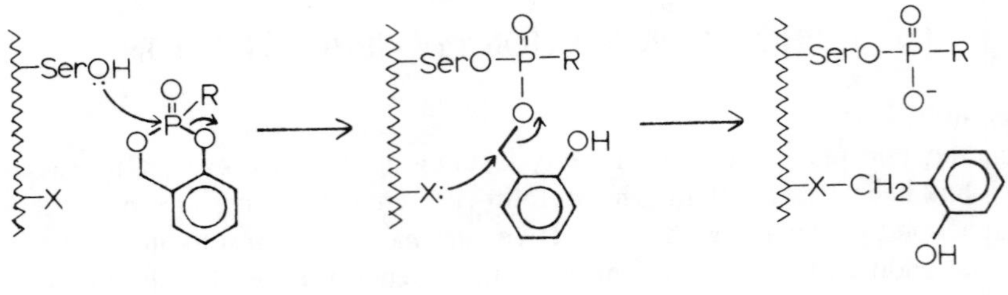

Scheme 3.   Containing Structural Formula 3.2.

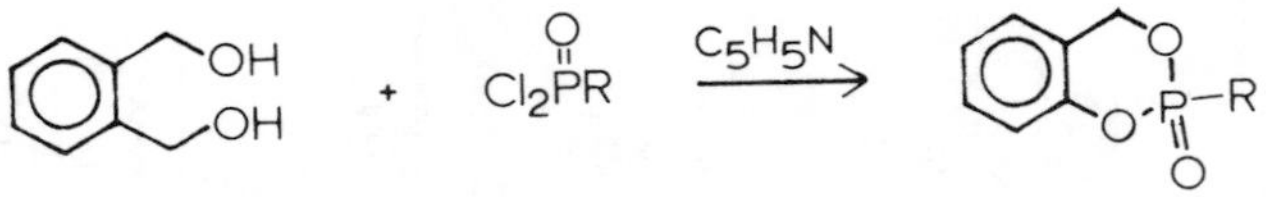

Scheme 4.

II. PHOSPHORYLATION /CARBOXYLATION/SUBSTITUTION

A. Phosphoenol-3-bromopyruvate

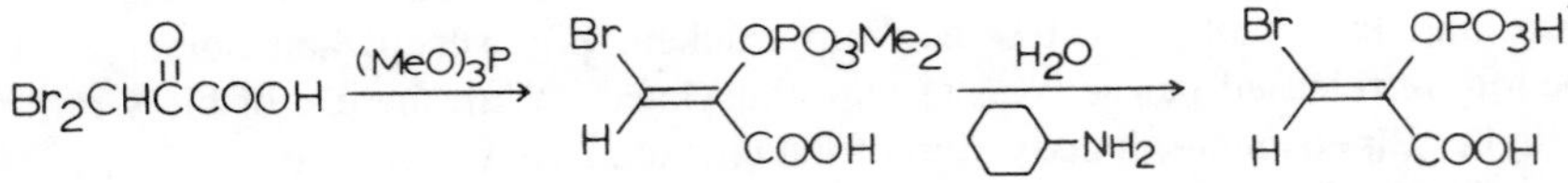

Scheme 5.   Containing Structural Formula 3.3.

Phosphoenol-3-bromopyruvate (Structural Formula 3.3) was originally synthesized by Stubbe and Kenyon;[5] by the sequence shown in Scheme 5. The NMR spectral data indicate exclusive formation of the Z isomer. This compound is a pseudo first-order, time-dependent inactivator of phosphoenolpyruvate carboxylase from maize when preincubation is in the presence of Mn(II) and bicarbonate.[6] The rate of inactivation is slower in the presence of phospholactate, a potent active site inhibitor of the enzyme. If the inactivation is carried out in the presence of $H^{14}CO_3^-$ and Mn(II) and the inactivated enzyme is reduced with sodium borohydride prior to dialysis, 0.6 equivalent of radioactivity is incorporated into the enzyme. If sodium borohydride treatment is omitted from the procedure, the enzyme is inactivated, but no radioactivity is incorporated. These results support the mechanism proposed by O'Leary and Diaz[6] (Scheme 6).

If carboxylation and dephosphorylation occur simultaneously, then this example could be categorized as a carboxylation reaction, in which case it is the sole example for that reaction type of inactivation.

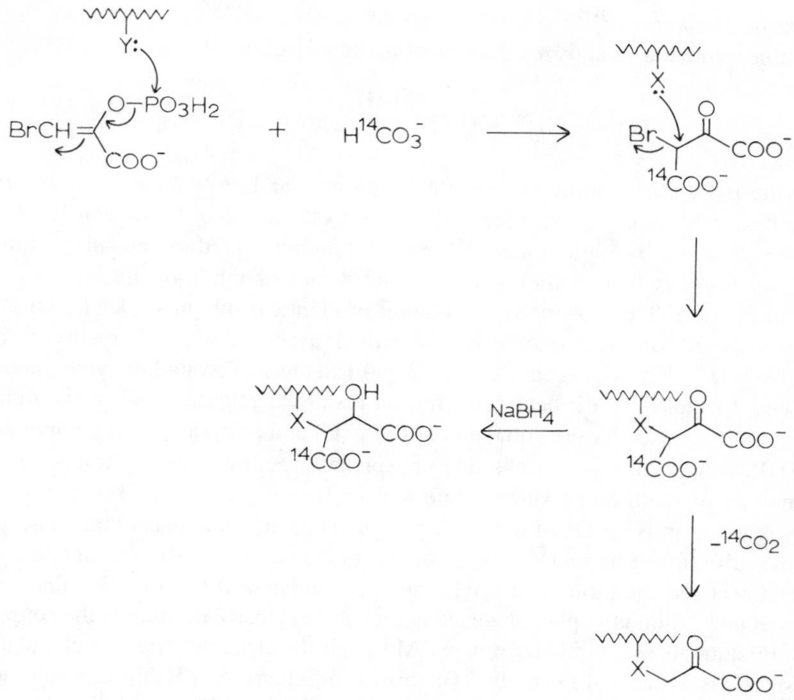

Scheme 6.

III. NONCOVALENT INACTIVATION

The examples described in this section involve noncovalent mechanism-based transition state analogues, i.e., transition-state analogues that are generated by enzyme-catalyzed phosphorylation.

A. Methionine Sulfoximine

When certain proteins rich in methionine are treated with nitrogen trichloride, the active ingredient in agene which was used commercially to treat flour, they become toxic and can produce convulsions. The toxic factor was found to be the formation of methionine sulfoximine (Structural Formula 3.4, Scheme 7), which is released upon digestion of the protein.[7] Methionine sulfoximine has been known to be a potent inhibitor of glutamine synthetase and other enzymes which utilize glutamine since 1952.[8] Because of the structural similarity of methionine sulfoximine to glutamate, it was originally suggested by Ronzio and Meister[9] that methionine sulfoximine binds to both the glutamate and ammonia binding sites (Structural Formula 3.5, Scheme 7).

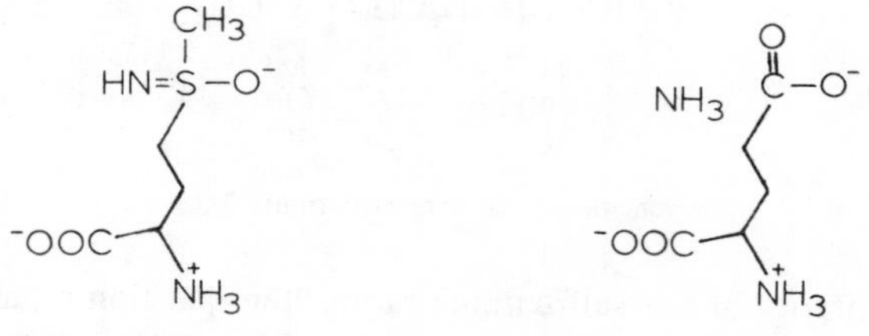

Scheme 7. Structural Formulas 3.4 and 3.5.

1. Glutamine Synthetase

Glutamine synthetase catalyzes the reaction shown below.

$$Glu + NH_3 + ATP \overset{M(II)}{\rightleftharpoons} Gln + ADP + P_i$$

The enzyme from sheep brain was shown by Meister and co-workers[9,10] to be irreversibly inhibited by methionine sulfoximine only when ATP and Mg(II) or Mn(II) were added; inhibition is blocked by L-glutamate. When [14C]methyl-L-methionine sulfoximine is used, radioactivity remains bound after gel filtration; 8 mol of inhibitor are bound per mole of inactivated enzyme. The enzyme is an octamer of identical subunits. Denaturation with acid or heat results in complete release of the radioactivity. When [14C]methyl-L-methionine sulfoximine and [32P]β,γ-ATP are used, the gel filtered inactivated enzyme contains 8 mol of [14C] and 16 mol of [32P]; heat denaturation results in release of all of the radioactivity. One compound released has an equal amount of [14C] and [32P]; a second compound released is [32P]ADP. Acid hydrolysis or alkaline phosphatase treatment of the first compound gives equal amounts of methionine sulfoximine and inorganic phosphate. These results indicate that phosphorylation is involved in the enzyme mechanism. It appears that a phosphorylated methionine sulfoximine and ADP bind so tightly to the enzyme as to prevent enzyme activity. The phosphorylated methionine sulfoximine was synthesized from methionine sulfoximine methyl ester and cyanoethyl phosphate using dicyclohexylcarbodiimide as the coupling agent, followed by aqueous ammonia treatment. Although the structure was not elucidated, it was hypothesized[9] as Structural Formula 3.6, shown in Scheme 8. Methionine sulfone also is a potent inhibitor of the enzyme, but is reversible.[10]

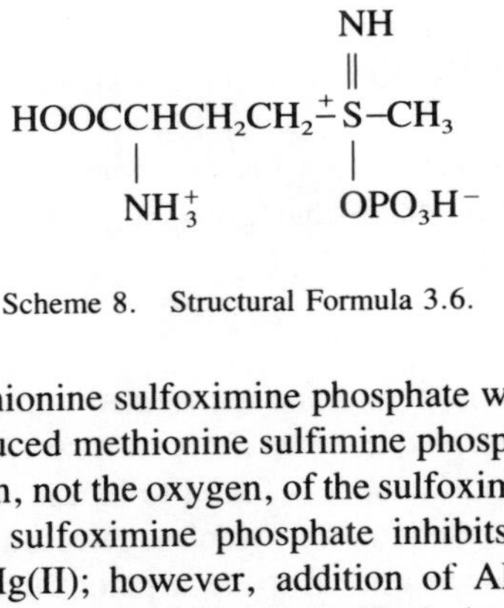

$$\overset{\displaystyle NH}{\underset{\displaystyle NH_3^+ \qquad OPO_3H^-}{HOOCCHCH_2CH_2 \overset{+}{-} \overset{\|}{S}-CH_3}}$$

Scheme 8.   Structural Formula 3.6.

The actual structure of methionine sulfoximine phosphate was elucidated by Rowe et al.[11] Reduction with Na_2S_2O_4 produced methionine sulfimine phosphate, indicating that the phosphate is attached to the nitrogen, not the oxygen, of the sulfoximine group (Structural Formula 3.7, Scheme 9). Methionine sulfoximine phosphate inhibits glutamine synthetase in the absence of added ATP or Mg(II); however, addition of ADP and Mg(II) increases the inhibition and tightness of binding of methionine sulfoximine phosphate.

$$\overset{+NPO_3^-}{\underset{NH_3^+ \qquad O}{^-OOCCHCH_2CH_2 - \overset{\|}{S}-CH_3}}$$

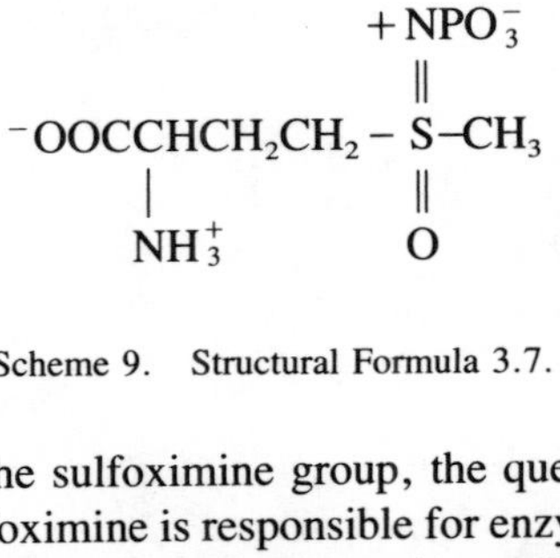

Scheme 9.   Structural Formula 3.7.

Because of the chirality of the sulfoximine group, the question regarding which diastereomer of L-methionine *R,S*-sulfoximine is responsible for enzyme inactivation was addressed by Manning et al.[12] L-Methionine *S*-sulfoximine and L-methionine *R*-sulfoximine are separable by ion exchange chromatography. The corresponding phosphates are separable as α-N-L-leucyl dipeptides by ion exchange chromatography. The methionine sulfoximine phosphate isolated from the enzyme following inactivation of glutamine synthetase by L-methionine R,S-sulfoximine and denaturation corresponds to the S-diastereomer (after conversion to the α-N-L-leucyl derivative). Furthermore, only L-methionine S-sulfoximine inactivates the enzyme; the R-diastereomer does not inhibit the enzyme at all.[12,13] Sixteen other brain enzymes are unaffected by the active isomer.[13] On the basis of these results and observations with Dreiding models, it was proposed that the sulfoximine nitrogen atom of the R-isomer cannot interact with the enzyme site required for phosphorylation.[12] However, the S-isomer can assume a conformation in which the tetrahedral sulfoximine is closely analogous to the tetrahedral intermediate produced during conversion of glutamate to glutamine. The results of studies with various substrates and inhibitors of glutamine synthetase were compiled and the three-dimensional coordinates were computer generated. From this work a hypothetical "active site" was mathematically derived.[14] Unlike the earlier hypothesis of Ronzio and Meister[9] that the sulfoximine nitrogen interacts with the ammonia binding site, the computer-generated results of Gass and Meister[14] suggest that the methyl groups of methionine sulfoximine and methionine sulfone interact with the ammonia binding site, indicating that unionized ammonia is involved in the normal reaction. The results also support the hypothesis[12] that methionine sulfoximine and methionine sulfone are analogues of the tetrahedral intermediate rather than of glutamate. If this is the case, methionine sulfoximine is a mechanism-based transition state analogue.

Because of some reported differences for glutamine synthetase from *E. coli* and sheep brain, the effect of methionine sulfoximine on the bacterial enzyme was determined; similar results for the two enzymes were obtained.[15] Only one isomer, the L-methionine-S-sulfoximine, is phosphorylated and inhibits the enzyme with incorporation of tightly bound phosphorylated methionine sulfoximine.

Several physical measurements were made on the complex formed upon addition of L-methionine R,S-sulfoximine to the *E coli* enzyme. Addition of the inactivator to the enzyme·Mg(II)·ATP complex results in a marked increase in the fluorescence; addition of L-glutamate has no further effect on the fluorescence intensity.[16] However, if the order of addition of L-glutamate and L-methionine R,S-sulfoximine is reversed, the usual fluorescence increase is observed with L-glutamate addition, followed by an additional small increase when L-methionine R,S-sulfoximine is added. This suggests that the inactivator displaces the L-glutamate and the additional small fluorescence increase may correspond to a higher fluorescence intensity of the inactivator·enzyme·ATP·Mg(II) complex than the L-glutamate·enzyme·ATP·Mg(II) complex. The fluorescence increase is much weaker in the absence of ATP, indicating a much weaker binding of the inactivator to the enzyme·Mg(II) complex (i.e., the unadenylated complex) than the enzyme·ATP·Mg(II) complex (the adenylated complex). However, the inactivator binding in the absence of ATP is still tighter than that of L-glutamate in the presence and absence of ATP. These results support a mechansim involving phosphorylation of the inactivator. On the basis of a study of the frequency dependence of the paramagnetic contribution to the longitudinal relaxation rate of solvent protons in the NMR spectrum of unadenylated *E. coli* W glutamine synthetase (in the Mn(II) form), when L-methionine R,S-sulfoximine is present, the number of rapidly exchanging water molecules at the high affinity Mn(II) site is reduced from 2 to 0.2.[17] The EPR spectrum of the enzyme-bound Mn(II) dramatically sharpens in the presence of L-methionine R,S-sulfoximine, suggesting reduced solvent accessibility to bound Mn(II) and conformational changes produced by binding of the inactivator. Binding is enhanced and additional conformational changes are seen in the EPR by ADP. These data further support a hypothesis that L-methionine R,S-sulfoximine is a transition state analogue which intimately interacts with enzyme-bound Mn(II).[17,18] Inactivator binding also appears to be influenced by the addition of the substrates ATP·Mg(II) and ammonia.[18] It is suggested that the enzyme may require the presence of all of the substrates to attain catalytic function.[18] Further EPR studies in the presence of L-methionine R,S-sulfoximine, show that the EPR linewidth of the bound Mn(II) narrows drastically, and forbidden transitions are observed.[19] Axial distortion of the high-affinity Mn(II) site decreases with increasing adenylation; the Mn(II) hyperfine coupling constant increases linearly with adenylation. If the inactivator mimics the transition state of the substrate reaction, then these changes at the high affinity Mn(II) sites may reflect corresponding changes near the γ-carboxyl group of L-glutamate whose activation is believed to be required for the catalytic reaction. The EPR results in the presence and absence of ATP suggest that the inactivator is phosphorylated.[20]

By observation of the UV difference spectra, it was found by Shrake et al.[21] that L-methionine R,S-sulfoximine produces tyrosine residue perturbations on binding to unadenylated, but not to adenylated, *E. coli* glutamine synthetase. Approximately one tyrosyl residue is buried per unadenylated subunit. It is believed that the perturbed tyrosyl residue is at the subunit site of adenylation because binding of L-methionine R,S-sulfoximine to the fully adenylated manganese enzyme produces a perturbation of the covalently bound 5'-adenylate groups without concomitant tyrosyl perturbation. Sedimentation velocity measurements show that unadenylated and fully adenylated Mn(II) enzymes have the same conformation in the absence of L-methionine R,S-sulfoximine, but binding of the inactivator produces small, but different, alterations in the hydrodynamic particle shape of the two enzymes. UV spectral studies of resolved L-S-and L-R-diastereomers of L-methionine R,S-sulfoximine show that the S-isomer binds reversibly to the unadenylated manganese enzyme with a stoichiometry of 1 equiv/subunit and with negative cooperativity.[22] The affinity of this enzyme complex is greatest for the S-isomer and 10 times lower for the R-isomer. The affinity of the S-isomer is enhanced greater than 35-fold by ADP and is decreased threefold by adenylation of the enzyme. The UV spectra perturbations that occur on binding of saturating levels of the S- and R-isomers are the same as those reported for the R,S-mixture[21]; therefore, the greater binding of the S-isomer relative to the R-isomer is not manifested in the spectral perturbations. Also, the subunit interactions that give rise to the observed negative cooperativity of binding are not reflected in the spectral changes. This suggests that the spectral perturbations are derived from a conformational change that marks the occupancy of the single subunit site by either isomer. The apparent, first-order rate constant for irreversible inactivation of *E. coli* glutamine synthetase by L-methionine S-sulfoximine progressively decreases from the first-order rate.[23] This, again, suggests negative cooperativity by the inactivated subunit on the reactivity of its neighboring subunits toward methionine sulfoximine and ATP. L-Methionine R-sulfoximine does not irreversibly inactivate the enzyme in the presence of ATP, but it does protect the enzyme from inactivation by the S-isomer. Changes in the protein fluoresence during binding of the S-isomer suggest an apparent negative cooperative binding isotherm, whereas the R-isomer shows an apparent positive cooperative pattern. All of these data indicate the existence of homologous subunit interactions among the 12 identical unadenylated subunits.

With either unadenylated or fully adenylated *E. coli* glutamine synthetase, L-methionine R,S-sulfoximine produces a 30-fold increase in the apparent association constant for Mn(II) binding to the high affinity (n_1) site of each subunit without affecting the Mn(II) binding to the n_2 site.[24] This provides evidence that bound inactivator interacts with Mn(II) bound to n_1 sites of the enzyme. It also was shown that the tight binding of Mg(II), L-methionine sulfoximine phosphate, and ADP to inactivated enzyme blocks enzyme-catalyzed adenylation of tyrosine residues at the adenylation sites of the enzyme.

Active site interactions of L-S- and L-R-diastereomers of L-methionine R,S-sulfoximine with *E. coli* glutamine synthetase were studied by Gorman and Ginsburg[25] using colorimetry techniques. Calculations were made of the number of proton equivalents either taken up or released for each binding reaction. These studies were carried out for the binding of the S-, R-, and R,S-isomers of L-methionine R,S-sulfoximine to unadenylated and adenylated manganese glutamine synthetase. The results indicate conformational differences between these enzyme forms, involving mainly pK perturbations of ionizable protein groups without major structural alterations.

E. coli glutamine synthetase, inactivated with L-methionine-S-sulfoximine and ATP, is completely reactivated at pH 3.5 to 4.6 in 1 *M* KCl and 0.4 *M* ammonium sulfate.[26] Both unadenylated and adenylated Mg(II) and Mn(II) enzymes are reactivated. Concomitant with reactivation is the release of one equivalent each of L-methionine-S-sulfoximine phosphate and ADP and two equivalents of Mn(II) from each subunit. When the pH of the reactivated solution is raised above pH 6 in the presence of L-methionine-S-sulfoximine phosphate, ADP, and Mn(II), complete reinactivation results, indicating the reversibility of this process.[26]

2. γ-Glutamylcysteine Synthetase

γ-Glutamylcysteine synthetase catalyzes a reaction similar to that of glutamine synthetase, as shown below.

$$\text{Glu} + \text{Cys} + \text{ATP} \overset{\text{M(II)}}{\rightleftharpoons} \gamma\text{-Glu} \cdot \text{Cys} + \text{ADP} + \text{P}_i$$

Only the L-methionine-S-sulfoximine isomer inhibits γ-glutamylcysteine synthetase and shows almost all of the properties described previously for glutamine synthetase.[27] One major difference is the tightness of binding of methionine sulfoximine phosphate. Although methionine sulfoximine phosphate is not released except by denaturation of glutamine synthetase, gel filtration in the absence of ATP and Mg(II) leads to dissociation of methionine sulfoximine phosphate from γ-glutamylcysteine synthetase. Addition of EDTA also releases the inhibitor. Since it has been postulated[14] that the methyl group of methionine sulfoximine binds to the ammonia binding site in glutamine synthetase, it was speculated by Richman et al.[27] that the absence of this site in γ-glutamylcysteine synthetase may account for the low binding of methionine sulfoximine to the latter enzyme. Inhibition of this enzyme may inhibit the transport in vivo of certain amino acids into the brain and this may also be related to the convulsant activity of methionine sulfoximine.

B. α-Alkyl Analogues of Methionine Sulfoximine

In order to determine if inhibition of glutamine synthetase or of γ-glutamylcysteine synthetase is responsible for the convulsant effect of methionine sulfoximine, α-methyl- and α-ethylmethionine sulfoximines were studied by Griffith and Meister.[28] Both of these compounds induce convulsions similar to those produced by methionine sulfoximine, but neither is metabolized to a significant extent in vivo. The implication, then, is that the convulsant effect is a result of these molecules themselves and not their metabolites. Both α-methyl- and α-ethylmethionine sulfoximine inactivate glutamine synthetase, but only the former inactivates γ-glutamylcysteine synthetase. Since they both induce convulsions, it suggests that inhibition of glutamine synthetase is the important block leading to convulsions.

C. Higher Homologues of Methionine Sulfoximine

A compound was desired that could inactivate γ-glutamylcysteine synthetase and not glutamine synthetase.[29] Since the methyl group of methionine sulfoximine was proposed to occupy the ammonia binding site,[14] a larger substituent should not bind to glutamine synthetase. However, γ-glutamylcysteine should have a larger binding site to accommodate its second substrate, cysteine (also γ-aminobutyrate), and therefore this enzyme should be inhibited by a compound with a larger S-substituent (see Structural Formula 3.8, Scheme 10).

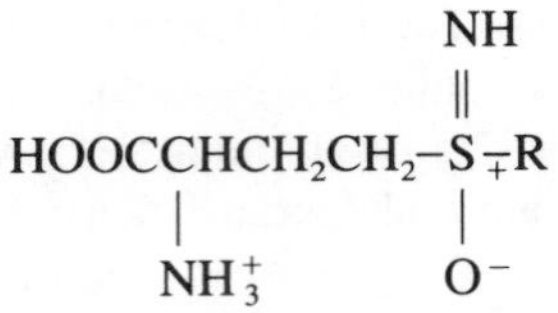

$$
\underset{\underset{NH_3^+}{|}}{HOOCCHCH_2CH_2\!-\!\underset{\underset{O^-}{|}}{S}\overset{\overset{NH}{\|}}{\mp}R}
$$

Scheme 10.   Structural Formula 3.8.

D,L-Prothionine *R,S*-sulfoximine (*S-n*-propylhomocysteine sulfoximine; **3.8**, R = Pr) and α-methyl prothionine sulfoximine (**3.8**, R = CH(CH₃)CH₂CH₃) are potent inhibitors of γ-glutamylcysteine synthetase that have little effect on glutamine synthetase; methionine sulfoximine is not as potent.[29] These results can be explained by hypothetical active sites of the two enzymes as shown in Scheme 11.

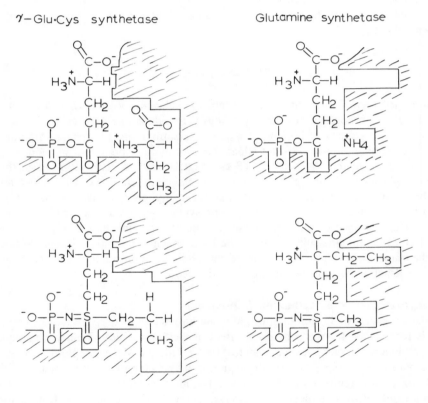

Scheme 11.

S-*n*-Butyl homocysteine sulfoximine (D,L-buthionine *R,S*-sulfoximine; **3.8**, R = Bu) and D,L-α-methylbuthionine *R,S*-sulfoximine (**3.8**, R = CH(CH₃)CH₂CH₂CH₃) are the most potent inhibitors of γ-glutamylcysteine synthetase thus far examined.[30] No detectable inhibition of glutamine synthetase is observed. The mechanism of inactivation of γ-glutamylcysteine synthetase by buthionine sulfoximine is the same as methionine sulfoximine, namely phosphorylation by ATP·Mg(II).[31] One equivalent of buthionine sulfoximine phosphate completely inhibits the enzyme. Pseudo first-order inactivation is observed. D-Buthionine *R,S*-sulfoximine, DL-buthionine *R,S*-sulfoxide, and DL-buthionine sulfone are not inactivators. Although longer alkyl chain amino acids are poor or nonsubstrates for the enzyme, pentathionine-, hexathionine-, and heptathionine sulfoximine are inactivators.[31]

The chemical ionization mass spectra of L-methionine-, DL-ethionine-, DL-prothionine-, DL-buthionine-, α-methyl-DL-methionine-, α-ethyl-DL-methionine-, and β-methyl-DL-methionine R,S-sulfoximine are unique among amino acids.[32]

D. 2-Amino-4-(methylphosphinyl)butanoic Acid (Phosphinothricin)

E. coli glutamine synthetase also is irreversibly inactivated by the phosphorus analogue of methionine sulfone, namely, 2-amino-4-(methylphosphinyl)butanoic acid (phosphinothricin, Structural Formula 3.9, Scheme 12) in the presence of ATP or adenylylimidodiphosphate, but not adenylyl(β,γ-methylene)diphosphonate.[33] When [γ-^{32}P]ATP was used, approximately 11 mol of ADP and [^{32}P] were bound per dodecamer. The inactivated enzyme can be reactivated with 50 mM acetate (pH 4.4), 1 M KCl, and 0.4 M ammonium sulfate with concomitant release of ADP, [^{32}P]phosphate, and phosphinothricin. All of these results are consistent with a mechanism of inactivation similar to that for methionine sulfoximine.

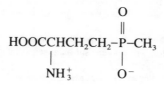

Scheme 12.   Structural Formula 3.9.

REFERENCES

1. **Bender, M. L. and Wedler, F. C.,** Phosphate and carbonate ester "aging" reactions with α-chymotrypsin. Kinetics and mechanism, *J. Am. Chem. Soc.,* 94, 2101, 1972.
2. **Hamilton, S. E., Dudman, N. P. B., deJersey, J., Stoops, J. K., and Zerner, B.,** Organophosphate inhibitors: the reactions of bis(*p*-nitrophenyl)methyl phosphate with liver carboxylesterases and α-chymotrypsin, *Biochim. Biophys. Acta,* 377, 282, 1975.
3. **Toia, R. F. and Casida, J. E.,** Phosphorylation, "aging" and possible alkylation reactions of saligenin cyclic phosphorus esters with α-chymotrypsin, *Biochem. Pharmacol.,* 28, 211, 1979.
4. **Eto, M. and Oshima, Y.,** Syntheses and degradation of cyclic phosphorus esters derived from saligenin and its analogues, *Agric. Biol. Chem.,* 26, 452, 1962.
5. **Stubbe, J. and Kenyon, G. L.,** Analogs of phosphoenolpyruvate. On the specificity of pyruvate kinase from rabbit muscle, *Biochemistry,* 10, 2669, 1971.
6. **O'Leary, M. H. and Diaz, E.,** Phosphoenol-3-bromopyruvate, a mechanism-based inhibitor of phosphoenolpyruvate carboxylase from maize, *J. Biol. Chem.,* 257, 14603, 1982.
7. **Bentley, H. R., McDermott, E. E., Moran, T., Pace, J., and Whitehead, J. K.,** Action of nitrogen trichloride on certain proteins. I. Isolation and identification of the toxic factor, *Proc. R. Soc. London Ser. B,* 137, 402, 1950.
8. **Pace, J. and McDermott, E. E.,** Methionine sulfoximine, *Nature (London),* 169, 415, 1952.
9. **Ronzio, R. A. and Meister, A.,** Phosphorylation of methionine sulfoximine by glutamine synthetase, *Proc. Natl. Acad. Sci. U.S.A.,* 59, 164, 1968.
10. **Ronzio, R. A., Rowe, W. B., and Meister, A.,** Studies on the mechanism of inhibition of glutamine synthetase by methionine sulfoximine, *Biochemistry,* 8, 1066, 1969.
11. **Rowe, W. B., Ronzio, R. A., and Meister, A.,** Inhibition of glutamine synthetase by methionine sulfoximine. Studies on methionine sulfoximine phosphate, *Biochemistry,* 8, 2674, 1969.
12. **Manning, J. M., Moore, S., Rowe, W. B., and Meister, A.,** Identification of L-methionine S-sulfoximine as the diastereoisomer of L-methionine S,R-sulfoximine that inhibits glutamine synthetase, *Biochemistry,* 8, 2681, 1969.
13. **Rowe, W. B. and Meister, A.,** Identification of L-methionine-S-sulfoximine as the convulsant isomer of methionine sulfoximine, *Proc., Natl. Acad. Sci. U.S.A.,* 66, 500, 1970.
14. **Gass, J. D. and Meister, A.,** Computer analysis of the active site of glutamine synthetase, *Biochemistry,* 9, 1380, 1970.

15. **Weisbrod, R. E. and Meister, A.,** Studies on glutamine synthetase from *Escherichia coli*. Formation of pyrrolidone carboxylate and inhibition by methionine sulfoximine, *J. Biol. Chem.*, 248, 3997, 1973.
16. **Timmons, R. B., Rhee, S. G., Luterman, D. L., and Chock, P. B.,** Mechanistic studies of glutamine synthetase from *Escherichia coli*. Fluorometric identification of a reactive intermediate in the biosynthetic reaction, *Biochemistry*, 13, 4479, 1974.
17. **Villafranca, J. J., Ash, D. E., and Wedler, F. C.,** Evidence for methionine sulfoximine as a transition-state analog for glutamine synthetase from NMR and EPR data, *Biochem. Biophys. Res. Commun.*, 66, 1003, 1975.
18. **Villafranca, J. J., Ash, D. E., and Wedler, F. C.,** Manganese(II) and substrate interaction with una-denylated glutamine synthetase (*Escherichia coli* W). II. Electron paramagnetic resonance and nuclear magnetic resonance studies of enzyme-bound manganese (II) with substrates and a potential transition-state analogue, methionine sulfoximine, *Biochemistry*, 15, 544, 1976.
19. **Hofmann, G. E. and Glaunsinger, W. S.,** EPR investigation of the Mn(II) binding sites in glutamine synthetase (*Escherichia coli* W). I. High-affinity binding sites, *J. Biochem.*, 83, 1769, 1978.
20. **Hofmann, G. E. and Glaunsinger, W. S.,** EPR investigation of the Mn(II) binding sites in glutamine synthetase (*Escherichia coli* W). II. Intermediate-affinity binding sites, *J. Biochem.*, 83, 1779, 1978.
21. **Shrake, S., Whitley, E. J., Jr., and Ginsburg, A.,** Conformational differences between unadenylated and adenylated glutamine synthetase from *Escherichia coli* on binding L-methionine sulfoximine, *J. Biol. Chem.*, 255, 581, 1980.
22. **Shrake, A., Ginsburg, A., Wedler, F. C., and Sugiyama, Y.,** On the binding of L-S- and L-R-diaster-eoisomers of the substrate analog L-methionine sulfoximine to glutamine synthetase from *Escherichia coli*, *J. Biol. Chem.*, 257, 8238, 1982.
23. **Rhee, S. G., Chock, P. B., Wedler, F. C., and Sugiyama, Y.,** Subunit interaction in unadenylated glutamine synthetase from *Escherichia coli*. Evidence from methionine sulfoximine inhibition studies, *J. Biol. Chem.*, 256, 644, 1981.
24. **Hunt, J. B. and Ginsburg, A.,** Mn^{2+} and substrate interactions with glutamine synthetase from *Escherichia coli*, *J. Biol. Chem.*, 255, 590, 1980.
25. **Gorman, E. G. and Ginsburg, A.,** Binding enthalpies for glutamine synthetase interactions with L-S- and L-R-diastereoisomers of the substrate analog L-methionine sulfoximine, *J. Biol. Chem.*, 257, 8244, 1982.
26. **Maurizi, M. R. and Ginsburg, A.,** Reactivation of glutamine synthetase from *Escherichia coli* after autoinactivation with L-methionine-S-sulfoximine, ATP, and Mn^{2+}, *J. Biol. Chem.*, 257, 4271, 1982.
27. **Richman, P. G., Orlowski, M., and Meister, A.,** Inhibition of γ-glutamylcysteine synthetase by L-methionine-S-sulfoximine, *J. Biol. Chem.*, 248, 6684, 1973.
28. **Griffith, O. W. and Meister, A.,** Differential inhibition of glutamine and γ-glutamylcysteine synthetases by α-alkyl analogs of methionine sulfoximine that induce convulsions, *J. Biol. Chem.*, 253, 2333, 1978.
29. **Griffith, O. W., Anderson, M. E., and Meister, A.,** Inhibition of glutathione biosynthesis by prothionine sulfoximine (S-n-propyl homocysteine sulfoximine), a selective inhibitor of γ-glutamylcysteine synthetase, *J. Biol. Chem.*, 254, 1205, 1979.
30. **Griffith, O. W. and Meister, A.,** Potent and specific inhibition of glutathione synthesis by buthionine sulfoximine (S-n-butyl homocysteine sulfoximine), *J. Biol. Chem.*, 254, 7558, 1979.
31. **Griffith, O. W.,** Mechanism of action, metabolism, and toxicity of buthionine sulfoximine and its higher homologs, potent inhibitors of glutathione synthesis, *J. Biol. Chem.*, 257, 13704, 1982.
32. **Cooper, A. J. L., Griffith, O. W., Meister, A., and Field, F. H.,** Chemical ionization mass spectra of L-methionine and L-methionine analogs, *Biomed. Mass Spectrom.*, 8, 95, 1981.
33. **Colanduoni, J. A. and Villafranca, J. J.,** Inhibition of *Escherichia coli* glutamine synthetase by phos-phinothricin, *Bioorg. Chem.*, 14, 163, 1986.

Chapter 4

ADDITION REACTIONS*,**

I. ADDITION/ADDITION

A. 5-Fluoro-2'-deoxyuridine Monophosphate (FdUMP)

Because the van der Waals radius of fluorine (1.35 Å) is similar to that of hydrogen (1.20 Å), 5-fluorouracil was originally prepared by Heidelberger et al.[1] as an antimetabolite of uracil. It was shown to be metabolized to 5-fluoro-2'-deoxyuridine monophosphate (Structural Formula 4.1, FdUMP; Scheme 1) and it was proposed that the site of action of FdUMP was the methylation of deoxyuridine monophosphate,[2] the last step in the *de novo* biosynthesis of thymidylate.

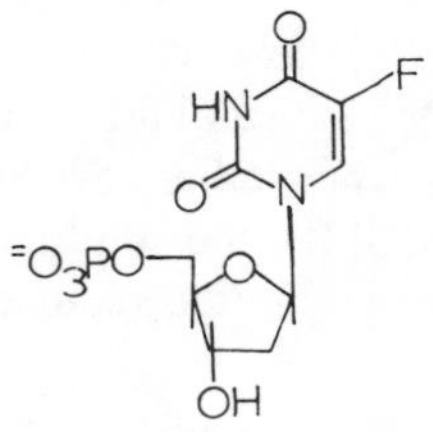

Scheme 1.   Structural Formula 4.1.

Early reports regarding the type of inhibition of thymidylate synthetase exhibited by FdUMP are in conflict. It was originally claimed that FdUMP was a competitive reversible inhibitor of thymidylate synthetase from Ehrlich ascites carcinoma cells,[3,4] but that it was an irreversible inactivator of the *E. coli* enzyme[5] and that of bacteriophage-infected *E. coli* B and 15$_T$.[6] Inactivation of the latter enzymes was shown to require preincubation of FdUMP. Thymidylate synthetase from T4 bacteriophage-infected *E. coli* is inactivated by FdUMP in the presence of the cofactor, 5,10-methylenetetrahydrofolate (CH_2H_4folate) to give a 1:1 stoichiometry with an average turnover number (two different experiments) of 113 molecules of tetrahydrofolate oxidized per minute per FdUMP binding site.[7] The FdUMP remains bound after gel filtration, but is released upon NaDodSO$_4$ denaturation. FdUMP, in the presence of CH_2H_4folate, reacts with T2 bacteriophage-induced thymidylate synthetase.[8] Increasing amounts of FdUMP results in a proportionate decrease in enzyme activity to a maximum of 2 mol FdUMP per mole of enzyme, at which amount the enzyme is completely inactive. The complex is stable to 6 *M* guanidine hydrochloride, but half of the FdUMP is removed by 6 *M* urea. Extensive dialysis against buffer results in loss of native enzyme activity and loss in its capacity to bind FdUMP; addition of 2-mercaptoethanol partially restores both properties. This suggests that a cysteine may be involved in FdUMP inactivation and catalysis. Thymidylate synthetase from *Streptococcus faecalis* can be titrated with FdUMP, resulting in irreversibly-inactivated enzyme.[9] Binding of FdUMP to either *E. coli*[10] or chick embryo[11] thymidylate synthetase requires the presence of CH_2H_4folate. The earlier reports[3,4] regarding reversibility of the inhibition were modified when it was shown that the inactivation of thymidylate synthetase from Ehrlich ascites carcinoma cells by [2-^{14}C]FdUMP produces a covalent adduct in the presence of CH_2H_4folate.[12] The complex, shown to contain the

---

* Most of the examples of addition reactions involve 5-fluoro 2'-deoxyuridylate and its analogues as inactivators of thymidylate synthetase.

** A list of abbreviations and shorthand notations can be found prior to Chapter 1.

cofactor by using [^{14}C]CH$_2$H$_4$folate, is stable to NaDodSO$_4$ denaturation and trichloroacetic acid precipitation. The structure of the proposed inactivator complex with FdUMP and CH$_2$H$_4$folate (Structural Formula 4.2, Scheme 2) is the same as that believed to be involved in the enzyme reaction with dUMP.

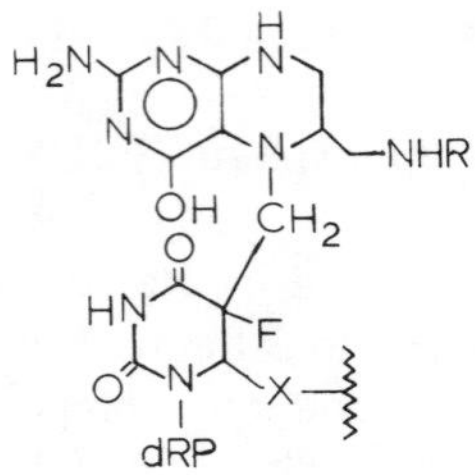

Scheme 2.   Structural Formula 4.2.

Because of the specificity of FdUMP as a covalent inactivator for thymidylate in the presence of CH$_2$H$_4$folate, [6-^{3}H]FdUMP is useful for the assay of enzyme activity in different tissues or crude homogenates, even when other reversible inhibitors are present.[13] After titration of the enzyme with [6-^{3}H]FdUMP, the solution to be assayed is poured through a nitrocellulose filter membrane which retains the radioactively labeled enzyme, but allows the free nucelotide to pass through.

Once it was found that methotrexate-resistant *Lactobacillus casei* is rich in thymidylate synthetase in a stable form,[14] this was used as the source for enzyme studies; and, unless otherwise stated, it will be the source of the thymidylate synthetase discussed henceforth. Santi and McHenry[15] showed that FdUMP is a time-dependent inactivator of thymidylate synthetase. No inactivation is observed in the absence of CH$_2$H$_4$folate; the rate and extent of inactivation depends upon the cofactor concentration. The corresponding nucleoside does not bind to nor inactivate the enzyme. Two moles of [6-^{3}H]FdUMP are required to inactivate one mole of enzyme dimer; the FdUMP-enzyme complex is not dissociated by denaturation of the enzyme, suggesting a covalent bond has formed. On the basis of model studies for the thymidylate synthetase-catalyzed reaction,[16,17] and the observation that the FdUMP chromophore at 269 nm is lost upon formation of the inactivated enzyme, a mechanism was proposed involving addition of an active site nucleophile across the pyrimidine 5,6-double bond[15] (Scheme 3). With the use of [6-^{3}H]- or [2-^{14}C]FdUMP as inactivator, 2 mol of inactivator are incorporated per mole of enzyme dimer.[18] Once the ternary complex (enzyme − CH$_2$H$_4$folate − [^{3}H]FdUMP) forms, it is stable for long periods, provided excess cofactor is present and temperatures remain below 37°C. Guanidine hydrochloride denaturation does not release radioactivity from the protein, and proteolytic digestion gives peptide fragments which contain the inactivator. Consistent with the loss of the pyrmidine chromophore[15,18] is a secondary tritium isotope effect (k_H/k_T) of 1.23 for the dissociation of [6−^{3}H]FdUMP from the complex, indicating a rehybridization of sp^3 to sp^2 during this process. Furthermore, upon inactivation, the optical spectrum of the CH$_2$H$_4$folate is altered. These data are consistent with a modified mechanism of inactivation (Scheme 4) which is believed to resemble the catalytic mechanism.

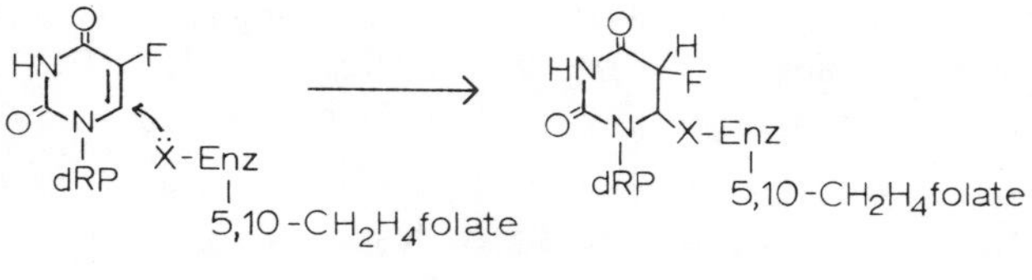

Scheme 3.

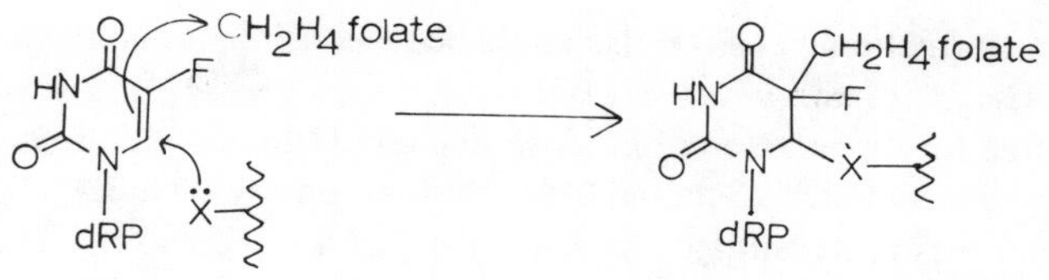

Scheme 4.

A model study for the inactivation of thymidylate synthetase by FdUMP was carried out by Kalman[19] with bisulfite as the nucleophile. An inverse isotope effect was observed ($k_H/k_D = 0.88$) when deuterium is in the C-6 position. The elimination reaction exhibits no secondary isotope effect; a primary isotope effect of 3.8 is observed when deuterium is at position 5. It was proposed that bisulfite adds rapidly in a Michael addition followed by rate determining tautomerization (Scheme 5). In the reverse reaction, a primary isotope effect at position 5 would occur, but since rehybridization follows the rate determining step, it is not observed. With regards to the enzyme-catalyzed inactivation, it was suggested that the rate-determining step is a conformational change of the enzyme prior to the nucleophilic addition of the active site cysteine to FdUMP.[19]

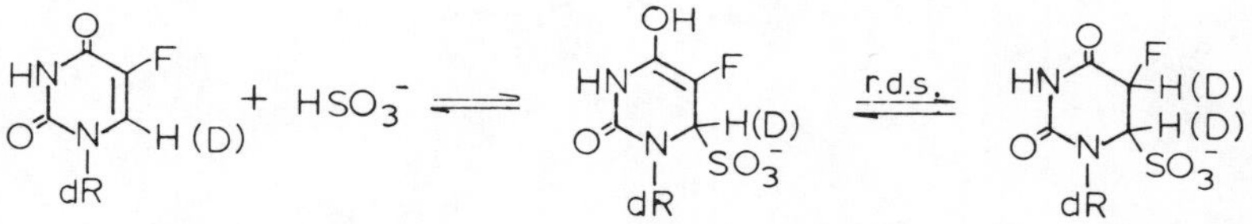

Scheme 5.

As suggested by the secondary tritium isotope effect experiment of Santi et al.,[18] inactivation of thymidylate synthetase by FdUMP is reversible. FdUMP substituted with deuterium at the C-6 position was prepared to investigate the mechanism of the reverse reaction.[19] Reversal of inactivation exhibits a secondary deuterium isotope effect of $k_H/k_D = 1.24$, indicating a change in hybridization at C-6 from sp^3 to sp^2 as was observed with [6-^{3}H]FdUMP.[18] With the use of a mixture of [6-^{3}H]FdUMP and [^{14}C]FdUMP for inactivation, no isotope discrimination was observed, suggesting that no secondary isotope occurs or that it occurs after the rate determining step of the reaction.[19] In accord with the principle of microscopic reversibility, fully active enzyme and FdUMP are obtained in the reverse reaction, provided reassociation of the inactivator is prevented by the addition of a high concentration of substrate (dUMP).[20] The reactivation is temperature dependent with an activation energy of 26.5 kcal/mol. Another study of the secondary α-hydrogen isotope effects on the formation and dissociation of the ternary complex, formed upon inactivation of thymidylate synthetase by FdUMP and CH$_2$H$_4$folate, was carried out.[21] Dissociation of [6-^{3}H]FdUMP from the complex again was shown to proceed with a secondary α-hydrogen kinetic isotope effect (k_H/k_T) of 1.23. There is no isotope effect on the formation of the complex, but isotopic equilibration occurs with an equilibrium isotope effect (K_H/K_T) of 1.24. Bruice and Santi[21] also conclude that the covalent bond formed by attack of the enzyme active site residue at C-6 of FdUMP is formed after the rate-determining step. By microscopic reversibility, it is cleaved prior to the slow step in the dissociation of the complex. Dissociation of [^{3}H]CH$_2$H$_4$folate from the complex occurs with a secondary kinetic isotope effect (k_H/k_T) of 1.03. It, again, is suggested that an enzyme conformational change occurs upon ternary complex formation; this perturbation may be the rate-determining step which occurs with, or shortly after, the initial reversible binding of CH$_2$H$_4$folate to the FdUMP-enzyme complex. This conformational change also could be responsible for conversion of the cofactor to 5-immonium CH$_2$H$_4$folate.

A kinetic scheme for the reaction of FdUMP and CH$_2$H$_4$folate with thymidylate synthetase was suggested by Santi et al.[18] (Scheme 6), the initial ternary complex is not covalent. The kinetic mechanism was reinvestigated by Danenberg and Danenberg.[22] The rate of recovery of enzyme activity is not affected by the concentration of added dUMP, suggesting first-order dissociation of FdUMP. Dissociation of [^{3}H]FdUMP from the complex occurs at the same rate as regeneration of enzyme activity in the absence of CH$_2$H$_4$folate, but the rate is decreased with increasing levels of cofactor linearly. Conversely, dissociation of [2-^{14}C] CH$_2$H$_4$folate is unaffected by increasing FdUMP concentrations. These results and other competitive studies indicate a sequential ordered mechanism for ternary complex formation and the enzymatic reaction in which binding of the nucleotide precedes binding of CH$_2$H$_4$folate (Scheme 7), a modification of the earlier proposal.[18]

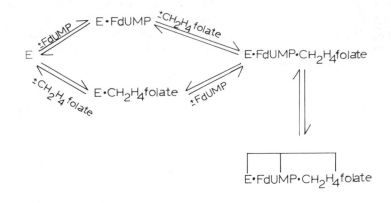

Scheme 6.

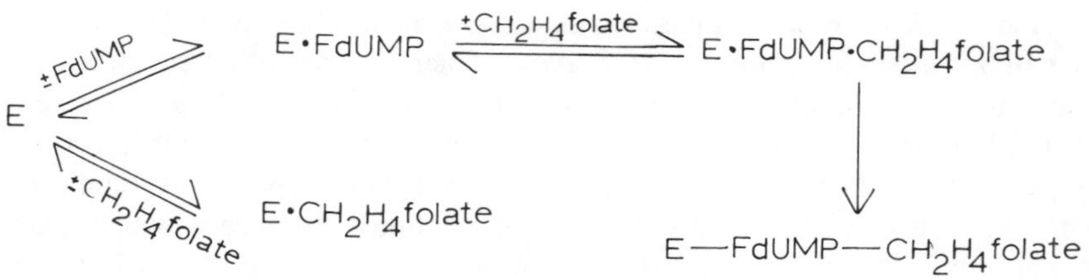

Scheme 7.

Since the 5,6-double bond of FdUMP appears to become saturated during inactivation,[15,18,23] the identity of the nucleophile that adds to the nucleotide was sought. A comparison of the total sulfhydryl content of thymidylate synthetase with that of the enzyme inactivated by FdUMP and CH$_2$H$_4$folate shows no difference when the enzyme is denatured with NaDodSO$_4$.[24] In the absence of NaDodSO$_4$, the inactivated enzyme has 1.5 fewer cysteine residues/enzyme than does native enzyme. It was concluded by Sommer and Santi,[25] however, that cysteine is **not** the nucleophile that attacks the C-6 position of substrate and inactivator. Inactivation of thymidylate synthetase with [6-^{3}H]FdUMP and CH$_2$H$_4$folate, followed by Pronase digestion and purification gives a peptide fragment containing the inactivator and CH$_2$H$_4$folate. The peptide contains 1 Glu, 2 Pro, 1 Ala, 1 Leu, 1 Thr, and 1 His. Since only Thr and His are nucleophilic, it was suggested that one of these amino acids is involved in the inactivation.

Inactivation of thymidylate synthetase with [6-^{3}H]FdUMP and [^{14}C]CH$_2$H$_4$folate was shown by Danenberg et al.[23] to give enzyme labeled with both [^{3}H] and [^{14}C] in a 1:1 stoichiometry. Tryptic digestion produces a peptide containing both [^{3}H] and [^{14}C]. The methylene group of CH$_2$H$_4$folate is required for covalent complex formation. Contrary to the earlier reports,[24,25] an active site cysteine residue was identified as being involved in the addition to the 5,6-double bond of FdUMP. When the enzyme is treated, in the absence of dUMP, with [^{14}C]iodoacetamide followed by acid hydrolysis, carboxymethylcysteine is formed; this is not the case in the presence of dUMP.[23] Raney nickel treatment of the ternary complex formed from [6-^{3}H]FdUMP and [^{14}C]CH$_2$H$_4$folate releases both isotopes simultaneously, indicating that the inactivator and cofactor are bound via a single sulfide linkage.[23] Further elegant studies by Danenberg and Heidelberger[26] with Raney nickel, which specifically reduces C–S bonds with no breakage of polypeptide chains, showed that the half-times for release of the [^{3}H] and [^{14}C] are identical. Chromatography of the released radioactivity indicated that both labels migrate in the same spot, intermediate between carrier FdUMP and CH$_2$H$_4$folate. In a very clever experiment in which *Lactobacillus casei* was grown in the presence of [^{35}S]cysteine to give [^{35}S]thymidylate synthetase, it was shown[26] that Raney nickel treatment of the ternary complex formed after inactivation of this [^{35}S] enzyme with [^{3}H]FdUMP and CH$_2$H$_4$folate results in release of both tritium and [^{35}S] at identical rates. All of these results support the involvement of a cysteine in the inactivation mechanism. The earlier conclusions of Santi and coworkers[24,25] regarding the identity of the active site nucleophile involved in the inactivation reaction were changed.[27] The structure of the peptide isolated by Pronase digestion of thymidylate synthetase which was inactivated by [^{3}H]FdUMP and CH$_2$H$_4$ folate was reinvestigated; one cysteine residue was indeed identified, and the isolated peptide was shown to have the structure Ala–Leu–Pro–Pro–(His,Cys)–Thr. The imidazole of the histidine can be modified without cleavage of the FdUMP; therefore, histidine is not attached to the FdUMP. The stability of the peptide fragment to air oxidation supports a hypothesis that attachment of CH$_2$H$_4$folate to FdUMP is through an N-5 methylene rather than an N-10 methylene linkage, since 5-substituted derivatives are known to be stable, whereas 10-substituted ones are not. Other experiments pertaining to the active site cysteine residue have been described. Studies with sulfhydryl modifying reagents indicate the loss of 1.4 to 1.8 sulfhydryls per enzyme dimer.[28] FdUMP was used to titrate thymidylate synthetase in the presence of CH$_2$H$_4$folate, and it was found that 1.6 molecules per enzyme dimer are bound. Loss of enzyme activity correlates with FdUMP binding and sulfhydryl modification. Experiments with sulfhydryl modifying reagents in the presence of FdUMP and CH$_2$H$_4$folate also suggest that thymidylate synthetase has an asymmetric arrangement of subunits.[29] Only one binding site is accessible on the free dimeric enzyme until the binding of both FdUMP (or dUMP) and CH$_2$H$_4$folate occurs. This, then, causes a conformational change which exposes the second binding site and the sulfhydryl group contained in it (a ligand-induced sequential model for subunit interactions). Alkylation of one of the four cysteines of thymidylate synthetase with either *N*-ethylmaleimide or iodoacetic acid prevents the binding of dUMP and the formation of the ternary complex of enzyme–FdUMP–CH$_2$H$_4$folate, but does not affect the binding of the cofactor.[30] This indicates that there are separate binding sites for FdUMP and CH$_2$H$_4$folate and that the enzyme can be inactivated in a way that does not prevent binding of the cofactor. Although four cysteines are present in native denatured thymidylate synthetase, only two cysteines were found in enzyme inactivated with FdUMP and CH$_2$H$_4$folate.[31] One equivalent of 5-mercurideoxyuridylate or *N*-ethylmaleimide is incorporated into native enzyme, but neither is incorporated when the enzyme is present as the ternary complex. FdUMP alone, however, does not prevent reactivity of the thiol reagents with cysteine. An explanation was offered by Galivan et al.[31] for the earlier results of McHenry and Santi[24] who observed no difference in the incorporation of *N*-ethylmaleimide into denatured native and ternary complex enzyme and, therefore, concluded that cysteine is not involved in the inactivation. *N*-Ethylmaleimide reacts with 1.5 additional residues other than cysteine (suggested to be histidine and lysine). These results also support the hypothesis that cysteine is involved in ternary complex formation.[31] Thymidylate synthetase, converted to a ternary complex with [2-^{14}C]FdUMP and CH_2H_4folate, then *S*-carboxymethylated with iodoacetate, was digested with cyanogen bromide to give four distinct peptide bands on gel electrophoresis.[32] The same results were obtained with native enzyme in which all four of its cysteine residues were *S*-carboxymethylated with iodo[1-^{14}C]acetate. In both cases only one peptide band was radiolabeled. Amino acid analyses of each reveal that they are almost identical except for the presence of two *S*-carboxymethylcysteine residues in the peptide from native enzyme and only one in the peptide from FdUMP-CH_2H_4folate-inactivated enzyme. The peptides were sequenced and shown to be identical except for the fifth residue, found to be cysteine in the native enzyme peptide, which could not be identified in the inactivated enzyme peptide. This again suggests that FdUMP is bound to a cysteine residue. The cyanogen bromide peptide which contains both of the cysteine residues was sequenced by Bellisario.[33] FdUMP reacts at Cys-198; the other cysteine residue is Cys-244.

Moore et al.[33a] used [2-^{14}C]FdUMP and [3',5',7,9-^{3}H]-(+)-CH_2H_4folate to inactivate the enzyme. Following trichloroacetic acid precipitation, cyanogen bromide fragmentation, and peptide analysis, a labeled pentapeptide was isolated. The sequence is Ala–Leu–Pro–Pro–Cys, where the cysteine residue, shown to be Cys-198, was associated with the labeled nucleotides and cofactor. The same active site peptide was isolated when [2-^{14}C]dUMP was used in place of FdUMP. These results show, conclusively, that a ternary complex between substrate (or inactivator), cofactor, and Cys-198 is involved in the enzyme reaction.

The trichloroacetic acid precipitation method was employed by Moore et al.[33b] to examine the binding the FdUMP, dUMP, and dTMP to thymidylate synthetase and the effect of various folates on binding. FdUMP covalently binds to the enzyme in the absence of added folates. The presence of folate, H_2folate, and H_4folate triggers increasing covalent binding enhancement of the nucleotides. HPLC isolation of the peptides formed from CNBr cleavage of FdUMP-, dUMP-, and dTMP-labeled thymidylate synthetase showed that the same pentapeptide, Ala·Leu·Pro·Pro·Cys–X, was obtained in each case. Cys-198 is attached to all of the nucleotides, supporting the concept that the same catalytic mechanism prevails for the three nucleotides.

The 94 MHz ^{19}F NMR spectrum of the peptide[27] isolated by Pronase digestion of thymidylate synthetase inactivated with [2,6-^{14}C]FdUMP in the presence of [6-^{3}H]CH_2H_4folate supports a structure in which the peptide is bound to C-6 of FdUMP and the CH_2H_4folate is bound to C-5 of FdUMP.[34] The configuration across the 5,6-bond is *trans* and pseudoequatorial (Structural Formula 4.3, Scheme 8).

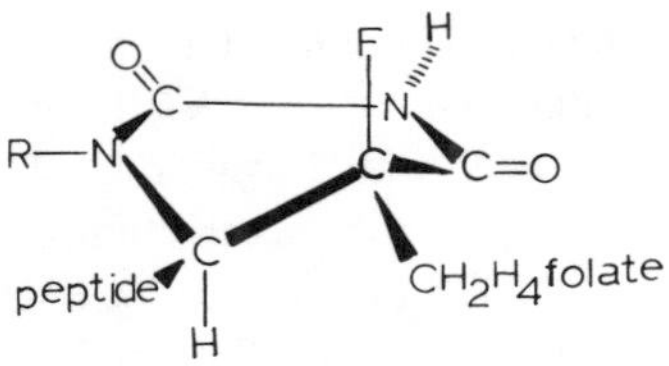

Scheme 8.   Structural Formula 4.3.

This suggests that if initial attack on the uracil ring occurs perpendicular to its plane, and CH_2H_4folate approach also is perpendicular, the ternary complex would have these groups *trans* and axial. In order for subsequent proton elimination to occur, an inversion must take place to put the C-5 proton axial (as shown for F above). James et al.[34] concluded, then, that elimination is syn. Other ^{19}F NMR spectroscopic studies of the ternary complex of thymidylate synthetase–CH$_2$H$_4$folate–FdUMP were carried out by Byrd et al.[35] The binding of FdUMP to form the ternary complex is accompanied by a 12.4 ppm upfield shift. With the use of model compounds, this shift was attributed to the nucleophilic attack of the pyrmidine ring followed by attachment of C-5 to the cofactor. Concomitant with this shift is the loss of H–F coupling. When the ternary complex is formed with CD$_2$H$_4$folate and ^{13}CD$_2$H$_4$folate, a ^{13}C–^{19}F coupling is observed, indicating that a covalent bond is formed between C-5 of FdUMP and the methylene of CH$_2$H$_4$folate. Denaturation of the ternary complex with NaDodSO$_4$ causes a 10.5 ppm shift downfield, yet the ^{13}C–^{19}F coupling remains intact. Comparison with model compounds suggests that upon denaturation, the C–F bond has moved relative to the anisotropy of the adjacent carbonyl group. Further spectral analyses indicate that in the native ternary complex one proton of the methylene of CH$_2$H$_4$folate is *trans* to the fluorine and the other is gauche. The C-6 proton of FdUMP and the fluorine are in a pseudo *trans* diequatorial relationship. Therefore, the cysteine and the methylene of CH$_2$H$_4$folate must be *trans* diaxial, a finding which is in disagreement with the results of James et al.,[34] who found that in the peptide fragment containing bound FdUMP and CH$_2$H$_4$folate, the cysteine and methylene groups are *trans* diequatorial. However, it is shown in the study by Byrd et al.[35] that upon denaturation, these two groups become *trans* diequatorial, which is in agreement with the peptide work of James et al.[34] The results in the native ternary complex, however, are the relevant data, and they are consistent with an approach of the cysteine residue which is perpendicular to the plane of the FdUMP ring followed by trapping of the incipient carbanion perpendicular to the ring. Therefore, it is not necessary to postulate ring inversion in the mechanism as previously proposed[34] based on their findings with the peptide. A ^{19}F NMR study of the reaction of FdUMP and folate derivatives with thymidylate synthetase was carried out by Lewis et al.[36] to determine what changes in the nucleotide might occur upon binding of CH$_2$H$_4$folate just prior to formation of the covalent bond between FdUMP and CH$_2$H$_4$folate. In all but one complex (methotrexate), the spectra indicate that the folates interact noncovalently with FdUMP in the thymidylate synthetase-FdUMP binary complex at essentially stoichiometric concentrations. The orientation of the N$_5$ substituent of the tetrahydropyrazine ring is the most striking structural feature which prohibits the strong interaction of folates with the binary complex. On the basis of these and other studies, a mechanism for ternary complex formation was proposed (Scheme 9).

The ^{19}F NMR spectrum of the intact ternary complex of thymidylate synthetase and CH$_2$H$_4$folate with 2′,5-difluoro-2′-deoxyuridylate, a good inhibitor of the enzyme, also was studied by Byrd et al.[37] This compound was shown to form two complexes in the ratio 1:1:1 and 2:2:1 (inhibitor:CH$_2$H$_4$folate:enzyme), as is the case with FdUMP (*vide infra*). The resonance due to the 2′-fluorine shifts upfield 2.4 ppm upon formation of the ternary complex as a result of changes in the microenvironment. The 5-fluorine resonance shifts upfield 11.9 ppm, as a result of a change in the bonding environment. In the ^{19}F spectroscopic study of James et al.[34] on a peptide containing the FdUMP and CH$_2$H$_4$folate, no upfield shift from that of FdUMP was observed.

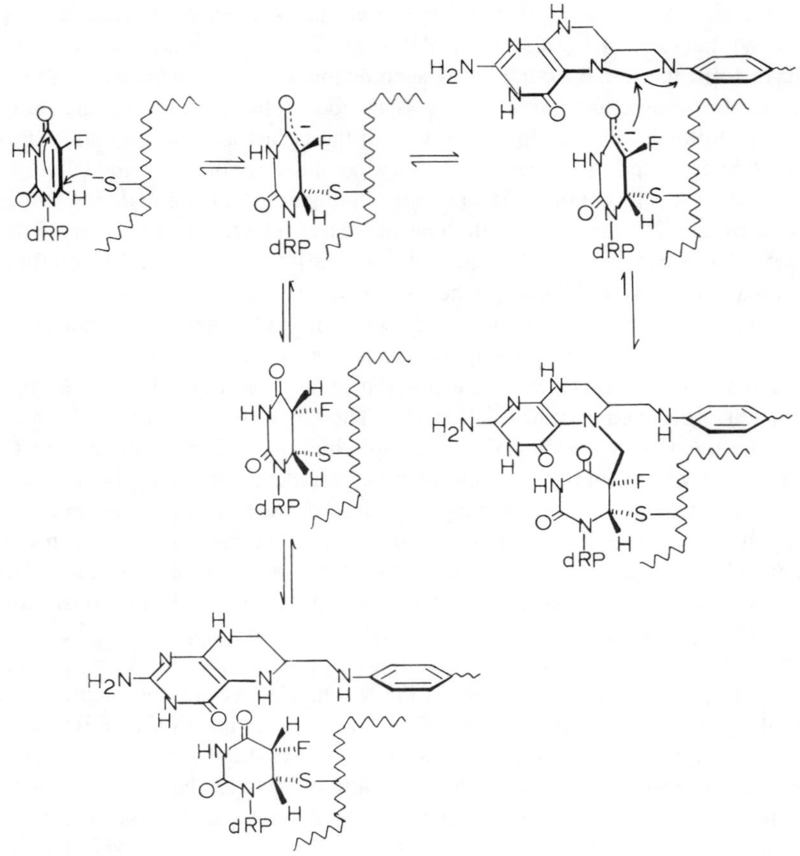

Scheme 9.

The nature of the ternary complexes formed with thymidylate synthetase, FdUMP, and CH₂H₄folate has been examined by several groups. Incubation of thymidylate synthetase with [6-³H]FdUMP and CH₂H₄[³H]folate results in the formation of two ternary complexes which are separable from each other and native enzyme by gel electrophoresis.[38,39] Spectral gel scans were used to show that the complexes are 1:1:1 and 2:2:1 mixtures of FdUMP:CH₂H₄folate:enzyme. The complexes are stable to gel filtration. This explains the differences in the stoichiometry of the ternary complex reported by other groups in the earlier literature. Santi and co-workers[15,18] and Galivan et al.[8] suggested a ratio of 2:1 for FdUMP:enzyme, but Heidelberger and collaborators[12,23] indicated a 1:1 stoichiometry. Titration of thymidylate synthetase with FdUMP in the presence of CH₂H₄folate gave two ternary complexes; when this titration is carried out with enzyme isolated at each step in its purification, 30% of the enzyme forms a 1:1:1 ternary complex and the remaining 70% forms a 2:2:1 ternary complex, regardless of the stage of enzyme purity.[28] This explains the controversy. Carboxypeptidase A treatment of thymidylate synthetase was shown by Galivan et al.[40] to prevent formation of the covalent ternary complex of (+)-CH₂H₄-folate, FdUMP, and enzyme, but not to prevent binding of (+)-CH₂H₄folate and FdUMP to the enzyme in the usual 2:2:1 ratio. Equilibrium dialysis experiments on the ternary complex formed from thymidylate synthetase, FdUMP, and (+)-CH₂H₄folate indicate that 2 mol of FdUMP are bound to 1 mol of enzyme in both phosphate and Tris-HCl buffers.[41] In the absence of cofactor, 2 mol of FdUMP are bound per mole of enzyme in Tris-HCl buffer, but only 1 mol is bound in potassium phosphate buffer. The phosphate ion, apparently, interferes with binding.

Other spectroscopic studies on inactivation of thymidylate synthetase by FdUMP have been reported. Quenching of tryptophan fluorescence of thymidylate synthetase occurs during titration with FdUMP in the presence of ($-$)-L-CH$_2$H$_4$folate.[42] Neither FdUMP nor ($-$)-L-CH$_2$H$_4$folate alone had an effect on enzyme fluorescence, but ($+$)-L-CH$_2$H$_4$folate alone causes quenching. This result was rationalized that during ternary complex formation with ($-$)-L-CH$_2$H$_4$folate, C-6 may invert its configuration and then produce quenching. A circular dichroism spectroscopic study of thymidylate synthetase in the presence of FdUMP and various folate analogues indicates that there are two binding sites for FdUMP and for each of the diastereomers of CH$_2$H$_4$folate.[43] The circular dichroism pattern produced by the ternary complex involving the active ($+$)-isomer of CH$_2$H$_4$folate is similar to that produced from the ($\pm$) mixture, indicating a binding preference of the active isomer. Titrations of thymidylate synthetase in the presence of FdUMP with either ($+$)- or ($\pm$)-CH$_2$H$_4$folate were monitored by absorption, fluorescence, and circular dichroic spectroscopy and polyacrylamide gel electrophoresis.[44] The ternary complex is formed primarily with the natural ($+$)-diastereomer of CH$_2$H$_4$folate, but the ($-$)-isomer is a weak competitor. The data suggest that CH$_2$H$_4$folate is bound at only one site on the enzyme; however, in the presence of excess FdUMP, there appear to be two binding sites for CH$_2$H$_4$folate. The laser-Raman spectra of the thymidylate synthetase ternary complex with FdUMP and ($+$)-L-CH$_2$H$_4$folate were studied by Sharma et al.[45] using an argon ion laser. A new band at 1618 cm^{-1} was assigned to a C=N stretching vibration, possibly due to formation of dihydrofolate or an immonium ion. Although the overall secondary structure of the enzyme does not change on ternary complex formation, local changes such as ionized carboxyl, tryptophan, and CH$_3$ and CH$_2$ deformation modes occur. The ultraviolet difference spectrum of the ternary inactivated enzyme complex, E–FdUMP–CH$_2$H$_4$folate, and E·CH$_2$H$_4$folate is similar to that for dihydrofolate minus CH$_2$H$_4$folate.[46] In order to determine if the cofactor has been oxidized during inactivation of the enzyme with FdUMP, [6-^{3}H]CH$_2$H$_4$folate was used in the inactivation. Dissociation of the bound cofactor shows that the tritium is still at the 6-position. If the cofactor had been oxidized during inactivation, the [^{3}H] would have been transferred to the methylene. Also, the same anomalous cofactor spectrum upon inactivation was observed using an analogue of the cofactor which is not susceptible to oxidation. Therefore, the cofactor is not oxidized during inactivation and the proposed structure is shown in Scheme 10.

On the basis of resonance Raman spectroscopy of the ternary complex between 5-FdUMP and *L casei* thymidylate synthase, two possible structures are proposed by Fitzhugh et al.[46a] (Structural Formulas 4.3a and 4.3b, Scheme 10A).

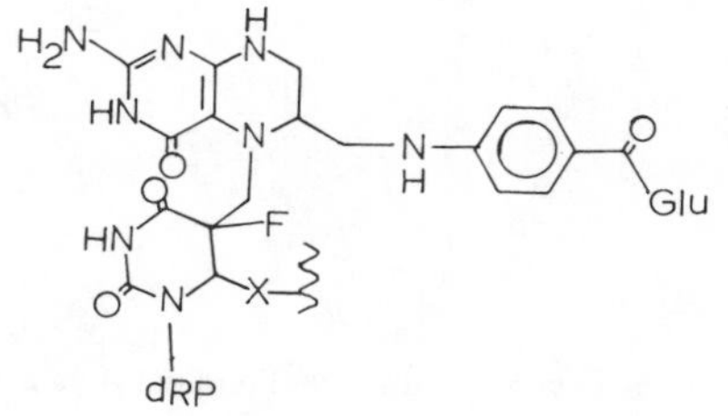

Scheme 10. Structural Formula 4.4.

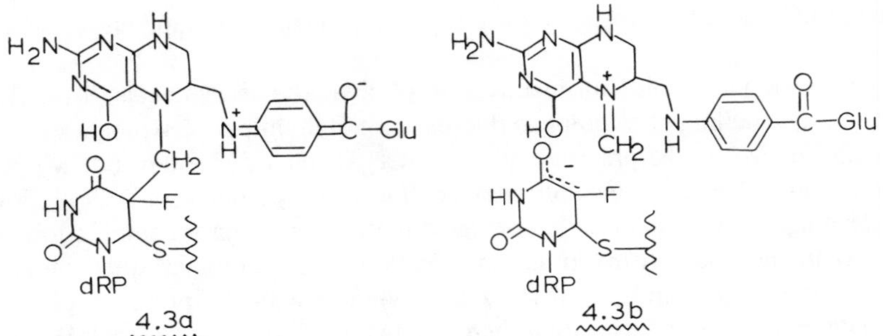

Scheme 10A.   Structural Formulas 4.3a and 4.3b.

No information regarding the site of attachment of FdUMP to CH$_2$H$_4$folate, i.e., at N-5 or N-10, is known except that the UV spectrum of a peptide containing FdUMP and CH$_2$H$_4$folate covalently bound is similar, but not identical, to that for 5-methyltetrahydrofolate[25] and that this peptide is stable to air.[27] A chemical approach to the identification of the site of attachment was taken by Pellino and Danenberg.[47] Degradative reactions were carried out on the ternary complex which are known to specifically cleave the C9-N10 bond. The two possible cleavage reactions are shown in Scheme 11.

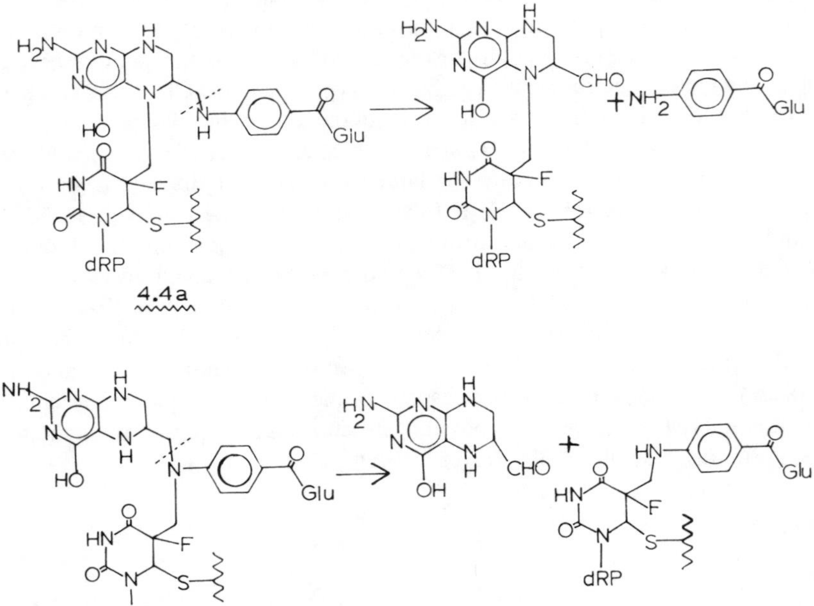

Scheme 11.   Containing Structural Formulas 4.4a and 4.4b.

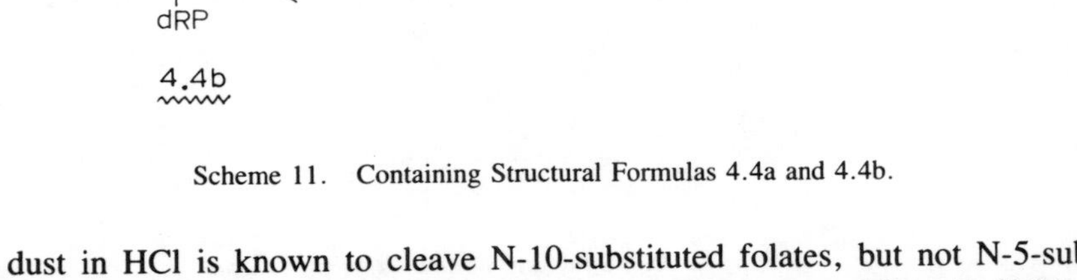

Zinc dust in HCl is known to cleave N-10-substituted folates, but not N-5-substituted ones. These conditions did not degrade the ternary complex prepared from [2-^{14}C,7,9,3′,5′-^{3}H] CH$_2$H$_4$folate, FdUMP, and thymidylate synthetase, thus supporting an N-5-substituted complex. The Bratton-Marshall reaction[48] of primary aromatic amines (nitrous acid treatment followed by conversion of the diazonium ion to a purple chromophoric species by reaction with α-naphthylenediamine) was carried out. In a control reaction it was shown that 5-methyltetrahydrofolate, but not 10-substituted tetrahydrofolates, is oxidized under the diazotization conditions to *p*-aminobenzoylglutamate, which is further diazotized and then converted to the purple chromophore with α-naphthylenediamine. Therefore, only N-5-substituted tetrahydrofolates give a positive result with the Bratton-Marshall reaction. This is the case with the ternary complex. Both of the experiments, then, suggest that an N-5-substituted folate complex (see Structural Formula 4.4a in Scheme 11) is formed during inactivation of thymidylate synthetase by FdUMP.

A ratio of 1.7 mol of FdUMP per mole of thymidylate synthetase from CCRF-CEM human lymphoblastic leukemia cells was observed only in the presence of CH_2H_4folate.[49] The ternary complex migrates intact on denaturing gels and can be precipitated with trichloroacetic acid. Urea (6 *M*), however, disrupts the ternary complex. Thymidylate synthetases from CCRF-CEM human lymphoblastic leukemic cells and from *L. casei* undergo substantial conformational changes upon inactivation with FdUMP in the presence of CH_2H_4folate.[50] The Stokes radius (and, therefore, the diffusion coefficient) of both enzymes decreases by 3.5% after formation of the ternary complex in spite of a 1.8% increase in molecular weight. The sedimentation coefficient also increases by 3.5%. There is a reduction in the frictional ratio upon ternary complex formation. Using the bacterial enzyme, it was found that 70% of the total conformational change occurs upon binding 1 mol of FdUMP and CH_2H_4folate per mole of enzyme. Even though the enzymes from the two sources have different amino acid compositions and dissimilar conformations, there are many similarities in the formation of the ternary complexes with FdUMP and CH_2H_4folate.

Studies on the binary complex of FdUMP with *L. casei* thymidylate synthetase in the absence of CH_2H_4folate also have been carried out. Calorimetric and gel filtration methods were utilized by Beaudette et al.[51] to determine the thermodynamic parameters ($\Delta H'$, $\Delta G'$, and $\Delta S'$) and the stoichiometry of binding of FdUMP to thymidylate synthetase. The calorimetric results suggest different binding energies for the two FdUMP binding sites of the dimeric enzyme. This binding can be written as 2 FdUMP + enzyme + rH^+ ⇌ Enzyme $-(FdUMP)_2-(H^+)_r$, where, at infinite ligand concentration, r > 1. This may explain why 2 mol of FdUMP bind to the enzyme in Tris buffer, but only 1 mol is bound in phosphate buffer;[41] the phosphate may be capable of binding preferentially to the phosphate binding site for FdUMP in only one of the subunits.

The interaction of FdUMP with the enzyme in the binary complex was examined by ^{19}F NMR spectroscopy by Lewis et al.[52] Two resonances are exhibited at 1.4 and 34.5 ppm upfield of free nucleotide, the first resulting from noncovalent association of the nucleotide to the enzyme and the second corresponding to a form covalently bound to the enzyme. Using the reaction of bisulfite with FdUMP as a model, the covalent species bound to the enzyme was identified as the 5,6-dihydro derivative of FdUMP linked to the active site cysteine. The enzyme stereoselectively produces a single isomer of the covalent binary complex.

In conjunction with ^{19}F NMR spectra studies, fluorescence, ultraviolet difference, and circular dichroism spectroscopy of the enzyme–FdUMP binary complex were examined.[53] All three light spectroscopic techniques reflect the interaction of FdUMP with thymidylate synthetase in the binary complex. Subtle alterations were proposed in the environments of certain tyrosine and tryptophan residues as observed by these chromophores when FdUMP associates with the enzyme.

Inactivation of thymidylate synthetase with [6-^{3}H]FdUMP in the presence of CH_2H_4folate followed by trichloroacetic acid precipitation gives a ternary complex with a binding ratio of 1.7 mol of FdUMP per mole of enzyme.[54] When the CH_2H_4folate is omitted, trichloroacetic acid precipitation gives a covalent binary complex with up to 0.9 mol of FdUMP bound per mole of enzyme. This provides chemical evidence for the existence of the covalent binary complex.

A nitrocellulose-filter binding assay for the binary complex was developed by Ahmed et al.[55] utilizing [6-³H]FdUMP. As measured by this method it takes a 600-fold excess of FdUMP to enzyme to reach saturation. At saturation, 0.5 to 0.6 equivalent of FdUMP is bound to the enzyme. If the enzyme complex is denatured prior to filtration, the stoichiometry is only 0.3 to 0.4.

2'-Deoxyuridylate hydroxymethylase from SPO1-infected *Bacillus subtilis* also is inactivated by FdUMP in the presence of CH₂H₄folate.[56] Stable radioactive complexes are formed when [6-³H]FdUMP and CH₂H₄folate or FdUMP and CH₂-[6-³H]H₄folate are used. Denaturation of the labeled enzyme does not result in dissociation of the inactivator. NaDodSO₄ gel electrophoresis demonstrates that both isotopes remain bound.

B. 1-(5-Phosphono-β-D-arabinofuranosyl)-5-fluorouracil

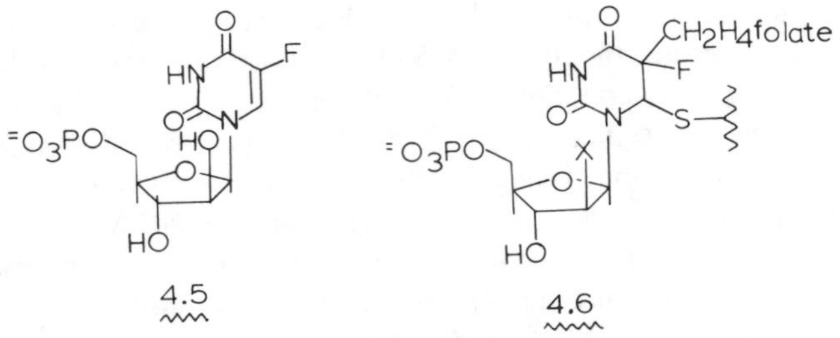

Scheme 12.   Structural Formulas 4.5 and 4.6.

1-(5-Phospho-β-D-arabinofuranosyl)-5-fluorouracil (Structural Formula 4.5, *ara*-FUMP; Scheme 12) in the presence of CH₂H₄folate, causes a time-dependent, pseudo first-order inactivation of thymidylate synthetase from methotrexate-resistant *L. casei*.[57] The rate of inactivation is about 40 times slower than that with FdUMP. The properties of the ternary complex with *ara*-FUMP and CH₂H₄folate are the same as FdUMP and CH₂H₄folate (Structural Formula 4.6, X = OH; Scheme 12). *ara*-FUMP is synthesized by phosphorylation and hydrolysis of *ara*-FU,[58] prepared as shown in Scheme 13.

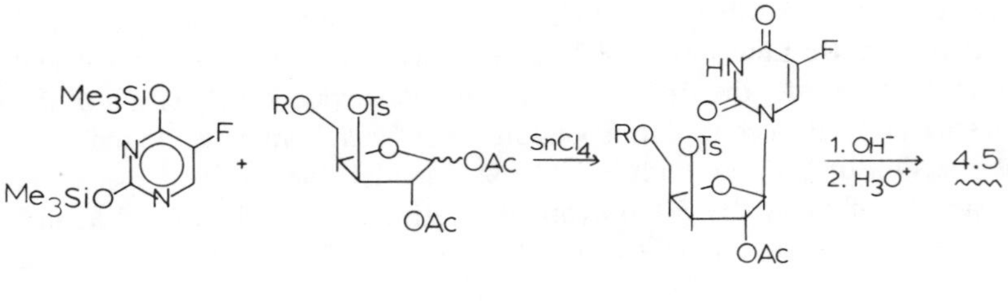

Scheme 13.

C. 2',5-Difluoro-1-arabinosyluridylate

2',5-Difluoro-1-arabinosyluridylate also is a time-dependent inactivator of *L. casei* thymidylate synthetase only in the presence of CH₂H₄folate.[59] When CH₂-[6-³H]H₄folate is used, 2 equivalents of [³H] remain bound to the dimeric enzyme after gel filtration. NaDodSO₄ denaturation does not release the radioactivity from the enzyme. The pyrimidine chromophore is bleached during inactivation, as is expected for a nucleophilic addition across the 5,6-double bond. Titration of the enzyme in the presence of CH₂H₄folate with the inactivator indicates that about 2 mol are bound per dimeric enzyme. The ternary complex formed was suggested by Coderre et al.[59] to be **4.6** (X = F). The inactivator is prepared by 2'-deoxythymidine kinase-catalyzed phosphorylation of 2',5-difluoro-1-arabinosyluracil.[60]

D. 5-Fluorocytosine

Santi et al.[61] proposed, from inspection of work described in the literature, that 5-fluorocytosine inactivates DNA-cytosine methyltransferase by the same mechanism that FdUMP inactivates thymidylate synthetase.

E. N⁴-Hydroxy-2'-deoxycytidylic Acid

In the presence of CH_2H_4-folate, N⁴-hydroxy-2-'-deoxycytidylic acid is a rapid, time-dependent inactivator of *L. casei* thymidylate synthetase.[62] With the use of [2-¹⁴C]-inactivator and CH_2-[6-³H]H_4folate, 2 mol of inactivator and 2 mol of CH_2H_2folate are bound per dimeric enzyme. The UV absorption for the pyrimidine is lost upon inactivation and does not reappear upon NaDodSO₄ denaturation. Upon denaturation, though, one-half of the CH_2H_4folate dissociates, but all of the inactivator remains bound. It was suggested that one subunit has both the cysteine residue and CH_2H_4folate bound to the inactivator (Structural Formula 4.7, Scheme 14) and the other subunit has only the cysteine bound (Structural Formula 4.8, Scheme 14).

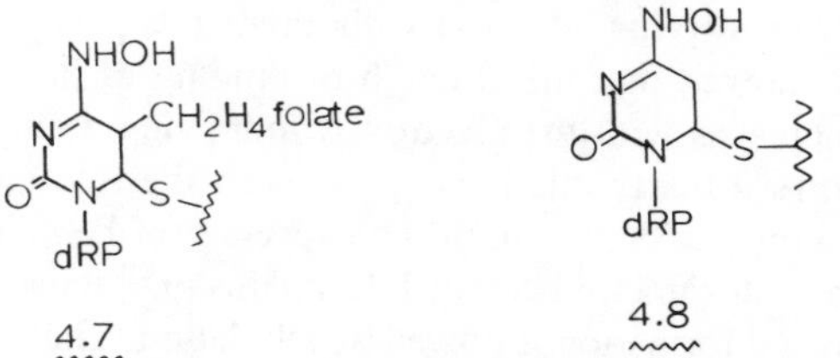

Scheme 14.   Structural Formulas 4.7 and 4.8.

F. 5-Bromouridine

E. coli alanyl-tRNA synthetase undergoes time-dependent inactivation by 5-bromouridine.[63] Similar results are obtained with *E. coli* isoleucyl-tRNA synthetase. No inactivation occurs with bromouracil, 5-bromo-2'-deoxyuridine, or 5'-bromouridine-5'-monophosphate. Inactivation by 5-bromouridine correlates with incorporation of [U-¹⁴C]bromouridine in a 1:1 stoichiometry. Denaturation of bromouridine-inactivated enzyme does not displace bound nucleoside, but dithiothreitol reactivates the enzyme. The nucleoside is bound to a protein segment required for tRNA interaction, and supports the hypothesis that tRNA synthetases transiently bind covalently to the uridine at position 8 in tRNAs. Since simple Michael adducts to 5-halogenated uracils are known to undergo further reactions to produce more stable complexes,[64] it was proposed by Starzyk et al.[63] that a further reaction of the initial Michael adduct takes place to inactivate the enzyme (Scheme 15). It is not clear if this proposed further reaction (e.g., a second amino acid attacking the 5-position to displace bromide) is necessary, or that a stable Michael addition product forms. If the latter is the case, then this compound is an affinity labeling agent.

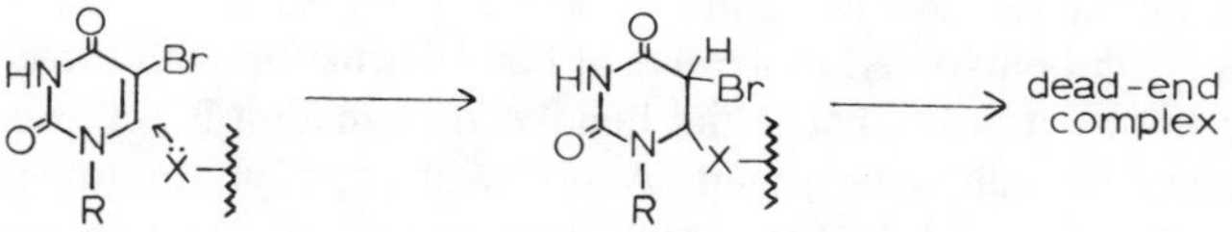

Scheme 15.

G. Halomethyl Ketones

Although halomethyl ketones are generally accepted as classical affinity labeling agents (see Chapter 1, Section III.B.) for serine proteases, Powers[64a] has suggested two alternative inactivation mechanisms which would classify these compounds as mechanism-based inactivators. One mechanism (Scheme 16A) involves hemiketal formation followed by active-site alkylation. The other possibility (Scheme 16B) involves hemiketal formation, followed by displacement of the halogen to give an enzyme-bound epoxide, which reacts with an active-site nucleophile. The mechanism shown in Scheme 16A was suggested originally by Kézdy et al.[64b] on the basis of their results on the inactivation of chymotrypsin by L-1-chloro-3-tosylamido-4-phenyl-2-butanone (TPCK), which indicate that the reaction proceeds by an adsorptive step showing the full specificity of α-chymotrypsin reactions. Further support for this mechanism comes from the findings of Weiner et al.[64c] that "anhydrochymotrypsin" (i.e., chymotrypsin in which the active site serine residue has been converted, by elimination, to a dehydroalanine residue) is not alkylated by TPCK; this suggests that His-57 alkylation by TPCK requires intact Ser-195. McMurray and Dyckes[64d] synthesized a series of model peptide ketones having the structure Lys–Ala–LysCH₂X, where X is a poor leaving group with varying electron-withdrawing capacity ($CH_2CO_2CH_3$, $COCH_3$, $OCOCH_3$, and F) and showed that the strength of binding of these competitive, reversible inhibitors to trypsin increases with the electron-withdrawing ability of X. A Hammett plot of $-\log K_i$ vs. σ_I shows a linear relationship between the free energy of binding and the electron-withdrawing ability of X, as would be expected for hemiketal formation. Since the K_i for the corresponding chloromethyl ketone falls on this line, it was suggested that reversible binding involves hemiketal formation followed by alkylation. The same conclusion regarding the mechanism of inactivation of the serine protease thermitase by peptide chloromethyl ketones was made by Fittkau et al.[64e] In this study it was shown that the reversible dissociation constant for binding of various peptide methyl ketones (K_i) is about 20-fold higher than that for the corresponding peptide chloromethyl ketones. The electron-withdrawing ability of the chlorine was suggested as the cause for stabilization of a hemiketal complex with the active-site serine prior to alkylation. Kinetic evidence was presented by Stein and Trainor[64f] in support of a mechanism-based inactivation mechanism for the inactivation of human leukocyte elastase by peptidyl chloromethyl ketones (see Scheme 16A). On the basis of the low K_i value, the solvent deuterium isotope effect on the inactivation step, and the proton inventory, a hemiketal intermediate was proposed.

X-ray data also can be used to support the mechanism shown in Scheme 16A. X-ray crystallography at 2.5 Å resolution of subtilisin inactivated by Phe–Ala–Lys–CH₂Cl indicates that attachment to the enzyme involves two covalent bonds.[64g] One bond is between the $N_\epsilon 2$ of His-64 and the methylene group of the inactivator. The other bond is between the oxygen of Ser-221 and ketone carbonyl of the inactivator. These data are used by Poulos et al.[64g] to support the inactivation mechanism shown in Scheme 16A. It is not clear, however, if hemiketal formation occurs prior to or subsequent to chloride displacement. *Streptomyces griseus* protease B is inactivated by two specific tripeptide chloromethyl ketones, Boc–L–Ala–Gly–L–PheCH₂Cl and Boc–Gly–L–Leu–L–PheCH₂Cl.[64h] X-ray crystallography of the inactivated enzyme indicates that covalent attachment occurs to both His-57 and to Ser-195. The Ser-195 is involved in hemiketal bond formation with the ketone carbonyl. It is suggested by James et al.[64h] that initial hemiketal formation is followed by nucleophilic attack by the histidine imidazole group. A series of peptidyl chloromethanes inactivates human thrombin;[64i] it is not clear if the mechanism is affinity labeling or mechanism-based inactivation.

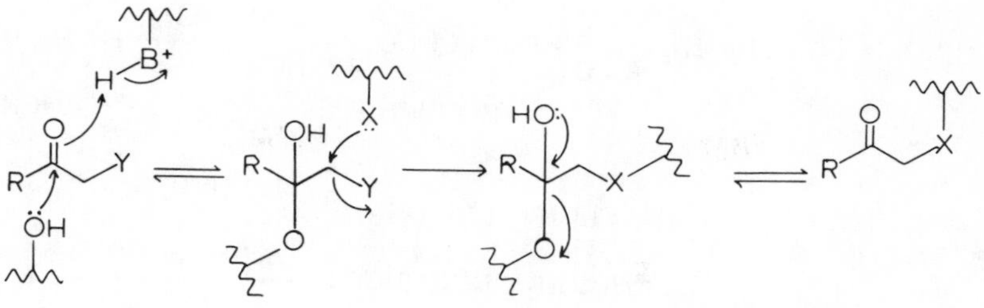

A

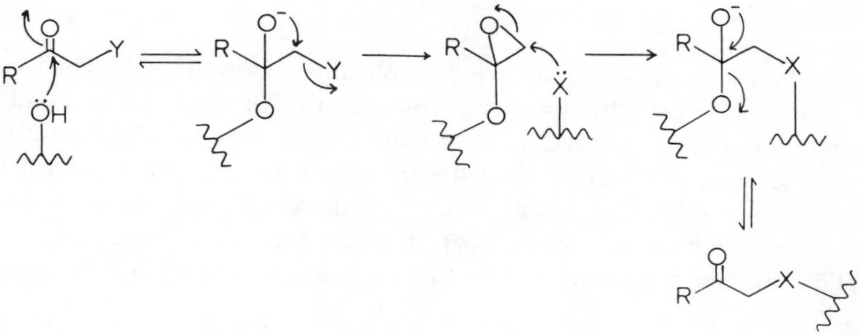

B

Scheme 16.   Two alternative inactivation mechanisms, each involving hemiketal formation followed by (A) active-site alkylation and (B) displacement of the halogen, giving an enzyme-bound epoxide which reacts with an active-site nucleophile.

Peptide fluoromethyl ketones were shown by Rasnick[64j] to inactivate the thiol protease, human cathepsin B. The K_I value for Cbz–Phe–AlaCH$_2$F is lower than that for the corresponding chloromethyl ketone, which is lower than that for the diazomethyl ketone. This corresponds to the decreasing electron-withdrawing ability of the substituents, and therefore was suggested as evidence to support a thiohemiketal intermediate during inactivation.

The dipeptide monofluoromethyl ketone **4.8a** (Scheme 17) is a time-dependent irreversible inactivator of α-chymotrypsin leading to covalent modification of a histidine residue.[64k] Inactivation is accompanied by fluoride ion release, as observed by ^{19}F NMR spectroscopy. The corresponding chloromethyl ketone reacts only 15 times faster with chymotrypsin than does **4.8a**, whereas the nonenzymatic reaction of *N*-acetylhistidine with typical chloromethyl ketones is > 200 times faster than with corresponding fluoromethyl ketones. Imperiali and Abeles,[64k] therefore, suggest that the epoxide mechanism shown in Scheme 16B may diminish the importance of the leaving group ability of the halide in the rate determining step and be relevant to this inactivation mechanism. The synthesis of **4.8a** is shown in Scheme 18.[64l]

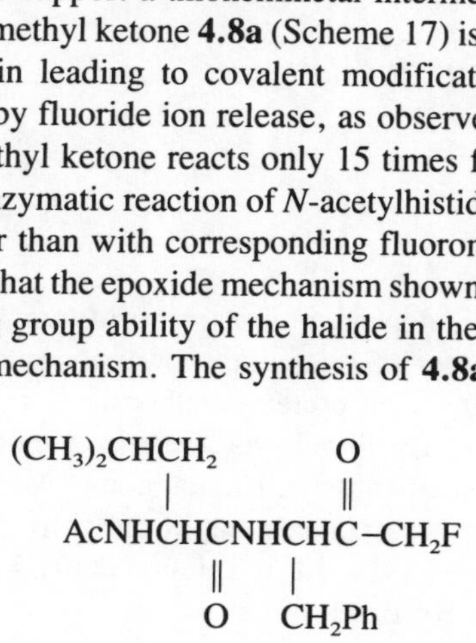

$$
\begin{array}{c}
(CH_3)_2CHCH_2 \qquad\quad O \\
| \qquad\qquad\quad || \\
AcNHCHCNHCHC{-}CH_2F \\
|| \quad\ | \\
O \quad\ CH_2Ph
\end{array}
$$

Scheme 17.   Structural Formula 4.8a.

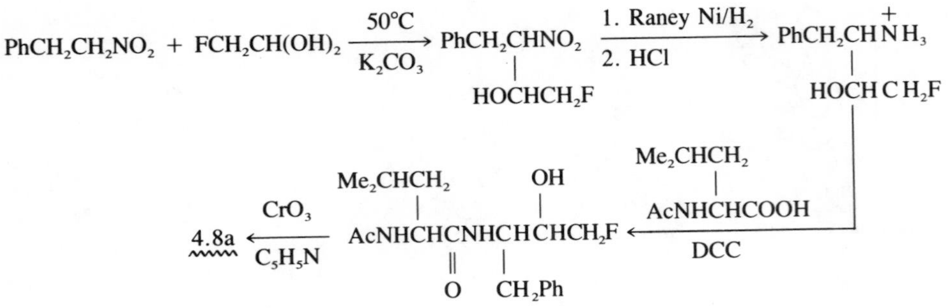

$$PhCH_2CH_2NO_2 + FCH_2CH(OH)_2 \xrightarrow[K_2CO_3]{50°C} \underset{HOCHCH_2F}{PhCH_2\overset{|}{C}HNO_2} \xrightarrow[2.\ HCl]{1.\ Raney\ Ni/H_2} \underset{HOCHCH_2F}{PhCH_2\overset{|}{C}H\overset{+}{N}H_3}$$

$$4.8a \underset{C_5H_5N}{\overset{CrO_3}{\longleftarrow}} \underset{O\quad CH_2Ph}{AcNHCHC\overset{Me_2CHCH_2}{\overset{|}{N}}HCH\overset{OH}{\overset{|}{C}}HCH_2F} \underset{DCC}{\overset{\underset{AcNHCHCOOH}{\overset{|}{Me_2CHCH_2}}}{\longleftarrow}}$$

Scheme 18.

α-Chymotrypsin is inactivated by the fluoromethyl ketone of Cbz-Phe; inactivation is slower than with the corresponding chloromethyl ketone.[64m] Inactivation with [³H]acetyl–Phe–CH₂F gives covalently bound radioactivity to the protein. Pig liver cathepsin B, a cysteine protease, is inactivated by Cbz–Phe–Phe–Ch₂F at a rate about half the rate of the corresponding chloromethyl ketone; Cbz–Phe–Ala–CH₂F also inactivates the enzyme. [³H]Acetyl–Phe–CH₂F inactivation of cathepsin B also leads to labeling of the enzyme.

The peptidylfluoromethanes are synthesized by Rauber et al.[64m] by the route show in Scheme 18A.

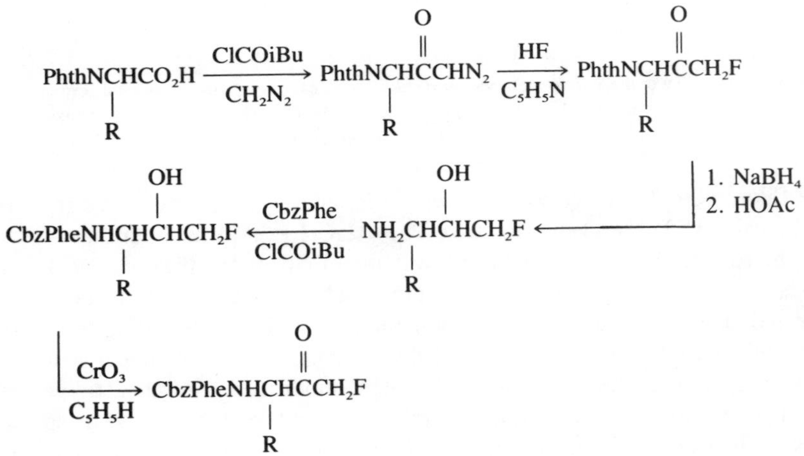

$$\underset{R}{\overset{|}{Phth}NCHCO_2H} \xrightarrow[CH_2N_2]{ClCOiBu} \underset{R}{\overset{|}{Phth}NCH\overset{O}{\overset{\|}{C}}CHN_2} \xrightarrow[C_5H_5N]{HF} \underset{R}{\overset{|}{Phth}NCH\overset{O}{\overset{\|}{C}}CH_2F}$$

$$\underset{R}{\overset{|}{Cbz}PheNHCH\overset{OH}{\overset{|}{C}}HCH_2F} \underset{ClCOiBu}{\overset{CbzPhe}{\longleftarrow}} \underset{R}{\overset{|}{N}H_2CH\overset{OH}{\overset{|}{C}}HCH_2F} \overset{1.\ NaBH_4}{\underset{2.\ HOAc}{\longleftarrow}}$$

$$\underset{C_5H_5H}{\overset{CrO_3}{\longrightarrow}} \underset{R}{\overset{|}{Cbz}PheNHCH\overset{O}{\overset{\|}{C}}CH_2F}$$

Scheme 18A.

Ala–Phe–Lys–CH₂F is a time-dependent inactivator of human plasmin and of trypsin; the rate of reaction is about an order of magnitude lower than the rate for the corresponding chloromethyl ketone.[64n] The cysteine protease cathepsin B from pig liver was inactivated by Ala–Phe–Lys–CH₂F and by Bz–Phe–Lys–CH₂F. In a model study of peptidyl fluoromethanes and peptidyl chloromethanes with glutathione, Angliker et al.[64n] found that the reactivity of the fluoromethane was 1/500th that of the corresponding chloromethane. Relative to the reaction of Ala–Phe–Lys–CH₂F in solution at pH 6.4, the corresponding reaction with cathepsin B at pH 6.4 is 10^8 times faster.

Lowe and Perham[86] suggested a mechanism for inactivation of thiamin pyrophosphate-dependent decarboxylase by bromopyruvate that is similar to that shown in Scheme 16A (see Section IV.A.1., Scheme 44, pathway a).

H. *(Z)*-2-Bromofumaric Acid

β-Methylaspartase from *Clostridium tetanomorphum* is irreversibly inactivated by (Z)-2-bromofumaric acid (Structural Formula 4.8b, Scheme 18B) in the presence of ammonia.[64o] A second aliquot of enzyme added after inactivation is complete is inactivated at the same rate as the initial aliquot of enzyme. No inactivation occurs in the absence of ammonia. The inactivation mechanism proposed by Akhtar et al.[64o] is shown in Scheme 18B.

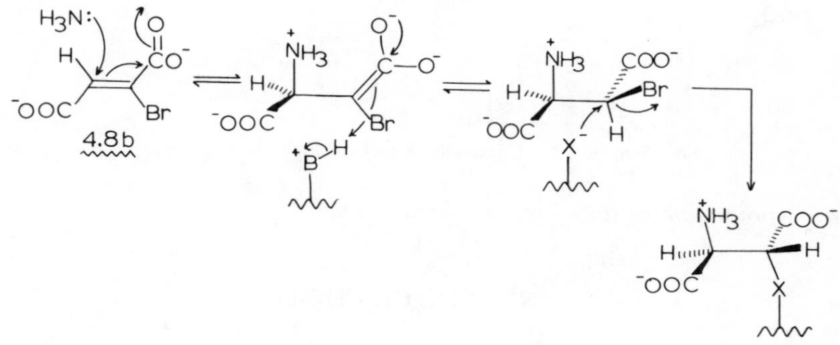

Scheme 18B.   Containing Structural Formula 4.8b.

II. ADDITION/ELIMINATION

A. Addition/Elimination

1. 5-Nitro-L-norvaline

5-Nitro-L-norvaline (Structural Formula 4.9, Scheme 20) is a time-dependent inactivator of three aminotransferases, alanine-, aspartate-, and 4-aminobutyrate aminotransferase.[65] Dialysis of aspartate aminotransferase inactivated by 5-nitro-L-norvaline results in complete restoration of enzyme activity, but with alanine- and 4-aminobutyrate aminotransferases, only 36% and 8%, respectively, of enzyme activity returns. The inactivation mechanism proposed is shown in Scheme 19. The inactivator can be synthesized from 2-amino-5-hydroxypentanoic acid[66] (Scheme 20).

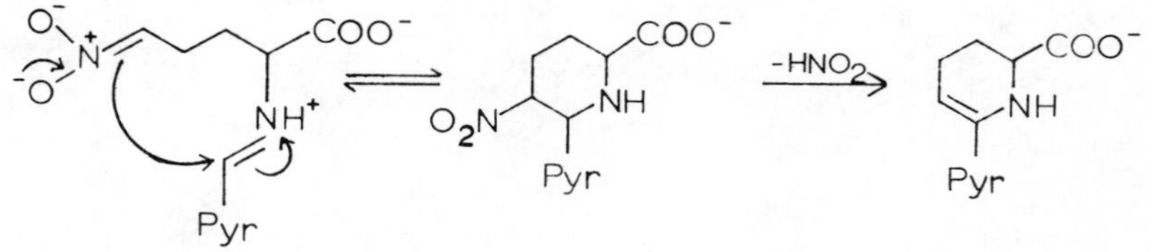

Scheme 19.

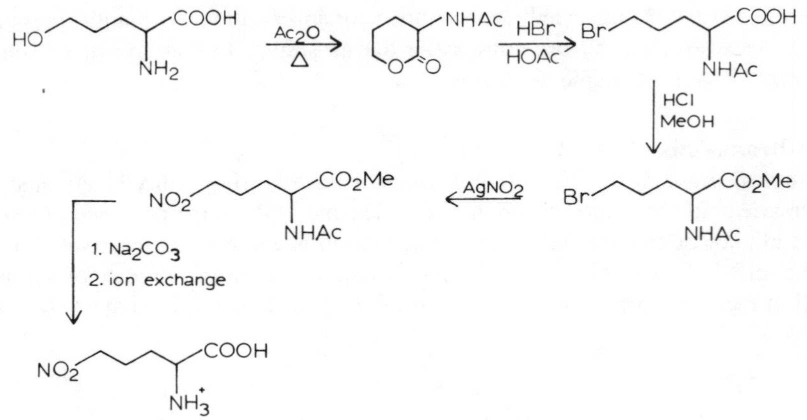

Scheme 20.   Containing Structural Formula 4.9.

2. 3-Nitro-1-propanamine and 4-Nitro-1-butanamine

$$\overset{+}{N}H_3(CH_2)_nCH_2CH_2NO_2$$

Scheme 20A.   Structural Formula 4.9a.

3-Nitro-1-propanamine (Structural Formula 4.9a ($n = 0$), Scheme 20A) is a substrate for pig brain GABA aminotransferase, but inactivates the enzyme from *Pseudomonas fluorescens*.[66a] Inactivated enzyme becomes reactivated upon dilution. Although acrolein is produced, it was shown that acrolein at low concentrations does not inactivate the enzyme. 4-Nitro-1-butanamine (Structural Formula 4.9a ($n = 1$), Scheme 20A) inactivates the enzyme from both sources, and no reactivation was observed upon dilution. The inactivation mechanism suggested by Alston et al.[66a] is related to that shown in Scheme 19.

B. Addition/Elimination/Addition
1. 1-Chloro-1-nitroethane

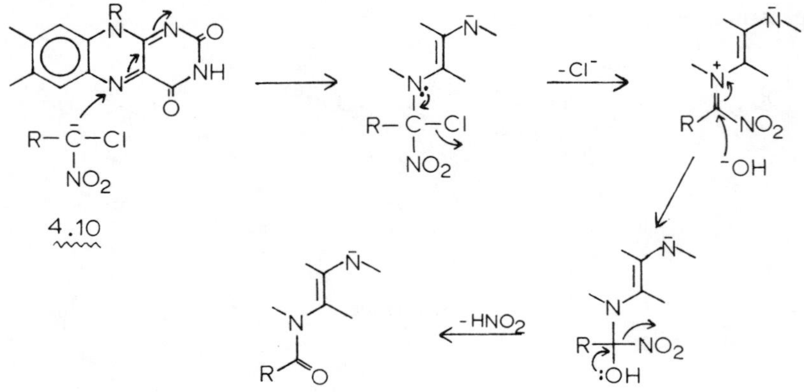

Scheme 21.   Containing Structural Formula 4.10.

1-Chloro-1-nitroethane, as the nitronate ion (Structural Formula 4.10, Scheme 21), inactivates porcine kidney D-amino acid oxidase.[67] Complete inactivation requires 1.5 equivalents of inactivator and produces chloride and nitrite, without the requirement of O_2. The additional 0.5 equivalent of inactivator is oxidized to acetate at about half the rate of inactivation. The flavin spectrum is bleached, and the flavin is converted to N^5-acetyl-1,5-dihydro FAD during inactivation. Scheme 21 shows the proposed mechanism of inactivation. Other flavoenzymes, e.g., lactate oxidase (*Mycobacterium smegmatis*), glycolate oxidase (spinach), glucose oxidase (*Aspergillus niger*), and succiniate dehydrogenase (bovine heart), are not inactivated by 1-chloro-1-nitroethane.

2. 5-Trifluoromethyl-2'-deoxyuridylate

Thymidylate synthetase from Ehrlich ascites carcinoma cells is irreversibly inactivated by 5-trifluoromethyl-2'deoxyuridylate (Structural Formula 4.11, Scheme 22); the kinetics change from competitive to noncompetitive with preincubation of inactivator and enzyme.[4] It was suggested that the enzyme becomes *N*-acylated.

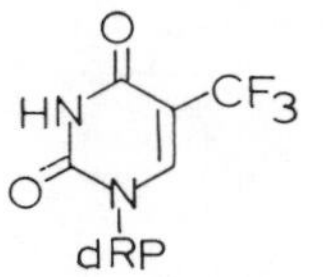

Scheme 22.   Structural Formula 4.11.

Model studies for the mechanism of inactivation of thymidylate synthetase by 5-trifluoromethyl-2'-deoxyuridylate were carried out by Santi and Sakai.[68] A kinetic analysis of the reaction of hydroxide ion with 1-methyl- and 1,3-dimethyl-5-trifluoromethyluracil indicates that the mechanism of hydrolysis to 1-methyl- and 1,3-dimethyl-5-carboxyuracil involves nucleophilic addition of hydroxide to the 6-position (Scheme 23). An intramolecular model compound for this mechanism also was prepared,[68] varying the length of the chain (Structural Formula 4.11a, Scheme 24). When n = 5, the same kinetics were observed as for 1-methyl-5-trifluoromethyluracil, but when n = 3, the rate of hydrolysis was 10^4 times faster (Scheme 25). This model supports a mechanism for inactivation as shown in Scheme 26. Contrary to these results, however, it was shown by Langenbach et al.[12] that [2-^{14}C]trifluoromethyl-2'-deoxyuridylate forms a complex with thymidylate synthetase from Ehrlich ascites carcinoma cells and CH_2H_4folate that is stable to dialysis, but which dissociates upon denaturation or acid precipitation; it was proposed that this compound does not form a covalent complex to the enzyme.

3. trans-5-(3,3,3-Trifluoro-1-propenyl)-2'-deoxyuridylate

trans-5-(3,3,3-Trifluoro-1-propenyl)-2'-deoxyuridylate (Structural Formula 4.12) can be prepared from dUMP by the route shown in Scheme 27.[69] Hydrolysis of this compound by hydroxide[69] occurs at a rate similar to that for 5-trifluoromethyl-2'-deoxyuridine,[68] giving the corresponding carboxylic acid. The nucleotide is a time-dependent inactivator of thymidylate synthetase in the presence or absence of CH_2H_4folate.

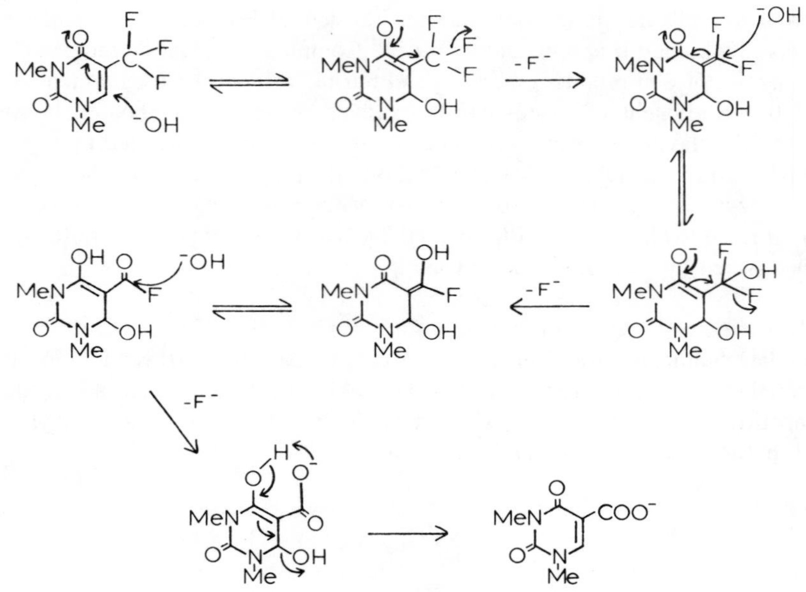

Scheme 23.

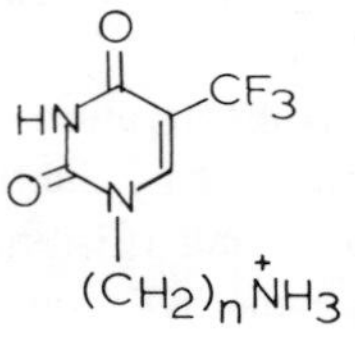

Scheme 24.   Structural Formula 4.11a.

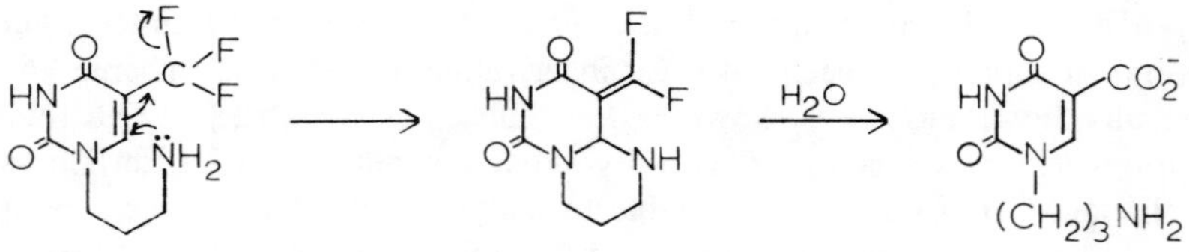

Scheme 25.

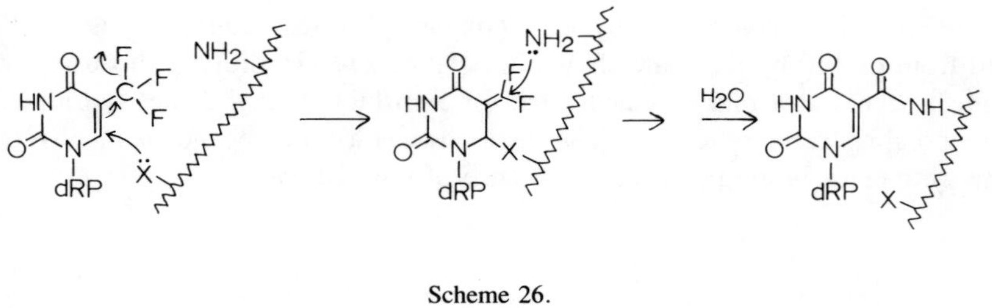

Scheme 26.

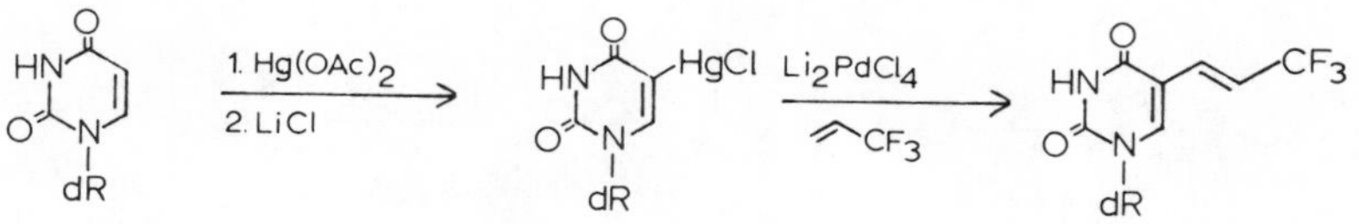

Scheme 27. Containing Structural Formula 4.12.

4. 3-Nitropropionate

The time-dependent inactivation of rat liver mitochondrial succinate dehydrogenase by 3-nitropropionate (Structural Formula 4.13) which is isoelectronic with succinate, was originally proposed by Alston et al.[70] to follow an addition/elimination/isomerization mechanism (Scheme 28). Centrifugation and washing of the pellet does not restore enzyme activity. This mechanism, however, was refuted by Coles et al.[71] 3-Nitropropionate inactivates succinate dehydrogenase from bovine heart mitochondria in stoichiometric amounts at a rate much slower than that for 3-nitroacrylate, the oxidation product which is the proposed inactivator. Spectral studies indicate that an N-5 alkylated flavin proposed by Alston et al.[70] is not compatible with the spectrum observed. Denaturation, even in the absence of oxygen, and proteolytic digestion abolishes the spectral change brought about by 3-nitropropionate. Unless the flavin adduct is very unstable, it suggests that no flavin adduct is formed. The spectrum of inactivated enzyme is similar to that brought about by oxalacetate, which is known to combine with an active site thiol. The modified mechanism proposed by Coles et al.,[71] therefore, is that shown in Scheme 29. The presence of dithiothreitol does not protect the enzyme from inactivation.

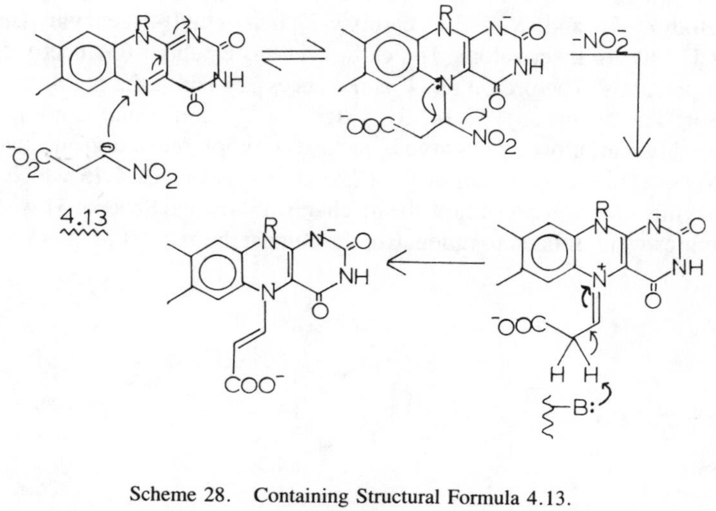

Scheme 28. Containing Structural Formula 4.13.

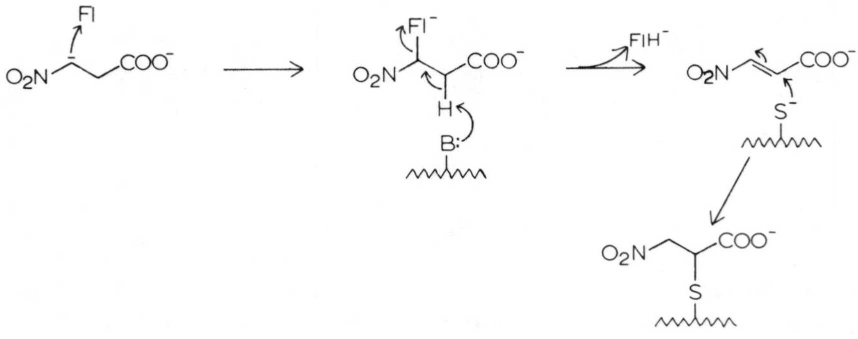

Scheme 29.

5. Amino Acid-Derived Hydantoins

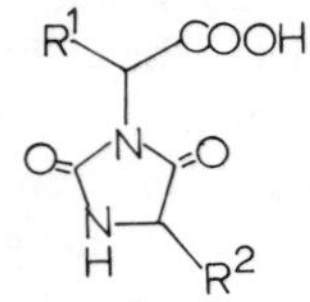

Scheme 30.   Structural Formula 4.14

Certain amino acid-derived hydantoins (Structural Formula 4.14, Scheme 30) were shown by Buntain et al.[72] to be inhibitors of dihydro-orotate dehydrogenase from *Zymobacterium (Clostridium) oroticum;* two derived from phenylalanine are time-dependent irreversible inactivators. Both (*S,S*)- and (*R,R*)-3-(1-carboxy-2-phenylethyl)-5-benzylhydantoins (**4.14**, $R^1 = R^2 = CH_2Ph$) are inactivators. The carboxylate is essential for inactivation as is the benzyl at both positions. The proton on N' is not necessary, since the N–Me compound also is an inactivator. The diisopropyl (**4.14**, $R^1 = R^2 = $ i–Pr) compound is not an inactivator, and is only weakly inhibitory. This argues against a simple acylation mechanism for the hydantoins. Nonenzymatic oxidation of the (*S,S*)-compound gives **4.15** which is an inactivator of the enzyme. On the basis of this, the mechanism shown in Scheme 31 was proposed.[72] Computer graphics were utilized to rationalize the comparable effects of the (*S,S*)- and (*R,R*)-compounds.

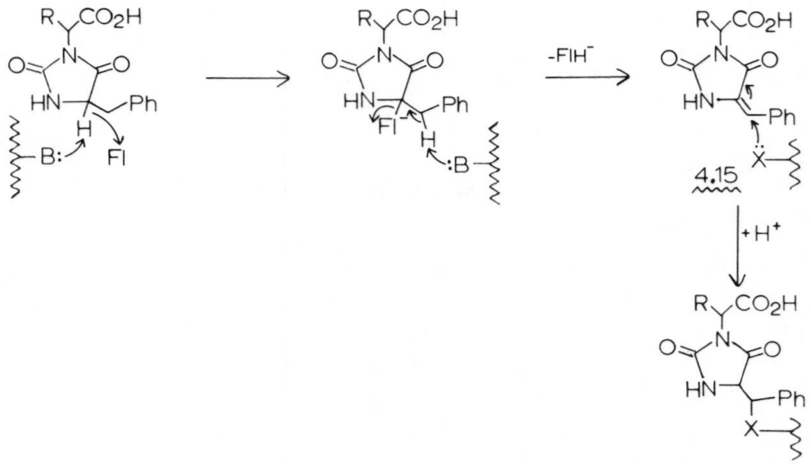

Scheme 31.   Containing Structural Formula 4.15.

6. *Monofluorofumarate and Difluorofumarate*

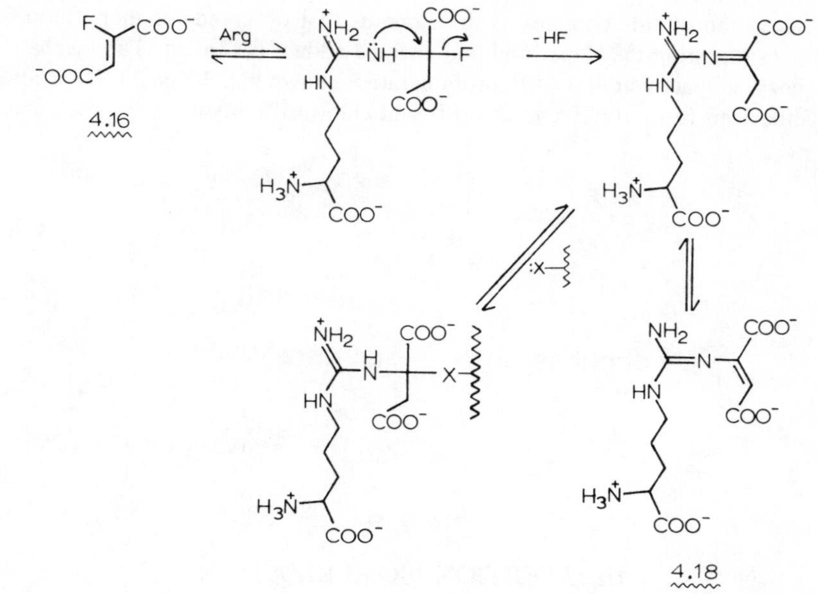

Scheme 32.   Containing Structural Formulas 4.16 and 4.18.

Monofluorofumarate (Structural Formula 4.16, Scheme 32) and difluorofumarate (Structural Formula 4.17, Scheme 33) are poor substrates, but potent time-dependent inactivators, in the presence of arginine, of the reverse reaction of bovine liver argininosuccinate lyase,[73] an enzyme that catalyzes the reversible degradation of argininosuccinate to arginine and fumarate. Dilution of inhibited enzyme into buffer containing argininosuccinate produces first-order reactivation. Slow passage through Sephadex G-25 also reactivates the enzyme. The proposed mechanism for inactivation by monofluorofumarate is shown in Scheme 32.

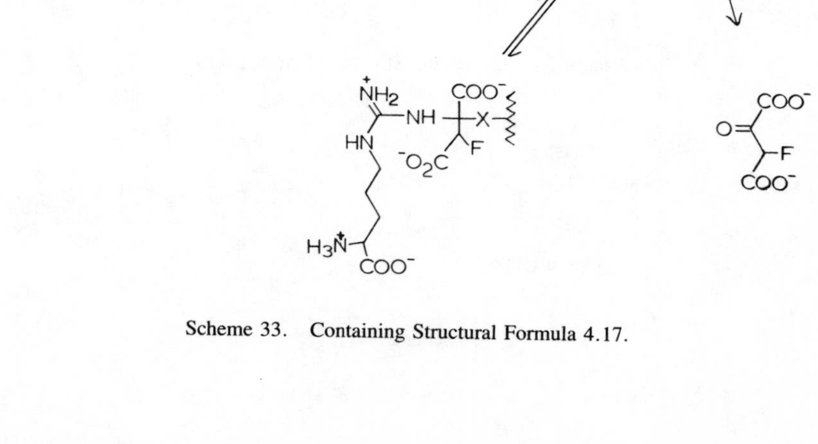

Scheme 33.   Containing Structural Formula 4.17.

On the basis of Structural Formula 4.18 (Scheme 32), acetylenedicarboxylic acid also was shown to inhibit the enzyme; addition of arginine to this compound should give **4.18**. The $t_{1/2}$ for reactivation of this complex is the same as that observed for monofluorofumarate inactivation, suggesting the same inhibited enzyme adduct is formed. The mechanism proposed for enzyme inactivation by difluorofumarate is shown in Scheme 33. Both compounds can be synthesized from vinylidene chloride and chlorotrifluoroethylene[74] (Scheme 34).

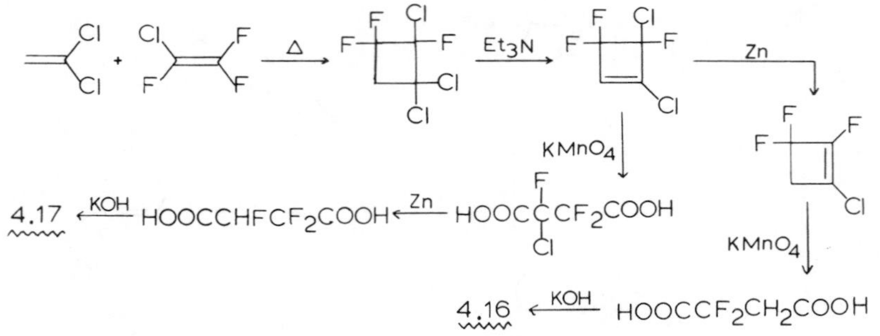

Scheme 34.

III. ADDITION/ISOMERIZATION

A. Addition/Isomerization

1. 5-p-Benzoquinonyl-2'-deoxyuridylate

5-p-Benzoquinonyl-2'-deoxyuridylate (Structural Formula 4.19) inactivates *L. casei* thymidylate synthetase.[75] A possible mechanism of inactivation is shown in Scheme 35. However, model studies with $HSCH_2CO_2Me$ show attack is at the quinone ring, suggesting that **4.19** does not undergo the reaction shown in Scheme 35, but rather is an affinity-labeling agent.[75a] Compound **4.19** can be synthesized by phosphorylation and oxidation[75] of the corresponding 5-(2,5-dimethoxyphenyl) nucleoside[76] (Scheme 36).

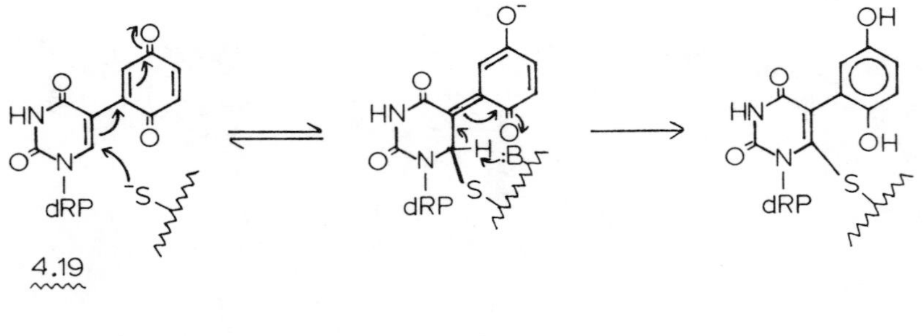

Scheme 35.   Containing Structural Formula 4.19.

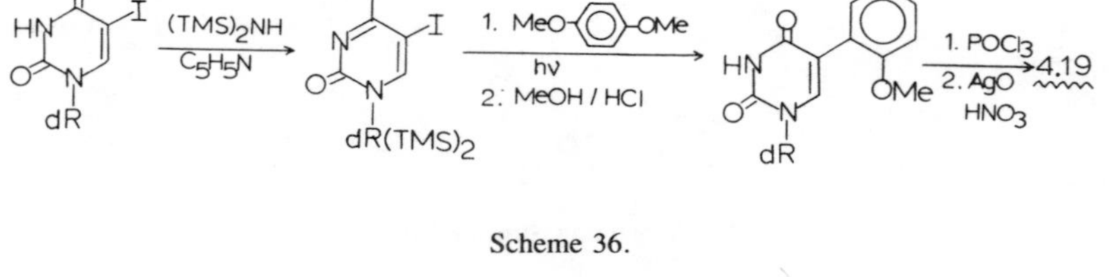

Scheme 36.

B. Addition/Isomerization/Addition

1. 5-Ethynyl-2'-deoxyuridylate

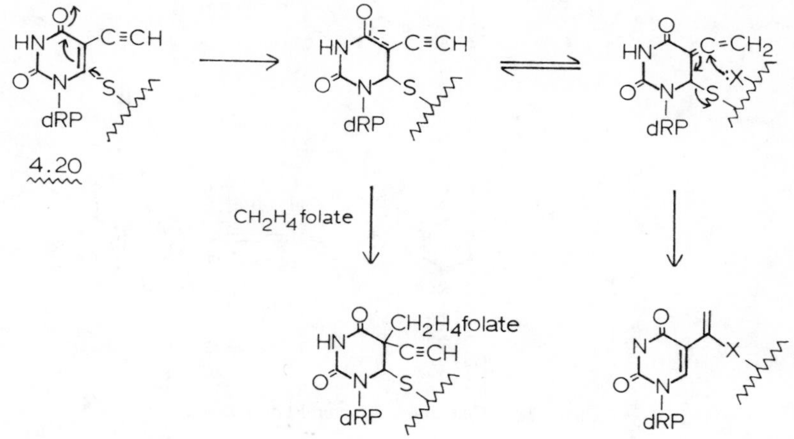

Scheme 37.   Containing Structural Formula 4.20.

5-Ethynyl-2'-deoxyuridylate (Structural Formula 4.20, Scheme 37) is a time-dependent inhibitor of thymidylate synthetase from *L. casei* in the presence and absence of CH_2H_4folate.[77] With the use of $[^{14}C]CH_2H_4$folate, a ternary complex was observed after gel filtration, but only 7% of the radioactivity is bound compared to when FdUMP is the inhibitor. Acid denaturation does not give any precipitated radioactivity. It was concluded by Danenberg et al.[77] that a covalent bond to the enzyme does not form. A different conclusion, however, was made by Santi and co-workers.[78] Several 5-alkynyl-2'deoxyuridylates were shown to be time-dependent inactivators of thymidylate synthetase in the presence of CH_2H_4folate; the 5-ethynyl analogue is the most potent. When $[2-^{14}C]$5-ethynyl-2'-deoxyuridylate and $CH_2[6-^3H]H_4$folate were used to inactivate the enzyme, the enzyme complex formed contained $[^{14}C]$ and $[^3H]$ in a 1:1 stoichiometry. In the absence of cofactor, the compound is a substrate. Two possible mechanisms of inactivation were offered (Scheme 37). The reason for the discrepancy in results by the two groups[77,78] may be that harsher conditions were used by Danenberg et al.[77] in their attempt to isolate the ternary complex. Also, this group included substrate in the dialysis buffer which is known[20] to cause reactivation of FdUMP-inactivated thymidylate synthetase and this may have been responsible for the observed lack of inactivation.

Detailed studies, including a chemical model study, of the reaction of 5-ethynyl-2'-deoxyuridylate with thymidylate synthetase were carried out by Barr et al.[79] In the model study the reaction of 5-ethynyl-2'-deoxyuridine with 2-mercaptoethanol in pH 7.4 buffer gave 5-[1-[(2-hydroxyethyl)thio]vinyl]-2'-deoxyuridine (Structural Formula 4.21) in a first-order reaction; the corresponding nucleotide reacts the same way at about one-half the rate. Scheme 38 depicts the proposed mechanism. This reaction has an inverse isotope effect (k_T/k_H) of 1.22 when $[6-^3H]$ 5-ethynyl-2'-deoxyuridine is used, thus supporting a nucleophilic attack at C-6 which results in conversion of the C-6 carbon to sp^3 hybridization. In the *absence* of CH_2H_4folate, the reaction of $[6-^3H]$5-ethynyl-2'-deoxyuridylate with 2-mercaptoethanol is catalyzed by thymidylate synthetase and the same inverse isotope effect is observed. These studies support the mechanisms shown in Scheme 37.

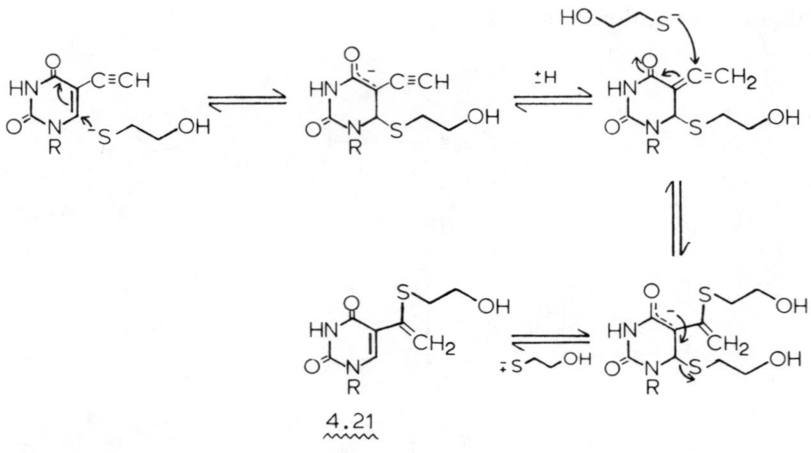

Scheme 38.   Containing Structural Formula 4.21.

5-Ethynyluracil[80,81] (**4.22**) and 5-ethynyl-2'-deoxyuridine (**4.22a**)[80-82] can be synthesized as shown in Schemes 39, 40, and 41.

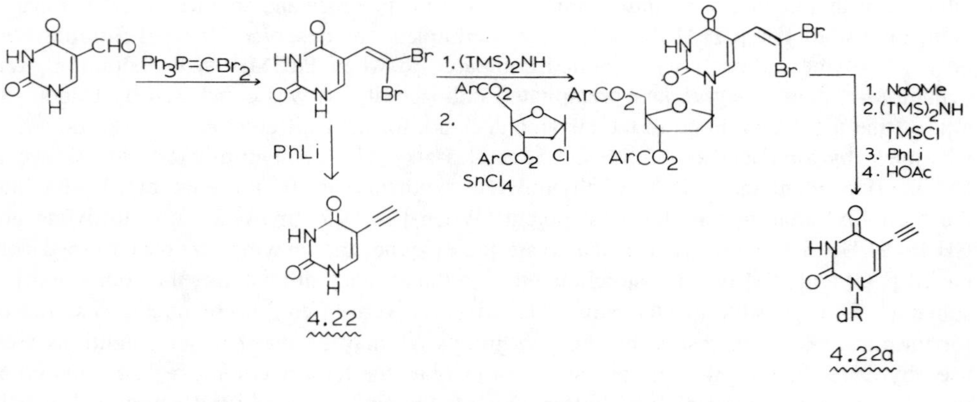

Scheme 39.   Containing Structural Formulas 4.22 and 4.22a.

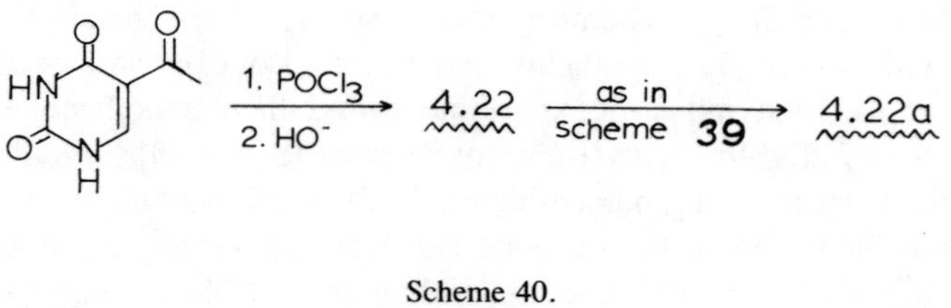

Scheme 40.

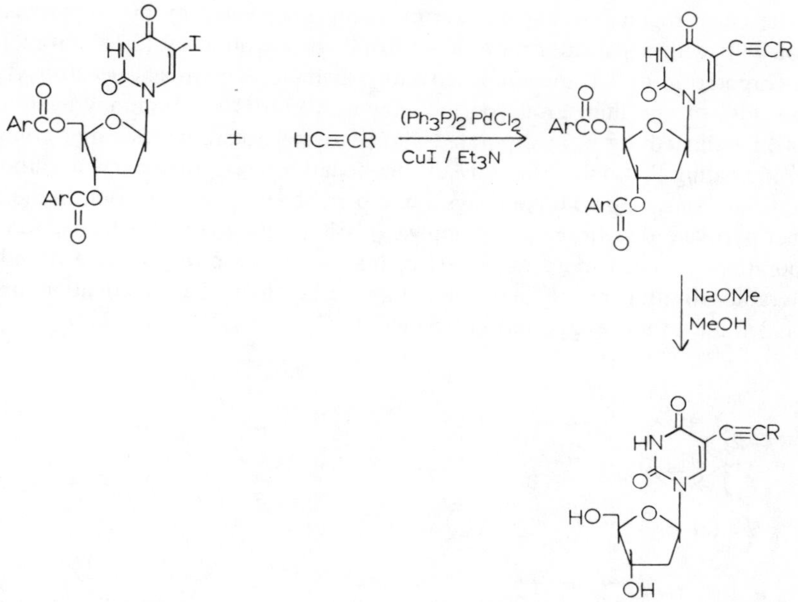

Scheme 41.

IV. ADDITION/DECARBOXYLATION

A. Addition/Decarboxylation/Addition

1. Halopyruvate

Bromopyruvate (Structural Formula 4.23, Scheme 42) is a time-dependent inactivator of *E. coli* pyruvate dehydrogenase complex;[83,84] enzyme activity does not return upon gel filtration.[83] Inactivation requires thiamin pyrophosphate (TPP), but bromopyruvate does not destroy the individual enzyme components of the enzyme complex by a TPP-dependent process. Although [2-^{14}C]bromopyruvate can lead to incorporation of radioactivity in a TPP-independent process, 3.9 times as much [^{14}C] is bound to the complex in the presence of TPP. Maldonado et al.[83] estimated that 2 mol of inactivator were bound per mole of lipoyl moieties.

$$\underset{\text{BrCH}_2\text{CCOOH}}{\overset{\displaystyle \text{O}}{\overset{\displaystyle \|}{}}}$$

Scheme 42.   Structural Formula 4.23.

Further studies by Apfel et al.[85] support the notion that bromopyruvate is not an affinity labeling agent; rather, it initially acts as a substrate for the first two steps of the pyruvate dehydrogenase complex followed by alkylation of the *S*-bromoacetyl dihydrolipoyl moieties of dihydrolipoyl transacetylase. Chemical degradation of the complex labeled with [2-^{14}C]bromopyruvate under conditions that would convert lipoyl groups to *S,S*-bis[^{14}C]carboxymethyl dihydrolipoic acid does generate that product. If these enzymes are considered to be separate entities, then this is an example of a multi enzyme-activated inactivator (see Chapter 1, Section III.D.). The mechanism of inactivation proposed is shown in Scheme 43. Lowe and Perham[86] found that the pyruvate decarboxylase activity of the *E.*

coli pyruvate dehydrogenase complex is irreversibly inactivated by bromopyruvate in the presence of TPP with a partition ratio of 40 to 60. Inactivation by [2-^{14}C]bromopyruvate leads to incorporation of 1.0 mol of inactivator per mole of pyruvate decarboxylase with concomitant loss of one thiol group on the enzyme. With [1-^{14}C]bromopyruvate only 0.4 to 0.7 mol of radioactivity is incorporated into pyruvate decarboxylase after complete inactivation, indicating that only 30 to 60% of the bound bromopyruvate has decarboxylated. These results are consistent with pathways a and b in Scheme 44 (X = Br). In the presence of the intact pyruvate dehydrogenase complex, [2-^{14}C]bromopyruvate inactivation leads to the incorporation of much more radioactivity than in the case of pyruvate decarboxylase alone. The radioactivity is mostly associated with the E2 chain. The inactivation mechanism proposed is similar to that suggested in Scheme 43.[85]

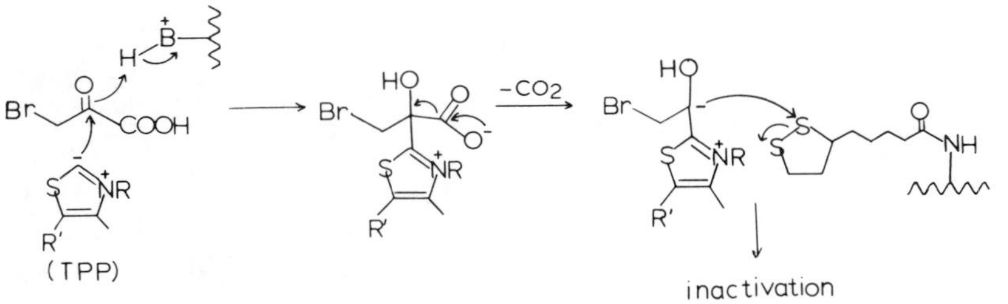

Scheme 43.

Chloropyruvate is neither a substrate nor an inactivator of the pyruvate dehydrogenase complex.[85] Fluoropyruvate is a substrate,[87] undergoing decarboxylation and elimination of fluoride via enolacetyl-TPP (Structural Formula 4.24, Scheme 44) to acetyl-TPP which hydrolyzes to acetate and TPP. A slow inactivation also occurs, possibly by Michael addition of an active site nucleophile to enolacetyl-TPP[85] (Scheme 44, pathway b, X = F).

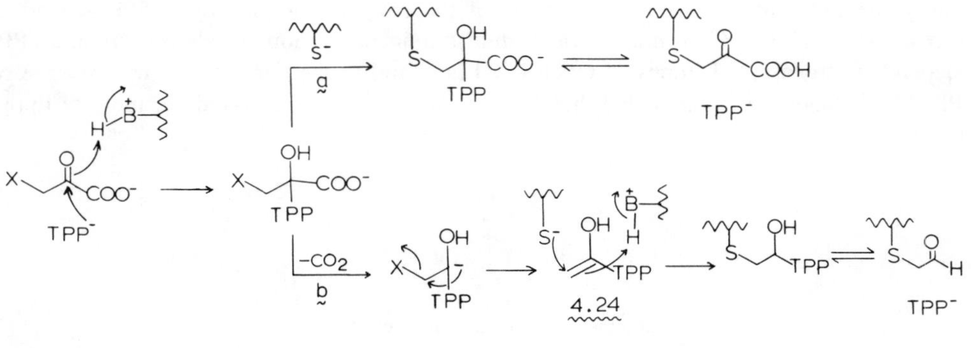

Scheme 44.   Containing Structural Formula 4.24.

V. NONCOVALENT INACTIVATION

A. Addition/Decarboxylation

1. α-Ketobutyric Acid

α-Ketobutyric acid causes pseudo first-order time-dependent inactivation of the pyruvate decarboxylating component of pigeon breast pyruvate dehydrogenase in the presence of thiamin pyrophosphate, Mg(II), and a hydrogen acceptor molecule.[88] [3-^{2}H$_2$]α-Ketobutyrate inactivates the enzyme with a k_H/k_D of 1.7. Pseudo first-order return of enzyme activity results when pyruvate is added to inactivated enzyme or if excess α-ketobutyrate is removed by gel filtration. It was suggested that the normal mechanism of oxidative decarboxylation occurs, but that product dissociation is slow.

B. Addition/Decarboxylation/Isomerization
1. (E)-4-(4-Chlorophenyl)-2-oxo-3-butenoic Acid and Related

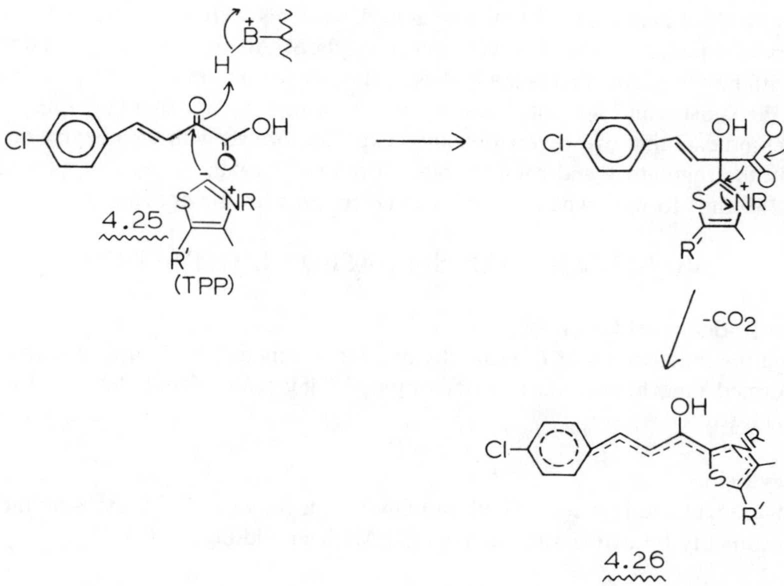

Scheme 45.   Containing Structural Formulas 4.25 and 4.26.

(E)-4-(4-Chlorophenyl)-2-oxo-3-butenoic acid (Structural Formula 4.25, Scheme 45) irreversibly inactivates brewers' yeast pyruvate decarboxylase in a time-dependent, biphasic, sigmoidal manner.[89] The substrate, pyruvate, also exhibits sigmoidal kinetics with this enzyme, a tetramer containing four thiamin pyrophosphates. No enzyme activity is restored upon dialysis. Pyruvamide, which is not a substrate, but is an allosteric activator, enhances the rate of inactivation by the inactivator; pyruvate protects the enzyme. These results suggest two binding sites, one catalytic and one regulatory. Pyruvamide only binds to the regulatory site and competes with the inactivator for that site, but not for the catalytic site, and therefore inactivation is enhanced. Pyruvate, however, competes at both sites. The kinetic expression for a two-site irreversible inhibition of an allosteric enzyme described in Chapter 1 (see Section VI.C) was used by Kuo and Jordan[89] to show that the kinetic behavior of the inactivator fits this expression for two-site binding. [1-^{14}C](E)-4-(4-Chlorophenyl)-2-oxo-3-butenoic acid inactivates the enzyme with concomitant release of $^{14}CO_2$. The partition ratio at pH 5.0 is 17; at pH 6.0, the partition ratio is 70. Glyoxalase I is not inhibited by this compound. The mechanism for inactivation proposed (Scheme 45) suggests that a highly-stabilized thiamin pyrophosphate adduct (that of Structural Formula 4.26 or some other transoid or cisoid isomer) is formed which is not released. Concomitant with inactivation is the appearance of a new absorption band at 440 nm that is attributed to the thiamin pyrophosphate-bound enamine product after decarboxylation.[90] Inactivation with [3-^{3}H] inactivator gives a stoichiometry of less than or equal to one inactivator per subunit. The observation of the enamine intermediate is used as evidence that decarboxylation and protonation of normal substrates is not concerted, that protonation and product release is concerted, that the anion formed after decarboxylation rather than hydroxyethyl thiamin pyrophosphate is the important intermediate, and that release of product is rate limiting. Brewers' yeast pyruvate decarboxylase was resolved into two isozymes by Kuo et al.[91] Compound **4.25** inactivates both isozymes at the same rate. When [3-^{3}H]-**4.25** was used, tritium became incorporated into both isozymes at the same rate and to the same extent. The inactivator can be synthesized by an aldol condensation of *p*-chlorobenzaldehyde with pyruvic acid.[92]

A series of (*E*)-4-aryl-2-oxo-3-butenoic acid derivatives were tested by Jordan et al.[93] as inactivators of brewers' yeast pyruvate decarboxylase. All of the phenyl analogues with electron-withdrawing substituents are biphasic, time-dependent inactivators; those with electron-donating substituents are not. Studies with the *m*-nitrophenyl analogue (the most potent inactivator) indicate that two reversible complexes are formed with the enzyme at two sites, presumably the regulatory and catalytic sites. Both the formation and disappearance of the enamine (440 nm) formed when this compound reacts with the enzyme were monitored.

VI. NOT MECHANISM-BASED INACTIVATION

A. 5-Nitro-2'-deoxyuridylate
Although the inactivation of *L. casei* thymidylate synthetase by 5-nitro-2'-deoxyuridylate has been termed a mechanism-based inactivation,[94-97] it is really affinity labeling by a Michael addition pathway.

B. 5-Azacytosine
When 5-azacytosine is incorporated into DNA, it inactivates DNA-cytosine methyltransferase, presumably by active site nucleophilic Michael addition to C-6.[98]

C. *N*-Hydroxyglutamate
Several PLP-dependent enzymes are inactivated by *N*-hydroxyglutamate which acts by forming an N-oxide Schiff base with the PLP.[99] Since PLP Schiff base formation has been defined as active-site binding in this book, this compound, then, is an affinity-labeling agent.

D. 5-*p*-Benzoquinonyl-2'-deoxyuridine 5'-phosphate and Related
As indicated in Section III.A.1. of this chapter, 5-*p*-benzoquinonyl-2'-deoxyuridine 5' phosphate was shown by Vadnere et al.[75a] to be an affinity labeling agent. In an attempt to change the mode of nucleophilic attack from the quinone to the nucleotide, the quinone ring was substituted (Structural Formula 4.27, R = H, R' = R'' = Me; Structural Formula 4.27, R = R'' = R'' = Me; and Structural Formula 4.28, all in Scheme 46).[100] Only **4.27** (R = H, R' = R'' = Me) was a time-dependent inactivator of *L. casei* thymidylate synthase. Al-Razzak et al.[100] suggest quinone attack as the most reasonable inactivation mechanism.

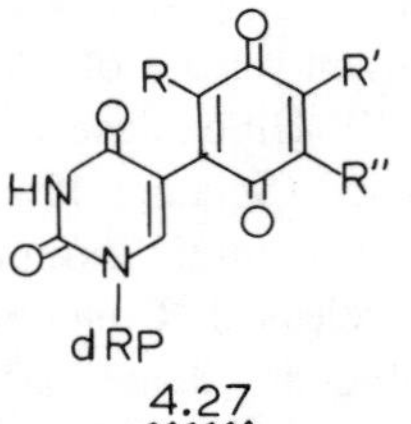

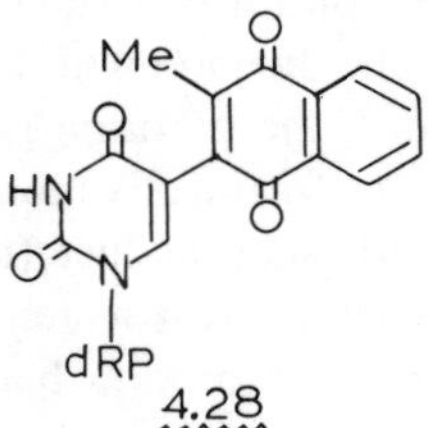

Scheme 46.   Structural Formulas 4.27 and 4.28.

REFERENCES

1. **Heidelberger, C., Chaudhuri, N. K., Danneberg, P., Mooren, D., Griesbach, L., Duschinsky, R., Schnitzer, R. J., Pleven, E., and Scheiner, J.,** Fluorinated pyrimidines, a new class of tumour-inhibitory compounds, *Nature (London),* 179, 663, 1957.
2. **Bosch, L., Harbers, E., and Heidelberger, C.,** Studies on fluorinated pyrimidines. V. Effects on nucleic acid metabolism *in vitro, Cancer Res.,* 18, 335, 1958.
3. **Hartmann, K.-U. and Heidelberger, C.,** Studies on fluorinated pyrimidines. XIII. Inhibition of thymidylate synthetase, *J. Biol. Chem.,* 236, 3006, 1961.
4. **Reyes, P. and Heidelberger, C.,** Fluorinated pyrimidines. XXVI. Mammalian thymidylate synthetase: its mechanism of action and inhibition by fluorinated nucleotides, *Mol. Pharmacol.,* 1, 14, 1965.
5. **Cohen, S. S., Flaks, J. G., Barner, H. D., Loeb, M. R., and Lichtenstein, J.,** The mode of action of 5-fluorouracil and its derivatives, *Proc. Natl. Acad. Sci. U.S.A.,* 44, 1004, 1958.
6. **Mathews, C. K. and Cohen, S. S.,** Inhibition of phage-induced thymidylate synthetase by 5-fluorodeoxyuridylate, *J. Biol. Chem.,* 238, 367, 1963.
7. **Capco, G. R., Krupp, J. R., and Mathews, C. K.,** Bacteriophage-coded thymidylate synthetase: characteristics of the T4 and T5 enzymes, *Arch. Biochem. Biophys.,* 158, 726, 1973.
8. **Galivan, J., Maley, G. F., and Maley, F.,** Purification and properties of T2 bacteriophage-induced thymidylate synthetase, *Biochemistry,* 13, 2282, 1974.
9. **Blakley, R. L.,** The biosynthesis of thymidylic acid. IV. Further studies on thymidylate synthetase, *J. Biol. Chem.,* 238, 2113, 1963.
10. **Lomax, M. I. S. and Greenberg, G. R.,** An exchange between the hydrogen atom of carbon 5 of deoxyuridylate and water catalyzed by thymidylate synthetase, *J. Biol. Chem.,* 242, 1302, 1967.
11. **Lorenson, M. Y., Maley, G. F., and Maley, F.,** The purification and properties of thymidylate synthetase from chick embryo extracts, *J. Biol. Chem.,* 242, 3332, 1967.
12. **Langenbach, R. J., Danenberg, P. V., and Heidelberger, C.,** Thymidylate synthetase: mechanism of inhibition of 5-fluoro-2′-deoxyuridylate, *Biochem. Biophys. Res. Commun.,* 48, 1565, 1972.
13. **Santi, D. V., McHenry, C. S., and Perriard, E. R.,** A filter assay for thymidylate synthetase using 5-fluoro-2′-deoxyuridylate as an active site titrant, *Biochemistry,* 13, 467, 1974.
14. **Crusberg, T. C., Leary, R., and Kisliuk, R. L.,** Properties of thymidylate synthetase from dichloromethotrexate-resistant *Lactobacillus casei, J. Biol. Chem.,* 245, 5292, 1970; **Leary, R. P. and Kisliuk, R. L.,** Crystalline thymidylate synthetase from dichloromethotrexate-resistant *Lactobacillus casei, Prep. Biochem.,* 1, 47, 1971.
15. **Santi, D. V. and McHenry, C. S.,** 5-Fluoro-2′-deoxyuridylate: covalent complex with thymidylate synthetase, *Proc. Natl. Acad. Sci. U.S.A.,* 69, 1855, 1972.
16. **Santi, D. V. and Brewer, C. F.,** Model studies of thymidylate synthetase. Neighboring-group facilitation of electrophilic substitution reactions of uracil furanosides, *J. Am. Chem. Soc.,* 90, 6236, 1968.
17. **Santi, D. V., Brewer, C. F., and Farber, D.,** Studies on the preparation and exchange reactions of 5-deuterated uracils, *J. Heterocycl. Chem.,* 7, 903, 1970.
18. **Santi, D. V., McHenry, C. S., and Sommer, H.,** Mechanism of interaction of thymidylate synthetase with 5-fluorodeoxyuridylate, *Biochemistry,* 13, 471, 1974.
19. **Kalman, T. I.,** Molecular aspects of the mechanism of action of 5-fluorodeoxyuridine, *Ann. N.Y. Acad. Sci.,* 255, 326, 1975.
20. **Kalman, T. I. and Yalowich, J. C.,** Survival and recovery: the reversibility of covalent drug-enzyme interactions, *Dev. Biochem.,* 6, 75, 1979.
21. **Bruice, T. W. and Santi, D. V.,** Secondary α-hydrogen isotope effects on the interaction of 5-fluoro-2′-deoxyuridylate and 5,10-methylenetetrahydrofolate with thymidylate synthetase, *Biochemistry,* 21, 6703, 1982.
22. **Danenberg, P. V. and Danenberg, K. D.,** Effect of 5,10-methylenetetrahydrofolate on the dissociation of 5-fluoro-2′-deoxyuridylate from thymidylate synthetase: evidence for an ordered mechanism, *Biochemistry,* 17, 4018, 1978.
23. **Danenberg, P. V., Langenbach, R. J., and Heidelberger, C.,** Structures of reversible and irreversible complexes of thymidylate synthetase and fluorinated pyrimidine nucleotides, *Biochemistry,* 13, 926, 1974.
24. **McHenry, C. S. and Santi, D. V.,** A sulfhydryl group is not the covalent catalyst in the thymidylate synthetase reaction, *Biochem. Biophys. Res. Commun.,* 57, 204, 1974.
25. **Sommer, H. and Santi, D. V.,** Purification and amino acid analysis of an active site peptide from thymidylate synthetase containing covalently bound 5-fluoro-2′-deoxyuridylate and methylenetetrahydrofolate, *Biochem. Biophys. Res. Commun.,* 57, 689, 1974.
26. **Danenberg, P. V. and Heidelberger, C.,** The effect of Raney nickel on the covalent thymidylate synthetase-5-fluoro-2′-deoxyuridylate-5,10-methylenetetrahydrofolate complex, *Biochemistry,* 15, 1331, 1976.

27. **Pogolotti, A. L., Jr., Ivanetich, K. M., Sommer, H., and Santi, D. V.,** Thymidylate synthetase: studies on the peptide containing covalently bound 5-fluoro-2'-deoxyuridylate and 5,10-methylenetetrahydrofolate, *Biochem. Biophys. Res. Commun.,* 70, 972, 1976.

28. **Plese, P. C. and Dunlap, R. B.,** Sulfhydryl group modification of thymidylate synthetase and its effect on activity and ternary complex formation, *J. Biol. Chem.,* 252, 6139, 1977.

29. **Danenberg, K. D. and Danenberg, P. V.,** Evidence for a sequential interaction of the subunits of thymidylate synthetase, *J. Biol. Chem.,* 254, 4345, 1979.

30. **Galivan, J. H., Maley, F., and Baugh, C. M.,** Demonstration of separate binding sites for the folate coenzymes and deoxynucleotides with inactivated *Lactobacillus casei* thymidylate synthetase, *Biochem. Biophys. Res. Commun.,* 71, 527, 1976.

31. **Galivan, J., Noonan, J., and Maley, F.,** Studies on the reactivity of the essential cysteine of thymidylate synthetase, *Arch. Biochem. Biophys.,* 184, 336, 1977.

32. **Bellisario, R. L., Maley, G. F., Galivan, J. H., and Maley, F.,** Amino acid sequence at the FdUMP binding site of thymidylate synthetase, *Proc. Natl. Acad. Sci. U.S.A.,* 73, 1848, 1976.

33. **Bellisario, R. L., Maley, G. F., Guarino, D. U., and Maley, F.,** The primary structure of *Lactobacillus casei* thymidylate synthetase. II. The complete amino acid sequence of the active site peptide, CNBr 4, *J. Biol. Chem.,* 254, 1296, 1979.

33a. **Moore, M. A., Ahmed, F., and Dunlap, R. B.,** Trapping and partial characterization of an adduct postulated to be the covalent catalytic ternary complex of thymidylate synthase, *Biochemistry,* 25, 3311, 1986.

33b. **Moore, M. A., Ahmed, F., and Dunlap, R. B.,** Evidence for the existence of covalent nucleotide-thymidylate synthase complexes, identification of site of attachment, and enhancement by folates, *J. Biol. Chem.,* 261, 12745, 1986.

34. **James, T. L., Pogolotti, A. L., Jr., Ivanetich, K. M., Wataya, Y., Lam, S. S. M., and Santi, D. V.,** Thymidylate synthetase: fluorine-19 nmr characterization of the active site peptide covalently bound to 5-fluoro-2'-deoxyuridylate and 5,10-methylenetetrahydrofolate, *Biochem. Biophys. Res. Commun.,* 72, 404, 1976.

35. **Byrd, R. A., Dawson, W. H., Ellis, P. D., and Dunlap, R. B.,** Elucidation of the detailed structures of the native and denatured ternary complexes of thymidylate synthetase via ^{19}F NMR, *J. Am. Chem. Soc.,* 100, 7478, 1978.

36. **Lewis, C. A., Jr., Ellis, P. D., and Dunlap, R. B.,** Fluorine-19 nuclear magnetic resonance characterization of ternary complexes of folate derivatives, 5-fluorodeoxyuridylate and *Lactobacillus casei* thymidylate synthetase, *Biochemistry,* 20, 2275, 1981.

37. **Byrd, R. A., Dawson, W. H., Ellis, P. D., and Dunlap, R. B.,** ^{19}F Nuclear magnetic resonance investigation of the ternary complex formed between native thymidylate synthetase, 5-fluoro-2'-deoxyuridylate, and 5,10-methylenetetrahydrofolate, *J. Am. Chem. Soc.,* 99, 6139, 1977.

38. **Aull, J. L., Lyon, J. A., and Dunlap, R. B.,** Gel electrophoresis as a means of detecting ternary complex formation of thymidylate synthetase, *Microchem. J.,* 19, 210, 1974.

39. **Aull, J. L., Lyon, J. A., and Dunlap, R. B.,** Separation, identification, and stoichiometry of the ternary complexes of thymidylate synthetase, *Arch. Biochem. Biophys.,* 165, 805, 1974.

40. **Galivan, J., Maley, F., and Baugh, C. M.,** Protective effect of the pteroylpolyglutamates and phosphate on the proteolytic inactivation of thymidylate synthetase, *Arch. Biochem. Biophys.,* 184, 346, 1977.

41. **Galivan, J. H., Maley, G. F., and Maley, F.,** Factors affecting substrate binding in *Lactobacillus casei* thymidylate synthetase as studied by equilibrium dialysis, *Biochemistry,* 15, 356, 1976.

42. **Sharma, R. K. and Kisliuk, R. L.,** Quenching of thymidylate synthetase fluorescence by substrate analogs, *Biochem. Biophys. Res. Commun.,* 64, 648, 1975.

43. **Galivan, J. H., Maley, G. F., and Maley, F.,** The effect of substrate analogs on the circular dichroic spectra of thymidylate synthetase from *Lactobacillus casei, Biochemistry,* 14, 3338, 1975.

44. **Donato, H., Jr., Aull, J. L., Lyon, J. A., Reinsch, J. W., and Dunlap, R. B.,** Formation of ternary complexes of thymidylate synthetase as followed by absorbance, fluorescence, and circular dichroic spectra and gel electrophoresis, *J. Biol. Chem.,* 251, 1303, 1976.

45. **Sharma, R. K., Kisliuk, R. L., Verma, S. P., and Wallach, D. F. H.,** Study of thymidylate synthetase-function by laser Raman spectroscopy, *Biochim. Biophys. Acta,* 391, 19, 1975.

46. **Santi, D. V., Peña, V. A., and Lam, S. S. M.,** On the structure of the cofactor in the complex formed with thymidylate synthetase, 5,10-methylenetetrahydrofolate and 5-fluoro-2'-deoxyuridylate, *Biochim. Biophys. Acta,* 438, 324, 1976.

46a. **Fitzhugh, A. L., Fodor, S., Kaufman, S., and Spiro, T. G.,** Resonance Raman spectroscopic evidence for alternative structures in the *native* ternary complex formed with thymidylate synthase, *J. Am. Chem. Soc.,* 108, 7422, 1986.

47. **Pellino, A. M. and Danenberg, P. V.,** Evidence from chemical degradation studies for a covalent bond from 5-fluoro-2'-deoxyuridylate to N-5 of tetrahydrofolate in the ternary complex of thymidylate synthetase-5-fluoro-2'-deoxyuridylate-5,10-methylenetetrahydrofolate, *J. Biol. Chem.,* 260, 10996, 1985.

48. **Bratton, A. C. and Marshall, E. K., Jr.,** A new coupling component for sulfanilamide determination, *J. Biol. Chem.*, 128, 537, 1939.

49. **Lockshin, A., Moran, R. G., and Danenberg, P. V.,** Thymidylate synthetase purified to homogeneity from human leukemic cells, *Proc. Natl. Acad. Sci. U.S.A.*, 76, 750, 1979.

50. **Lockshin, A. and Danenberg, P. V.,** Hydrodynamic behavior of human and bacterial thymidylate synthetases and thymidylate synthetase-5-fluoro-2'-deoxyuridylate-5,10-methylenetetrahydrofolate complexes. Evidence for large conformational changes during catalysis, *Biochemistry*, 19, 4244, 1980.

51. **Beaudette, N. V., Langerman, N., Kisliuk, R. L., and Gaumont, Y.,** A calorimetric study of the binding of 2'-deoxyuridine-5'-phosphate and 5-fluoro-2'-deoxyuridine-5'-phosphate to thymidylate synthetase, *Arch. Biochem. Biophys.*, 179, 272, 1977.

52. **Lewis, C. A., Jr., Ellis, P. D., and Dunlap, R. B.,** Fluorine-19 nuclear magnetic resonance investigation of the noncovalent and covalent binary complexes of 5-fluorodeoxyuridylate and *Lactobacillus casei* thymidylate synthetase, *Biochemistry*, 19, 116, 1980.

53. **Lewis, C. A., Jr., Hopper, W. E., Jr., Wheeler, M. R., and Dunlap, R. B.,** Light spectroscopic studies of the binary complex of *Lactobacillus casei* thymidylate synthetase and 5-fluoro-2'-deoxyuridylate, *J. Biol. Chem.*, 256, 7347, 1981.

54. **Moore, M. A., Ahmed, F., and Dunlap, R. B.,** Isolation of the covalent binary complex of 5-fluorodeoxyuridylate and thymidylate synthetase by trichloroacetic acid precipitation, *Biochem. Biophys. Res. Commun.*, 124, 37, 1984.

55. **Ahmed, F., Moore, M. A., and Dunlap, R. B.,** A nitrocellulose-filter assay for the binary complex of 5-fluorodeoxyuridylate and *Lactobacillus casei* thymidylate synthetase, *Anal. Biochem.*, 145, 151, 1985.

56. **Kunitani, M. G. and Santi, D. V.,** On the mechanism of 2'-deoxyuridylate hydroxymethylase, *Biochemistry*, 19, 1271, 1980.

57. **Nakayama, C., Wataya, Y., Santi, D. V., Saneyoshi, M., and Ueda, T.,** Interaction of 1-(5-phospho-β-D-arabinofuranosyl)-5-substituted uracils with thymidylate synthetase: mechanism-based inhibition by 1-(5-phospho-β-D-arabinosyl)-5-fluorouracil, *J. Med. Chem.*, 24, 1161, 1981.

58. **Saneyoshi, M., Inomata, M., and Fukuoka, F.,** Synthetic nucleosides and nucleotides. XI. Facile synthesis and antitumor activities of various 5-fluoropyrimidine nucleosides, *Chem. Pharm. Bull.*, 26, 2990, 1978.

59. **Coderre, J. A., Santi, D. V., Matsuda, A., Watanabe, K. A., and Fox, J. J.,** Mechanism of action of 2',5-difluoro-1-arabinosyluracil, *J. Med. Chem.*, 26, 1149, 1983.

60. **Watanabe, K. A., Reichman, U., Hirota, K., Lopez, C., and Fox, J. J.,** Nucleosides. 110. Synthesis and antiherpes virus activity of some 2'-fluoro-2'deoxyarabinofuranosylpyrimidine nucleosides, *J. Med. Chem.*, 22, 21, 1979.

61. **Santi, D. V., Garrett, C. E., and Barr, P. J.,** On the mechanism of inhibition of DNA-cytosine methyltransferases by cytosine analogues, *Cell*, 33, 9, 1983.

62. **Goldstein, S., Pogolotti, A. L., Jr., Garvey, E. P., and Santi, D. V.,** Interaction of N⁴-hydroxy-2'-deoxycytidylic acid with thymidylate synthetase, *J. Med. Chem.*, 27, 1259, 1984.

63. **Starzyk, R. M., Koontz, S. W., and Schimmel, P.,** A covalent adduct between the uracil ring and the active site of an aminoacyl tRNA synthetase, *Nature (London)*, 298, 136, 1982.

64. **Wataya, Y., Negishi, K., and Hayatsu, H.,** Debromination of 5-bromo-2'-deoxyuridine by cysteine. Formation of deoxyuridine and S-[5-(2'-deoxyuridyl)]cysteine, *Biochemistry*, 12, 3992, 1973.

64a. **Powers, J. C.,** Haloketone inhibitors of proteolytic enzyme, *Chem. Biochem. Amino Acids Pept. Proteins*, 4, 65, 1977.,

64b. **Kézdy, F. J., Thomson, A., and Bender, M. L.,** Studies on the reaction of chymotrypsin and L-1-chloro-3-tosylamido-4-phenyl-2-butanone, *J. Am. Chem. Soc.*, 89, 1004, 1967.

64c. **Weiner, H., White, W. N., Hoare, D. G., and Koshland, D. E., Jr.,** The formation of anhydrochymotrypsin by removing the elements of water from the serine at the active site, *J. Am. Chem. Soc.*, 88, 3851, 1966.

64d. **McMurray, J. S. and Dyckes, D. F.,** Evidence for hemiketals as intermediates in the inactivation of serine proteinases with halomethyl ketones, *Biochemistry*, 25, 2298, 1986.

64e. **Fittkau, S., Smalla, K., and Pauli, D.,** Thermitase: a thermostable serine protease. IV. Kinetic studies on the binding of N-acylated peptide ketones as substrate analogous inhibitors, *Biomed. Biochim. Acta*, 43, 883, 1984.

64f. **Stein, R. and Trainor, D. A.,** Mechanism of inactivation of human leukocyte elastase by a chloromethyl ketone: kinetic and solvent isotope effect studies, *Biochemistry*, 25, 5414, 1986.

64g. **Poulos, T. L., Alden, R. A., Freer, S. T., Birktoft, J. J., and Kraut, J.,** Polypeptide halomethyl ketones bind to serine proteases as analogs of the tetrahedral intermediate. X-ray crystallographic comparison of lysine- and phenylalanine-polypeptide chloromethyl ketone-inhibited subtilisin, *J. Biol. Chem.*, 251, 1097, 1976.

64h. **James, M. N. G., Brayer, G. D., Delbaere, L. T. J., Sielecki, A. R., and Gertler, A.,** Crystal structure studies and inhibition kinetics of tripeptide chloromethyl ketone inhibitors with *Streptomyces griseus* Protease B., *J. Mol. Biol.,* 139, 423, 1980.

64i. **Walker, B., Wikstrom, P., and Shaw, E.,** Evaluation of inhibitor constants and alkylation rates for a series of thrombin affinity labels, *Biochem. J.,* 230, 645, 1985.

64j. **Rasnick, D.,** Synthesis of peptide fluoromethyl ketones and the inhibition of human cathepsin B, *Anal. Biochem.,* 149, 461, 1985.

64k. **Imperiali, B. and Abeles, R. H.,** Inhibition of serine proteases by peptidyl fluoromethyl ketones, *Biochemistry,* 25, 3760, 1986.

64l. **Imperiali, B. and Abeles, R. H.,** A versatile synthesis of peptidyl fluoromethyl ketones, *Tetrahedron Lett.,* 27, 135, 1986.

64m. **Rauber, P., Angliker, H., Walker, B., and Shaw, E.,** The synthesis of peptidyl-fluoromethanes and their properties as inhibitors of serine proteinases and cysteine proteinases, *Biochem. J.,* 239, 633, 1986.

64n. **Angliker, H., Wikstrom, P., Rauber, P., and Shaw, E.,** The synthesis of lysylfluoromethanes and their properties as inhibitors of trypsin, plasmin and cathepsin B, *Biochem. J.,* 241, 871, 1987.

64o. **Akhtar, M., Cohen, M. A., and Gani, D.,** Enzymatic synthesis of 3-halogenoaspartic acids using β-methylaspartase: inhibition by 3-bromoaspartic acid, *J. Chem. Soc. Chem. Commun.,* 1290, 1986.

65. **Alston, T. A. and Bright, H. J.,** Inactivation of pyridoxal 5′-phosphate-dependent enzymes by 5-nitro-L-norvaline, an analog of L-glutamate, *FEBS Lett.,* 126, 269, 1981.

66. **Maurer, B. and Keller-Schierlein, W.,** Synthese des Ferrichroms, 1. Teil: (S)-α-amino-δ-nitrovalerian-säure (δ-nitro-L-norvalin), *Helv. Chim. Acta,* 52, 388, 1969.

66a. **Alston, T. A., Porter, D. J. T., and Bright, H. J.,** Inactivation of GABA amino-transferase by 3-nitro-1-propanamine, *J. Enz. Inhib.,* 1, 215, 1987.

67. **Alston, T. A., Porter, D. J. T., and Bright, H. J.,** Suicide inactivation of D-amino acid oxidase by 1-chloro-1-nitroethane, *J. Biol. Chem.,* 258, 1136, 1983.

68. **Santi, D. V. and Sakai, T. T.,** Thymidylate synthetase. Model studies of inhibition by 5-trifluoromethyl-2′-deoxyuridylic acid, *Biochemistry,* 10, 3598, 1971.

69. **Wataya, Y., Matsuda, A., Santi, D. V., Bergstrom, D. E., and Ruth, J. L.,** *trans*-5-(3,3,3-Trifluoro-1-propenyl)-2′-deoxyuridylate: a mechanism-based inhibitor of thymidylate synthetase, *J. Med. Chem.,* 22, 339, 1979.

70. **Alston, T. A., Mela, L., and Bright, H. J.,** 3-Nitropropionate, the toxic substance of *Indigofera*, is a suicide inactivator of succinate dehydrogenase, *Proc. Natl. Acad. Sci. U.S.A.,* 74, 3767, 1977.

71. **Coles, C. J., Edmondson, D. E., and Singer, T. P.,** Inactivation of succinate dehydrogenase by 3-nitropropionate, *J.. Biol. Chem.,* 254, 5161, 1979.

72. **Buntain, I. G., Suckling, C. J., and Wood, H. C. S.,** Irreversible inhibition of dihydro-orotate dehydrogenase by hydantoins derived from amino acids, *J. Chem. Soc. Chem. Commun.,* 242, 1985.

73. **Garrard, L. J., Mathis, J. M., and Raushel, F. M.,** Substrate-induced inactivation of arginosuccinate lyase by monofluorofumarate and difluorofumarate, *Biochemistry,* 22, 3729, 1983.

74. **Raasch, M. S., Miegel, R. E., and Castle, J. E.,** Mono- and difluorobutenedioic acids, *J. Am. Chem. Soc.,* 81, 2678, 1959.

75. **Maggiora, L., Chang, C. T. C., Hasson, M. E., Bigge, C. F., and Mertes, M. P.,** 5-*p*-Benzoquinonyl-2′-deoxyuridine-5′-phosphate: a possible mechanism-based inhibitor of thymidylate synthetase, *J. Med. Chem.,* 26, 1028, 1983.

75a. **Vadnere, M. K., Maggiora, L., and Mertes, M. P.,** Thiol addition to quinones: model reactions for the inactivation of thymidylate synthase by 5-*p*-benzoquinonyl-2′-deoxyuridine 5′-phosphate, *J. Med. Chem.,* 29, 1714, 1986.

76. **Bigge, C. F. and Mertes, M. P.,** A palladium-catalyzed coupling reaction and a photolytic reaction for the direct synthesis of 5-arylpyrimidine nucleotides, *J. Org. Chem.,* 46, 1994, 1981.

77. **Danenberg, P. V., Bhatt, R. S., Kundu, N. G., Danenberg, K., and Heidelberger, C.,** Interaction of 5-ethynyl-2′-deoxyuridylate with thymidylate synthetase, *J. Med. Chem.,* 24, 1537, 1981.

78. **Barr, P. J., Nolan, P. A., Santi, D. V., and Robbins, M. J.,** Inhibition of thymidylate synthetase by 5-alkynyl-2′-deoxyuridylates, *J. Med. Chem.,* 24, 1385, 1981.

79. **Barr, P. J., Robins, M. J., and Santi, D. V.,** Reaction of 5-ethynyl-2′-deoxyuridylate with thiols and thymidylate synthetase, *Biochemistry,* 22, 1696, 1983.

80. **Perman, J., Sharma, R. A., and Bobek, M.,** Synthesis of 1-(2-deoxy-β-D-*erythro*-pentofuranosyl)-5-ethynyl-1,2,3,4-tetrahydropyrimidine-2,4-dione(5-ethynyl-2′-deoxyuridine), *Tetrahedron Lett.,* 2427, 1976.

81. **Barr, P. J., Jones, A. S., Serafinowski, P., and Walker, R. T.,** The synthesis of nucleosides derived from 5-ethynyluracil and 5-ethynylcytosine, *J. Chem. Soc. Perkin Trans 1,* 1263, 1978.

82. **Robins, M. J. and Barr, P. J.,** Nucleic acid related compounds, 31. Smooth and efficient palladium-copper catalyzed coupling of terminal alkynes with 5-iodouracil nucleosides, *Tetrahedron Lett.,* 22, 421, 1981.

83. **Maldonado, M. E., Oh, K.-J., and Frey, P. A.,** Studies on *Escherichia coli* pyruvate dehydrogenase complex. I. Effect of bromopyruvate on the catalytic activities of the complex, *J. Biol. Chem.,* 247, 2711, 1972.

84. **Bisswanger, H.,** Substrate specificity of the pyruvate dehydrogenase complex from *Escherichia coli, J. Biol. Chem.,* 256, 815, 1981.

85. **Apfel, M. A., Ikeda, B. H., Speckhard, D. C., and Frey, P. A.,** *Escherichia coli* pyruvate dehydrogenase complex. Thiamin pyrophosphate-dependent inactivation by 3-bromopyruvate, *J. Biol. Chem.,* 259, 2905, 1984.

86. **Lowe, P. N. and Perham, R. N.,** Bromopyruvate as an active-site-directed inhibitor of the pyruvate dehydrogenase multienyzme complex from *Escherichia coli, Biochemistry,* 23, 91, 1984.

87. **Leung, L. S. and Frey, P. A.,** Fluoropyruvate: an unusual substrate for *Escherichia coli* pyruvate dehydrogenase, *Biochem. Biophys. Res. Commun.,* 81, 274, 1978.

88. **Hübner, G., Schellenberger, A., Bernhardt, R., Khailova, L. S., and Severin, S. E.,** Inactivation of the pyruvate dehydrogenase component from pigeon breast muscle pyruvate dehydrogenase complex by α-ketobutyric acid, *FEBS Lett.,* 84, 179, 1977.

89. **Kuo, D. J. and Jordan, F.,** Active-site directed irreversible inactivation of brewers' yeast pyruvate decarboxylase by the conjugated substrate analogue (E)-4-(4-chlorophenyl)-2-oxo-3-butenoic acid: development of a suicide substrate, *Biochemistry,* 22, 3735, 1983.

90. **Kuo, D. J. and Jordan, F.,** Direct spectroscopic observation of a brewer's yeast pyruvate decarboxylase-bound enamine intermediate produced from a suicide substrate. Evidence for nonconcerted decarboxylation, *J. Biol. Chem.,* 258, 13415, 1983.

91. **Kuo, D. J., Dikdan, G., and Jordan, F.,** Resolution of brewers' yeast pyruvate decarboxylase into two isozymes, *J. Biol. Chem.,* 261, 3316, 1986.

92. **Datta, A. K. and Daniels, T. C.,** Antitubercular activity of some aromatic aldehyde and ketone derivatives, *J. Pharm. Sci.,* 52, 905, 1963.

93. **Jordan, F., Adams, J., Farzami, B., and Kudzin, Z. H.,** Conjugated α-keto acids as mechanism-based inactivators of brewer's yeast pyruvate decarboxylase: electronic effects of substituents and detection of a long-lived intermediate, *J. Enzyme Inhib.,* 1, 139, 1986.

94. **Matsuda, A., Wataya, Y., and Santi, D. V.,** 5-Nitro-2′-deoxyuridylate: A mechanism-based inhibitor of thymidylate synthetase, *Biochem. Biophys. Res. Commun.,* 84, 654, 1978.

95. **Wataya, Y., Matsuda, A., and Santi, D. V.,** Interaction of thymidylate synthetase with 5-nitro-2′-deoxyuridylate, *J. Biol. Chem.,* 255, 5538, 1980.

96. **Mertes, M. P., Chang, C. T. C., De Clercq, E., Huang, G. F., and Torrence, P. F.,** 5-Nitro-2′-deoxyuridine 5′-monophosphate is a potent irreversible inhibitor of *Lactobacillus caesi* thymidylate synthetase, *Biochem. Biophys. Res. Commun.,* 84, 1054, 1978.

97. **Maggiora, L., Chang, C. T. C., Torrence, P. F., and Mertes, M. P.,** 5-Nitro-2′-deoxyuridine 5′-phosphate: a mechanism-based inhibitor of thymidylate synthetase, *J. Am. Chem. Soc.,* 103, 3192, 1981.

98. **Santi, D. V., Norment, A., and Garrett, C. E.,** Covalent bond formation between a DNA-cytosine methyltransferase and DNA containing 5-aza-cytosine, *Proc. Natl. Acad. Sci. U.S.A.,* 81, 6993, 1984.

99. **Cooper, A. J. L. and Griffith, O. W.,** N-Hydroxyamino acids, irreversible inhibitors of pyridoxal 5′-phosphate enzymes and substrates of D- and L-amino acid oxidase, *J. Biol. Chem.,* 254, 2748, 1979.

100. **Al-Razzak, L. A., Schwepler, D., Decedue, C. J., Balzarini, J., De Clercq, E., and Mertes, M. P.,** 5-Quinone derivatives of 2′-deoxyuridine 5′-phosphate: inhibition and inactivation of thymidylate synthase, antitumor cell, and antiviral studies, *J. Med. Chem.,* 30, 409, 1987.

Chapter 5

ACYLATION REACTIONS*

I. INTRODUCTION

Since acylation reactions are covalent in nature, the inactivation pathways described in this chapter actually are initiated by affinity labeling of the enzyme. However, in order for them to be classified as mechanism-based inactivations, additional steps must be required after the acylation step.

II. ACYLATION/ADDITION

A. 6-Acetylmethylenepenicillanic Acid

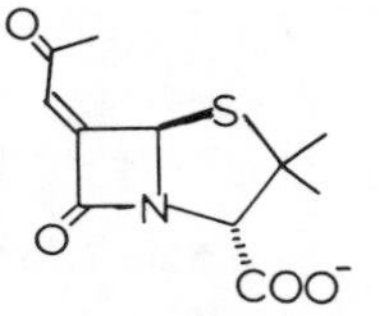

Scheme 1.   Structural Formula 5.1.

6-Acetylmethylenepenicillanic acid (Structural Formula 5.1, Scheme 1) was shown by Arisawa and Then[1] to be a time-dependent inactivator of β-lactamases from a variety of bacteria. Inactivation occurs at much lower concentrations than two other β-lactamase inactivators, clavulanic acid (see Section IV.C.1.) or sulbactam (see Section IV.B.1.c.). With purified TEM 1 β-lactamase, no hydrolysis of the inactivator was detected prior to inactivation. One mol of β-lactamase is inactivated by 1.04 mol of 6-acetylmethylenepenicillanic acid, whereas 110 and 2900 mol of clavulanic acid and sulbactam, respectively, are required. Inactivation is second order when equimolar amounts of enzyme and inactivator are used.[2] Isoelectric focusing of the enzyme after a short duration of inactivation at 10°C gives two new bands; after longer incubation times, one band disappears, leaving only the second new band. These results suggest that two inhibited forms of the enzyme result; one that is transient and another irreversible.[2] Several derivatives of 6-acetylmethylenepenicillanic acid were prepared by Arisawa and Adam;[3] the ketone in the C-6 side chain is required for inhibition. Changes in the absorption spectrum during inactivation resemble that observed when excess hydroxylamine is added to the inactivator. Incubation of inactivated enzyme at 37°C leads to slow reactivation; isoelectric focusing indicates at least two new forms of inhibited enzyme are formed during reactivation. A product is generated during reactivation which was identified as Structural Formula 5.2 (Scheme 2). A model study was carried out[3] in which 6-acetylmethylenepenicillanic acid was treated with sodium methoxide and the methyl ester of **5.2** was isolated. The mechanism for formation of **5.2** is shown in Scheme 3. On the basis of this mechanism, the enzyme inactivation mechanism shown in Scheme 4 was proposed.[3]

---

* A list of abbreviations and shorthand notations can be found prior to Chapter 1.

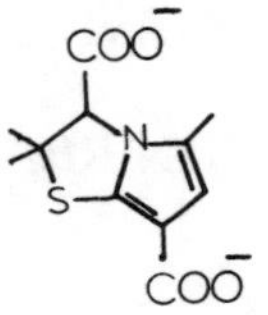

Scheme 2.   Structural Formula 5.2.

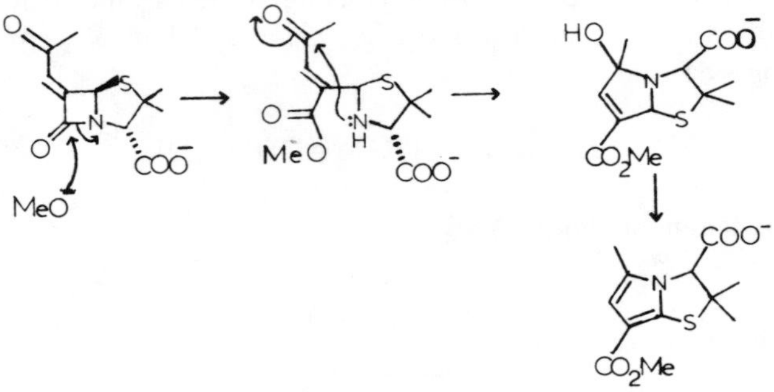

Scheme 3.

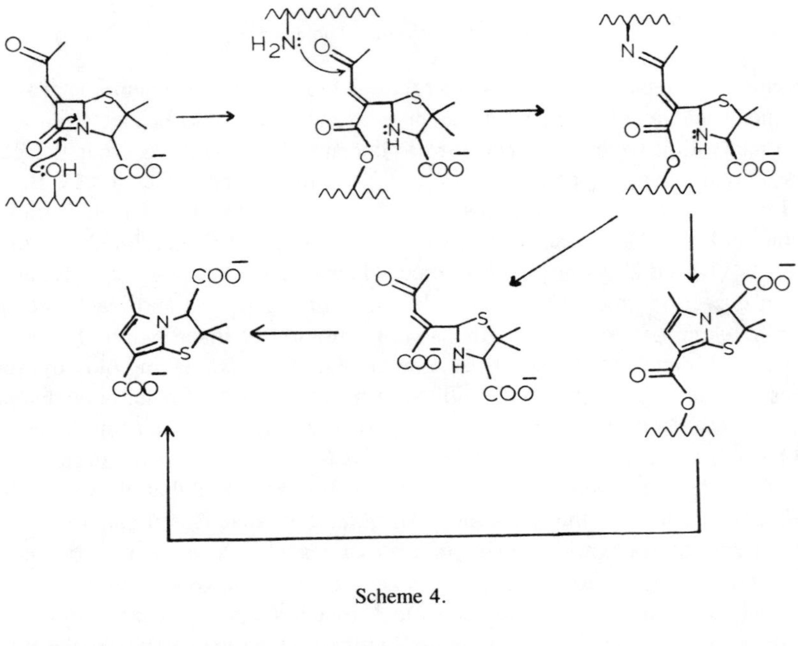

Scheme 4.

B. (Z)-6-(Methoxymethylene)penicillanic Acid

(Z)-6-(Methoxymethylene)penicillanic acid (**5.3**) is an inhibitor and inactivator of *Escherichia coli* RTEM β-lactamase, but the (*E*)-isomer is not a substrate, inhibitor, or inactivator.[4] The mechanism proposed by Brenner and Knowles[4] for inactivation is shown in Scheme 5. These compounds can be synthesized by the route shown in Scheme 6.[5] The (Z)-isomer is the first compound in a new class of penams having a heteroatom-substituted methylene group at the 6-position conjugated to the β-lactam carbonyl.

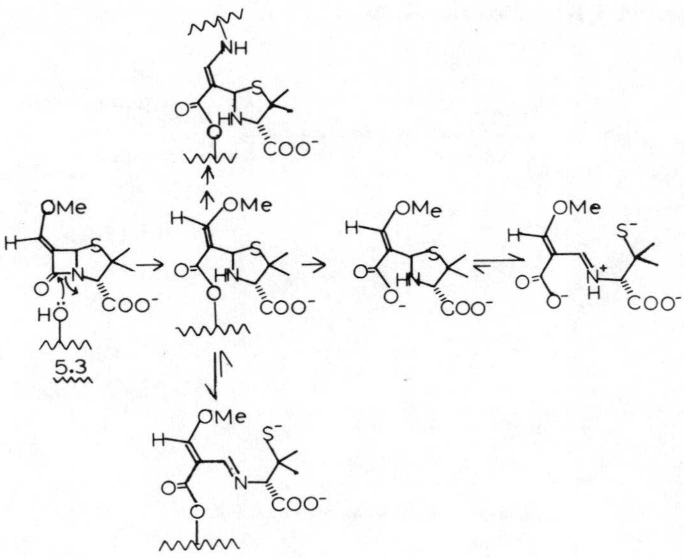

Scheme 5.   Containing Structural Formula 5.3.

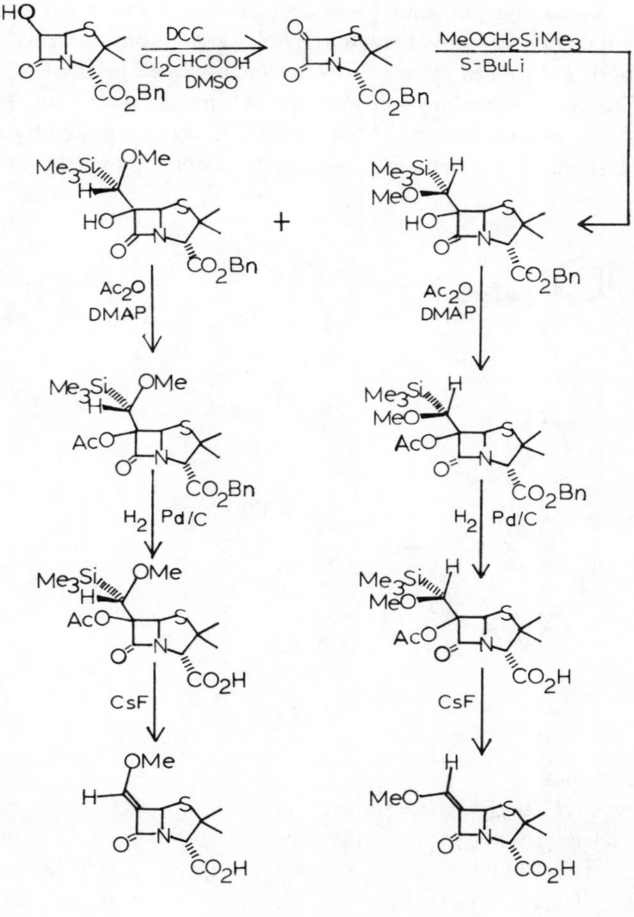

Scheme 6.

C. 2-Bromomethyl-3,1-benzoxazin-4-one

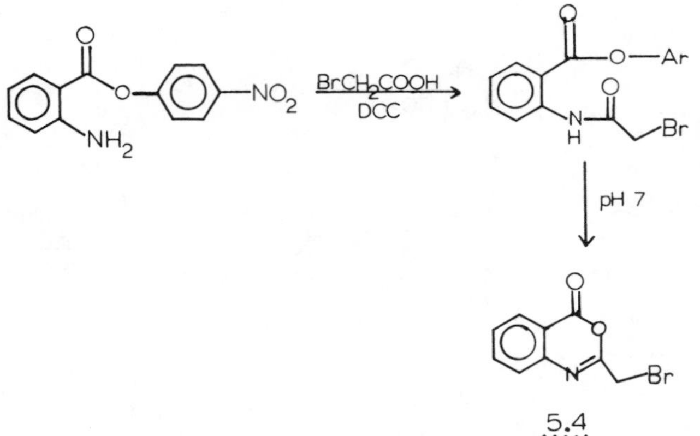

Scheme 7.   Containing Structural Formula 5.4.

The synthesis of 2-bromomethyl-3,1-benzoxazin-4-one (**5.4**)[6] is shown in Scheme 7. This compound irreversibly inactivates α-chymotrypsin, but not trypsin, at neutral pH in a time-dependent process.[7] Inactivation occurs with a change in absorbance for the α-chymotrypsin-proflavin complex, suggesting formation of an acyl enzyme. Using *p*-nitrophenyl *N*-(bromo[2-¹⁴C]acetyl)anthranilate (a compound known to hydrolyze to 2-bromomethyl-3,1-benzoxazin-4-one at neutral pH), 1.2 mol of radioactivity are incorporated per mole of enzyme. Amino acid analyses of native α-chymotrypsin indicate two methionines, but inactivated α-chymotrypsin shows only one methionine. Methionine-192 was suggested by Alazard et al.[7] to be the alkylated residue. The mechanism proposed is depicted in Scheme 8.

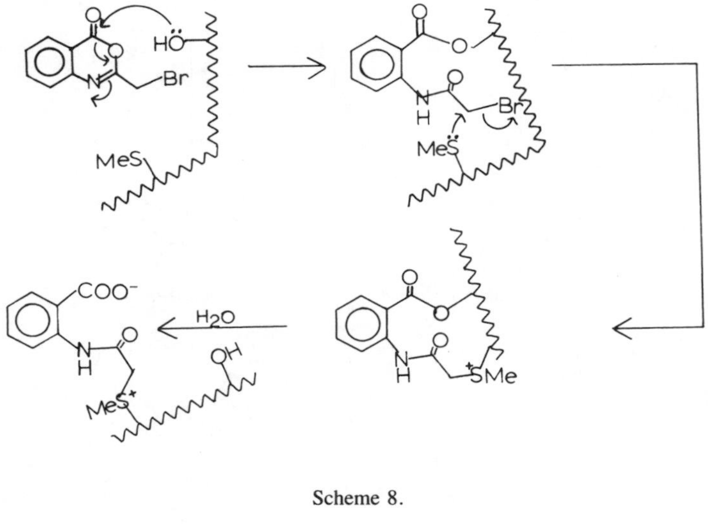

Scheme 8.

D. Halo Enol Lactones

1. (E)-5-(1-Halomethylidene)dihydro-3-phenyl-2(3H)-furanone Derivatives

On the basis of a suggestion by Rando[8] that halo enol lactones would be useful mechanism-based inactivators, new synthetic routes to these compounds were devised by Krafft and Katzenellenbogen.[9] The best general route is halolactonizaton of acetylenic acids (Scheme 9; X = Cl, Br, I). Other routes also are successful (Schemes 10A to D).

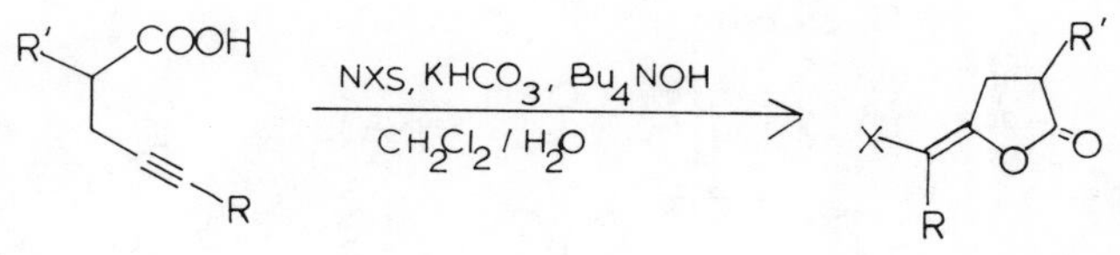

Scheme 9.

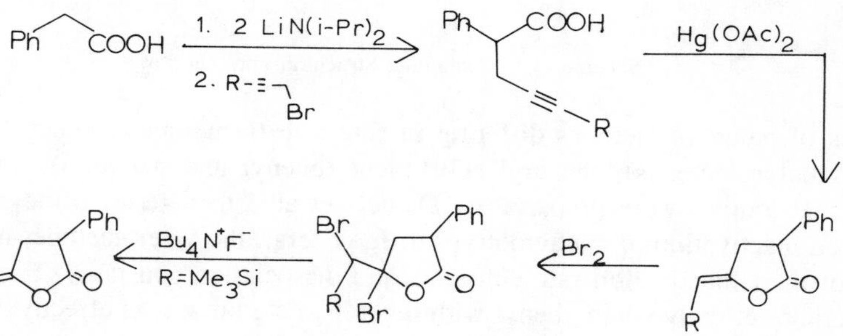

Scheme 10A.

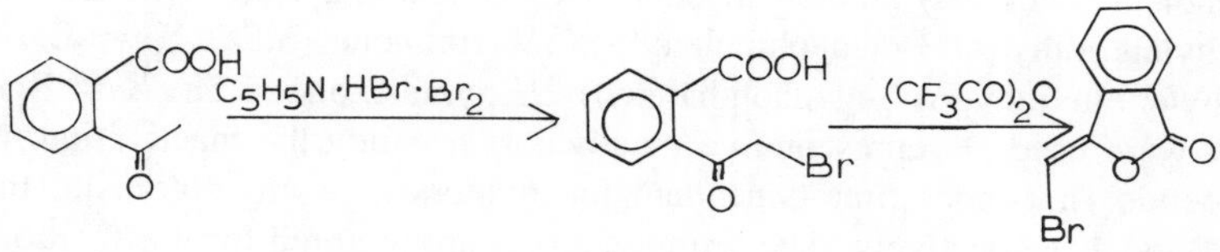

Scheme 10B.

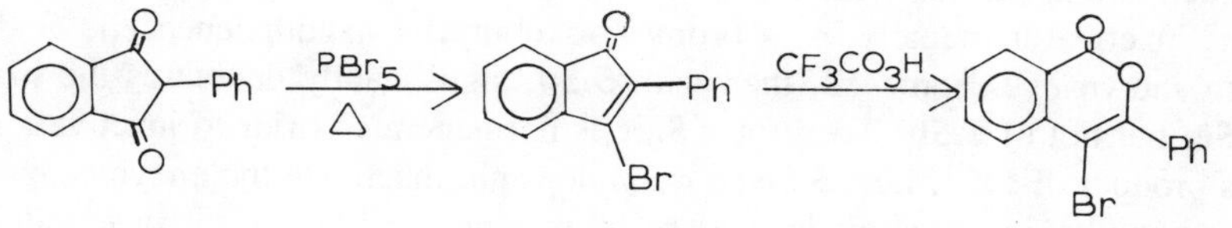

Scheme 10C.

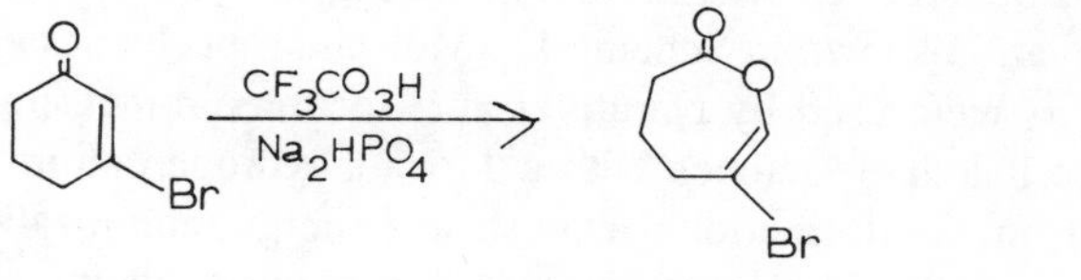

Scheme 10D.

3-Phenyl-(5*E*)-(1-bromoethylidene)dihydro-2(3*H*)-furanone (**5.5**, X = Br, R = Me) and 3-phenyl-(5*E*)-(iodomethylidene)dihydro-2(3*H*)-furanone (**5.5**, X = I, R = H) were shown by Chakravarty et al.[10] to inactivate α-chymotrypsin at a pseudo-first order time-dependent rate. The inactivation is irreversible to gel filtration. The halogen is required for inactivation, but the hydrolysis products of the halo enol lactones are not inactivators, indicating that enzyme-catalyzed reaction of the lactone carbonyl is necessary prior to alkylation as shown in Scheme 11.

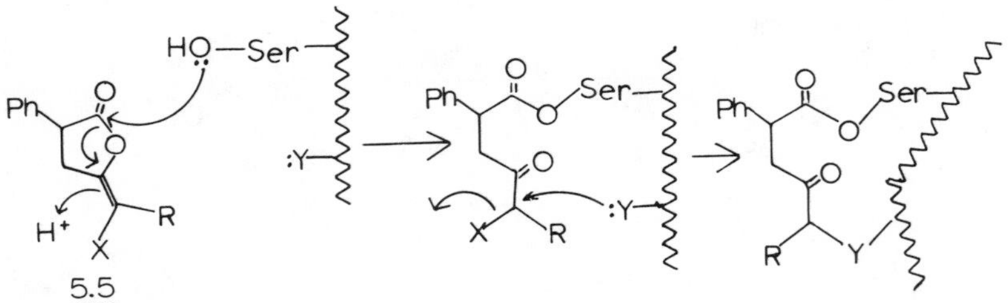

5.5

Scheme 11.   Containing Structural Formula 5.5.

A series of haloenol lactones differing in ring size (6-membered valerolactones and 5-membered butyrolactones), the aryl substituent (phenyl and naphthyl), and the halogen (bromine and iodine) were prepared by Daniels et al.[11] to determine the effects of these variables on inactivation of α-chymotrypsin. In general, the valerolactones are more potent in terms of their inactivation rate constants and dissociation constants. The naphthyl substituent is more effective than phenyl with regards to K_I, but not as effective in terms of the catalytic inactivation rate constant. The halogen has little effect on either K_I or k_{inact}. The rates of spontaneous hydrolysis of the compounds are low ($t_{1/2}$ = 73 to 710 min). The mechanism of inactivation of α-chymotrypsin by α-aryl-substituted five- and six-membered ring halo enol lactones was studied in detail by Daniels and Katzenellenbogen.[11a] (*E*)-5-(Bromomethylene)dihydro-3-(1-naphthalenyl)-2(3*H*)-furanone (**5.5a**, X = Br) and (*E*)-6-(bromomethylene)tetrahydro-3-(1-naphthalenyl)-2*H*-pyran-2-one (**5.5b**, X = Br), shown in Scheme 12, were used as representative inactivators to study the inactivation mechanism. Both are pseudo first-order time-dependent inactivators of α-chymotrypsin; the partition ratios are 11 and 4, respectively. The bromine atoms are required for inactivation, since the corresponding methylene analogues (**5.5a** and **5.5b**, X = H) are not inactivators. Evidence that the reactive species is not released prior to inactivation is as follows: (1) there is no lag time for inactivation; (2) methylamine, which reacts much more rapidly with the hydrolyzed form of the inactivator, namely, α-(4-bromo-3-oxobutyl)-1-naphthaleneacetic acid (**5.5c**, n = 2), also shown in Scheme 12, than with **5.5b**, significantly decreases the inactivation rate of **5.5c**, but not of **5.5b**. Therefore, **5.5c** is not generated prior to inactivation; (3) the hydrolysis product of **5.5a**, i.e., **5.5c** (n = 1) does not inactivate the enzyme significantly. In order to show that inactivation does not result by simple acylation, the inactivated enzyme was treated with hydrazine; no return of enzyme activity resulted. Chymotrypsin inactivated with [2-^{14}C]-**5.5b** (X = I) gives a 1:1 stoichiometry of [^{14}C] to enzyme. All of the radioactivity remains bound after denaturation in boiling ethanol. These results support the mechanism of inactivation shown in Scheme 11. Molecular mechanics calculations and computer graphics analysis were used by Naruto et al.[12] for three-dimensional mapping of the interactions of three haloenol lactones with a 115 nonhydrogen atom model for the active site of α-chymotrypsin. Conformational analysis and energy minimization of the three key stages of interactions, namely, Michaelis complex formation, acylation of Ser-195, and the alkylation of His-57 to give the bis-adduct, were modeled. The results of these studies support the mechanism of inactivation previously proposed by Daniels et al.[11] Estimates of van der Waals interaction energies correlated with the experimental K_I values, and estimates of enthalpy of the alkylation reaction parallel k_{inact} values. This computational approach may be useful in formulating mechanisms and in guiding the synthesis of new inactivators. Phenyl and naphthyl groups were shown by minimization calculations to occupy the hydrophobic pocket of the enzyme in the Michaelis complex. The acylation reaction is similar to that with peptide substrates. Alkylation of Met-192, instead of His-57, was modeled; the complex leading to alkylation of His-57 is 10 kcal/mol more stable.

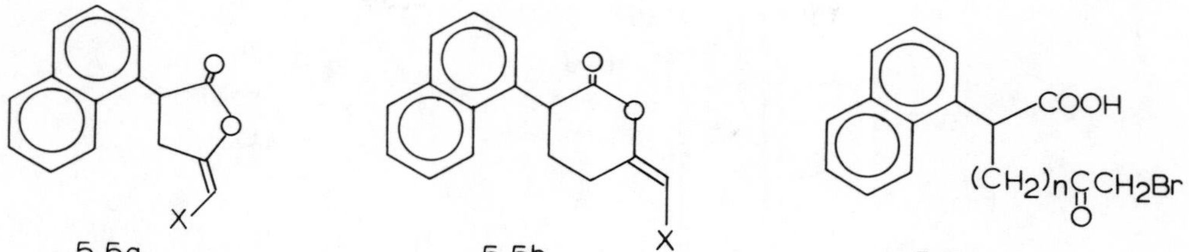

Scheme 12. Structural Formulas 5.5a, b, and c.

2. *Halo Enol Lactone Analogues of Amino Acids*

In order to achieve more selective inactivators of proteases, syntheses of α-amino halo enol lactone analogues of amino acids were developed by Sofia et al.[13] The approach would then be extended to incorporation of the α-amino halo enol lactones into oligopeptides at the scissile bond (Structural Formula 5.6, Scheme 13). This way, specificity for selective proteases could be achieved by appropriate choice of the carrier peptide. The halo enol lactone of glycine can be prepared as shown in Scheme 14A and of phenylglycine in Scheme 14B.

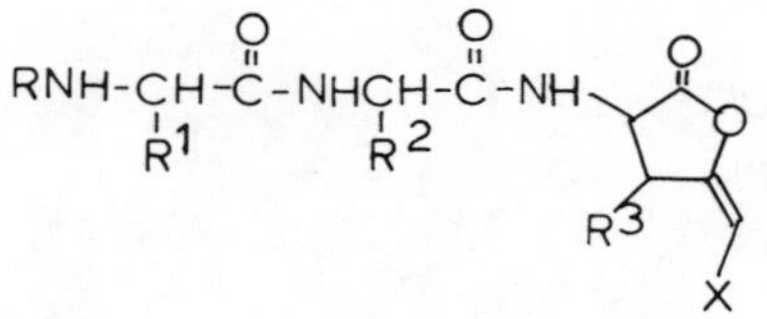

Scheme 13. Structural Formula 5.6.

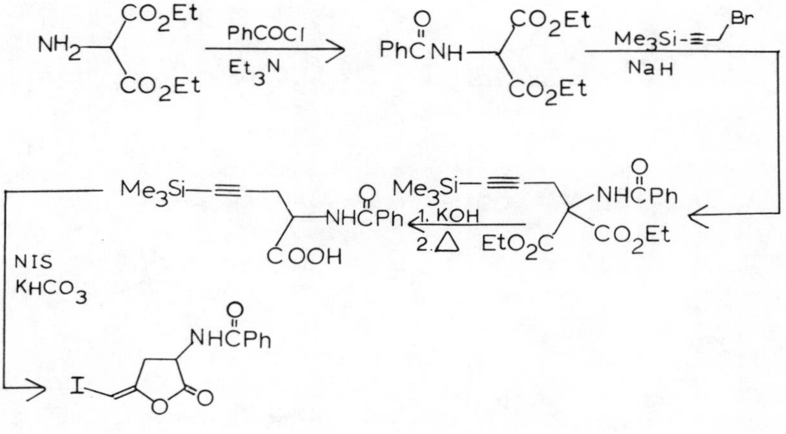

Scheme 14A.

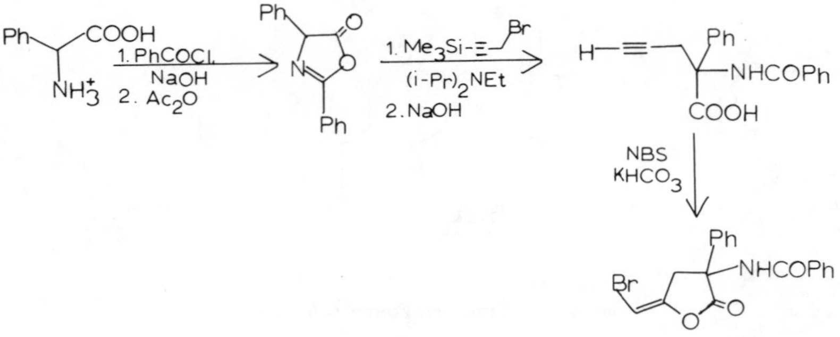

Scheme 14B.

Syntheses of α-propargyl amino acids were required in order to obtain the halo enol lactones of other amino acids; these are shown in Schemes 15A to C. α-Propargylphenylalanine was used to prepare 3-amino-3-benzyl lactones (Scheme 15D).

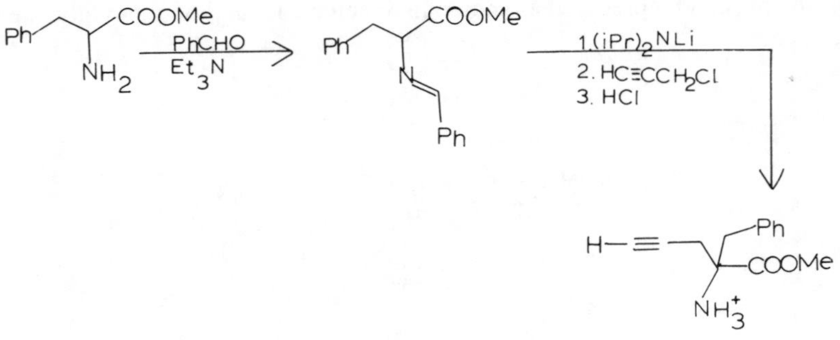

Scheme 15A.

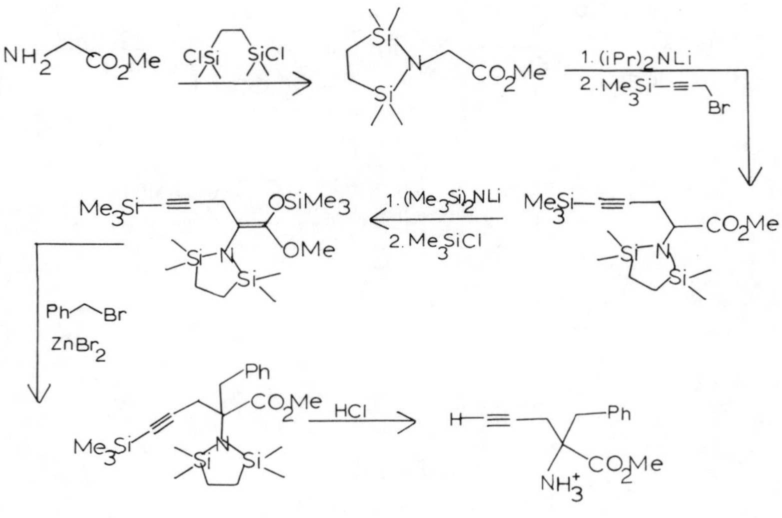

Scheme 15B.

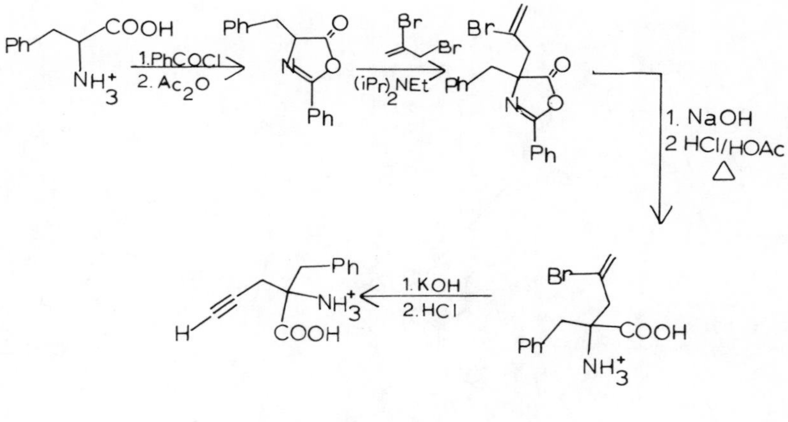

Scheme 15C.

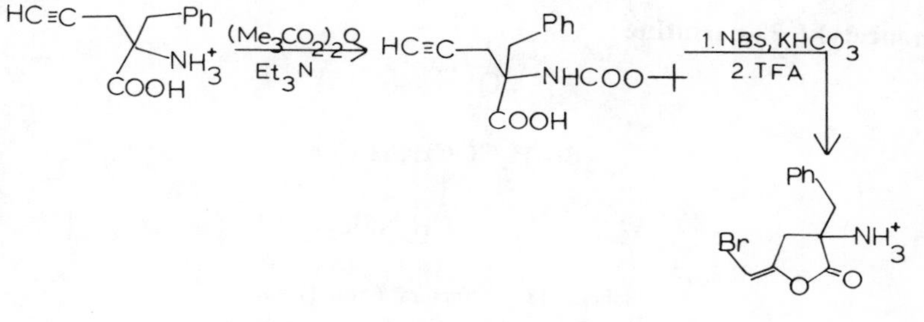

Scheme 15D.

E. Ynenol Lactones

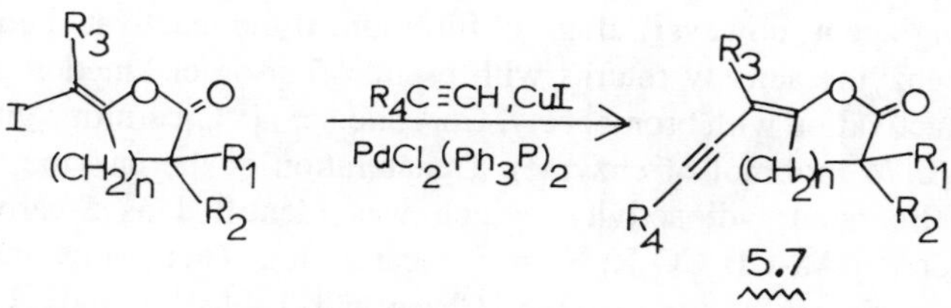

Scheme 16.   Containing Structural Formula 5.7.

Halo enol lactones synthesized by Chakravarty et al.[10] were utilized as starting materials for ynenol lactones (**5.7**) by Tam et al.[14] (Scheme 16). Several of the (*E*)-ynenol lactones are time-dependent inactivators of human leukocyte elastase;[14] gel filtration does not regenerate enzyme activity. The ratios of k_{inact}/K_I for inhibition of human leukocyte elastase, porcine pancreatic elastase, and bovine trypsin (all serine proteases) are 22,000: 730: 17 for **5.7** (R_1 = PhCH$_2$, R_2 = R_3 = R_4 = H) and 23,000: 260: 43 for **5.7** (R_1 = *n*-Bu, R_2 = R_3 = R_4 = H). The mechanism proposed is shown in Scheme 17 (see Section V.B.1).

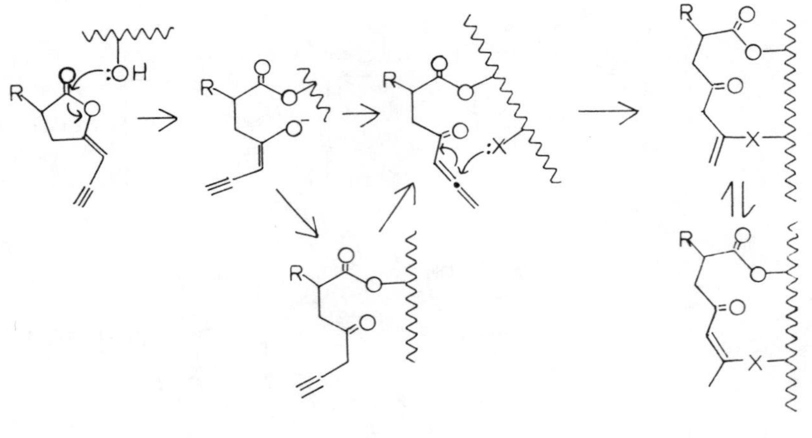

Scheme 17.

F. Bromoacetyl-(−)-carnitine

$$
\underset{\underset{CH_2NMe_3}{\overset{|}{\underset{+}{}}}}{\overset{O}{\overset{\|}{BrCH_2COCHCH_2COO^-}}}
$$

Scheme 18.   Structural Formula 5.8.

Bromoacetyl-(−)-carnitine (Structural Formula 5.8, Scheme 18) plus CoA or bromoacetyl CoA plus (−)-carnitine rapidly inactivate pigeon breast muscle carnitine acetyltransferase in a pseudo first-order process.[15] In both cases 1 mol of inactivator binds per mole of enzyme, and enzyme activity does not return upon gel filtration. It may be thought that this is simply an affinity-labeling agent, however, after gel filtration, if the inactivated enzyme is allowed to stand at 4°C, enzyme activity returns with pseudo first-order kinetics and a half-life of about 15 days. Inactivation with bromoacetyl CoA and (±) [^{14}C]carnitine gives radioactively labeled enzyme (0.74 mol/mol of enzyme). Denaturation of the enzyme was followed by isolation of the liberated radioactivity, which was identified as S-carboxymethyl CoA (−)carnitine ester (CoAS-CH$_2$CO$_2$R; R = (−)carnitine). This compound is a very slow time-dependent inactivator of the enzyme. Chase and Tubbs[15] proposed that a bridge is formed between the binding sites for CoASH and carnitine as shown in Scheme 19. These results support a catalytic mechanism involving ternary enzyme-substrate complex formation.

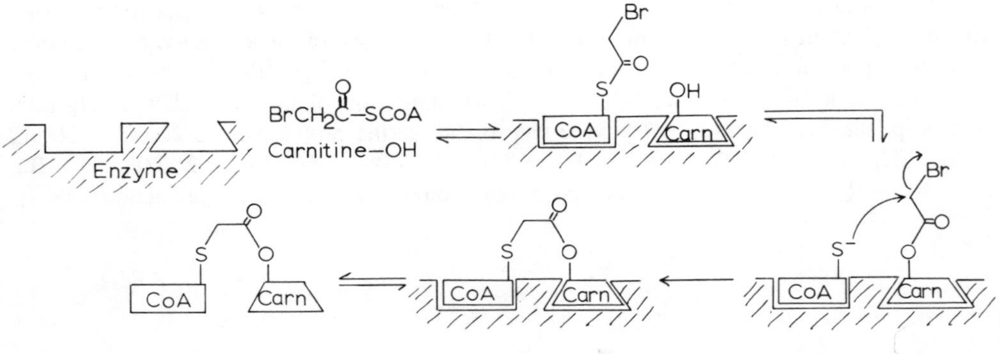

Scheme 19.

G. 3-Keto-4-pentenoyl CoA

The following metabolites of 4-pentenoic acid were tested by Zhong et al.[16] as inactivators of carnitine acetyltransferase from pigeon breast muscle: 2,4-pentadienoyl CoA, 3-keto-4-pentenoyl CoA, and acryloyl CoA. Only 3-keto-4-pentenoyl CoA (Structural Formula 5.9, Scheme 20) is a time-dependent inactivator; inactivation is much more rapid in the presence of the substrate L-carnitine than in its absence. At first sight, this compound also could act simply as an affinity-labeling agent, since it is an α,β-unsaturated ketone. However, the following evidence supports a mechanism in which 3-keto-4-pentenoyl CoA is converted to 3-keto-4-pentenoyl-carnitine prior to inactivation: D-Carnitine, a competitive inhibitor for L-carnitine, does not stimulate inactivation; 3-keto-4-pentenoyl CoA is a substrate for the enzyme in the presence of L-carnitine with concomitant release of a sulfhydryl-containing compound shown to be CoASH; no CoASH is formed in the absence of L-carnitine; less than 5 mol of inactivator are required to inactivate 1 mol of enzyme. No return of enzyme activity is evident after gel filtration or dialysis. The mechanism proposed is shown in Scheme 20.

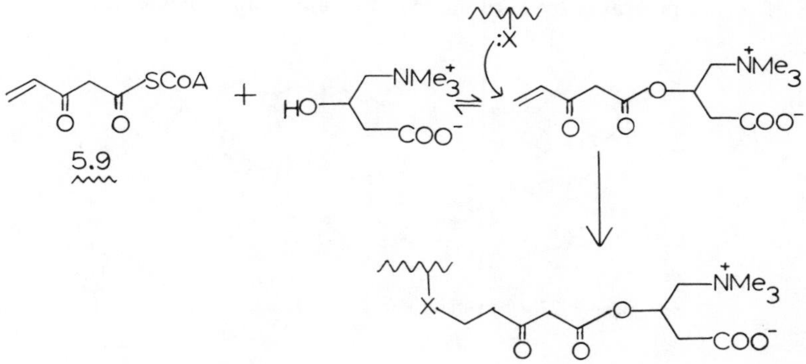

Scheme 20.   Containing Structural Formula 5.9.

III. ACYLATION/ACYLATION

A. Acylation/Acylation
1. 3-Chloroisocoumarin, 3,3-Dichlorophthalide, and Related

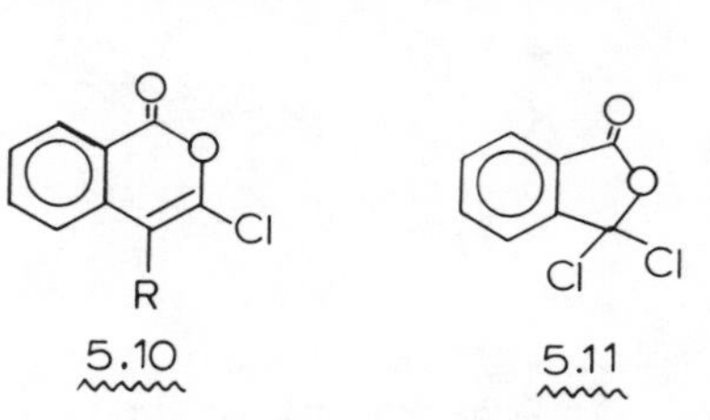

Scheme 21.   Structural Formulas 5.10 and 5.11.

3-Chloroisocoumarin (3-chloro-1*H*-2-benzopyran-1-one) (**5.10**) and 3,3-dichlorophthalide (**5.11**), shown in Scheme 21, are potent mechanism-based inactivators of the serine proteases, human leukocyte elastase, porcine pancreatic elastase, human leukocyte cathepsin G, and chymotrypsin.[17] Neither compound inactivates trypsin nor papain. 3-Chlorophthalide does not inactivate any of the enzymes. Inactivation of the elastases by **5.10** leads to stable adducts, but inactivation of all four enzymes by **5.11** produces an inhibited enzyme that gains activity upon standing. Treatment of the elastases inactivated by **5.10** with hydrazine results in a rapid return of enzyme activity. The partition ratios for inactivation of human leukocyte elastase and porcine pancreatic elastase by **5.10** are greater than 15 and greater than 4, respectively. The values are given as lower limits because of the nonenzymatic hydrolytic lability of the compound. Chymotrypsin A$_\gamma$ becomes 90% inactivated with only 1 equivalent of **5.10**, concomitant with the release of 0.9 equivalent of protons. Further studies by Harper et al.[18] showed that **5.10**, **5.11**, 3,4-dichloroisocoumarin (**5.12**, Scheme 22) and 3-acetoxyisocoumarin (**5.13**, Scheme 22) inactivate, to different extents, human leukocyte elastase, porcine pancreatic elastase, cathepsin G, chymotrypsin, *Streptomyces griseus* protease A, *Staphylococcus aureus* protease V-8, rat mast cell protease I, and rat mast cell protease II. Pseudo first-order plots were obtained, but enzyme inactivation was not irreversible. Enzyme activity returns slowly upon standing, after dialysis, or more rapidly in the presence of hydroxylamine. Inactivation of chymotrypsin is stoichiometric with **5.10** and **5.12**. With human leukocyte elastase, **5.10** and **5.12** turnover 15 and 3.1 times, respectively, per inactivation event. There are 0.92 and 1.25 equivalent of protons released during inactivation of chymotrypsin with **5.10** and **5.12**, respectively. The mechanism for inactivation of serine proteases by **5.10** and **5.12** is shown in Scheme 23.

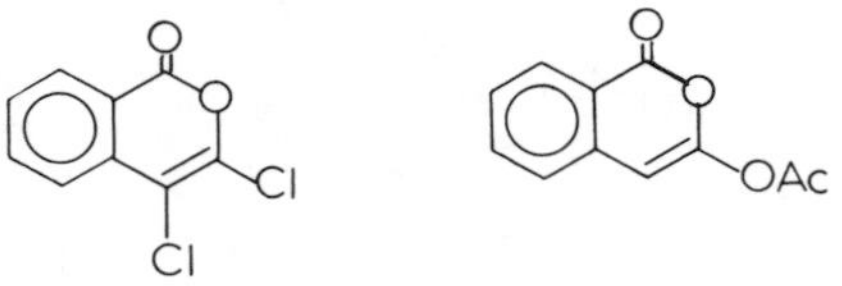

Scheme 22.   Structural Formulas 5.12 and 5.13.

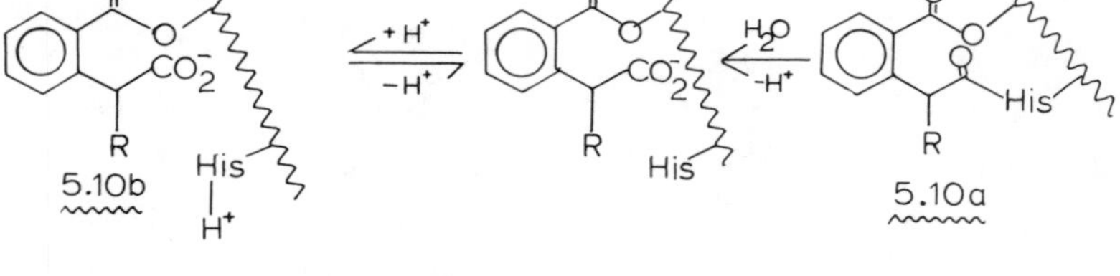

Scheme 23.   Containing Structural Formulas 5.10a and 5.10b.

Since only 1 equivalent of protons is released during inactivation, **5.10a** is most likely the inactivation product, although **5.10b** also would be consistent with the data. If only acylation occurs, then these compounds are affinity-labeling agents; if **5.10a** is involved, then they are mechanism-based inactivators. 3-Acetoxyisocoumarin **5.13** must inactivate the enzymes by a different mechanism, since it is possible to produce **5.10a** from inactivation by **5.13**, yet the rate of reactivation is much faster than for enzyme inactivated by **5.10**. Similar experiments with **5.11** and chymotrypsin A$_\gamma$ result in the release of 2.7 equivalents of protons with 92% loss of enzyme activity. These results are consistent with the mechanism shown in Scheme 24.

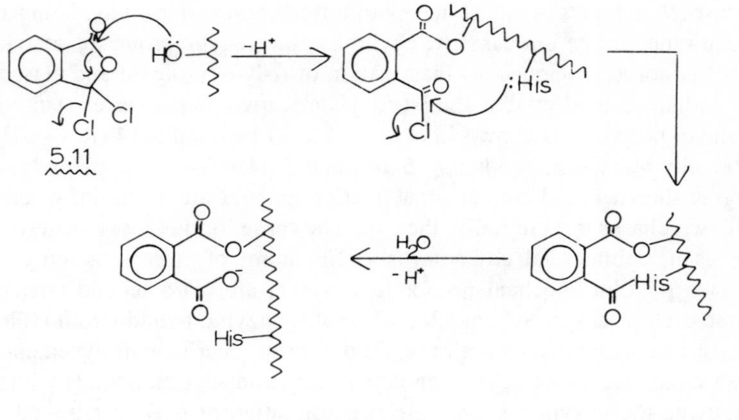

Scheme 24.

Compounds **5.10**[19] and **5.13**[20] can be synthesized from homophthalic acid with $PCl_5/POCl_3$ (**5.10**) or with acetyl chloride/tri-elamine (**5.13**). Chlorination of **5.10** followed by tri-ethylamine elimination of HCl gives **5.12**.[21] Compound **5.11** is prepared from phthalic anhydride[22] (Scheme 25).

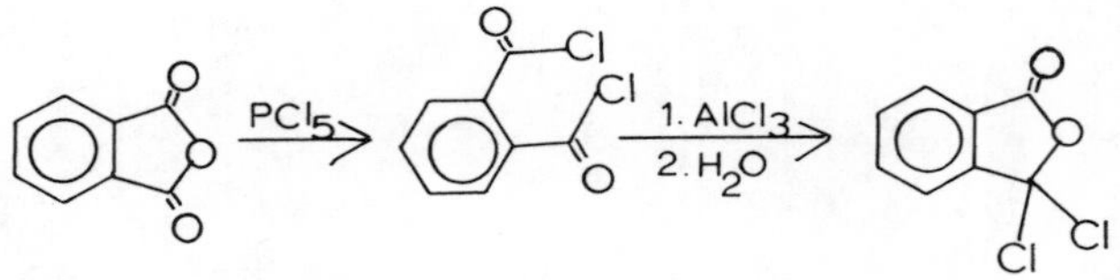

Scheme 25.

2. 6-Chloro-2-pyrone Analogues

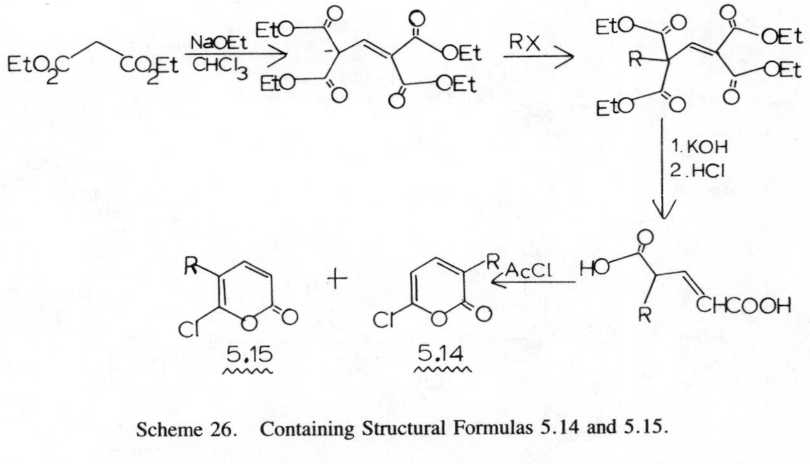

Scheme 26.   Containing Structural Formulas 5.14 and 5.15.

3-Substituted- and 5-substituted-6-chloro-2-pyrones (**5.14** and **5.15**, respectively) were synthesized by Westkaemper and Abeles[23] by the route in Scheme 26 (R = $PhCH_2$ or CH_3). The 3-benzyl derivative (**5.14**, R = CH_2Ph) has a structural similarity to specific substrates for α-chymotrypsin and the 3-methyl compound resembles substrates for α-lytic protease and elastase. Neither trypsin, which acts on substrates with charged side chains, nor carboxypeptidase, which acts on substrates with a free carboxy terminus, is inactivated. Specificity among other serine proteases (α-chymotrypsin, α-lytic protease, acetylcholinsterase, and elastase) is not as pronounced. Inactivation of α-lytic protease and elastase is, indeed, more rapid with the 3-methyl- than the 3-benzyl derivative. Papain is resistant to inactivation. The mechanism proposed is shown in Scheme 27. Compound **5.14** (R = CH$_2$Ph) also is a substrate for chymotrypsin, producing **5.14a** and **5.14b**. These, apparently, come from a partitioning as shown in Scheme 28. Inactivation kinetics are biphasic, where the partition ratio for the two phases is 14 initially, then 40. The cause for this is two different catalytically active species in commercial chymotrypsin. No return of enzyme activity occurs by gel filtration. Two possible mechanisms for inactivation are (1) a second nucleophile attacks one of the acid chlorides in Scheme 28, while still enzyme bound (from CPR models His-57 is the most reasonable nucleophile) or (2) the serine ester is hydrolyzed and the half acid chloride can either escape or acylate an active site residue. Compounds without a chlorine do not inactivate the enzyme, suggesting that formation of α,β-unsaturated acyl enzymes alone is not responsible for inactivation; another activated group is needed. The optical spectrum of inactivated enzyme shows that the pyrone ring is not present.

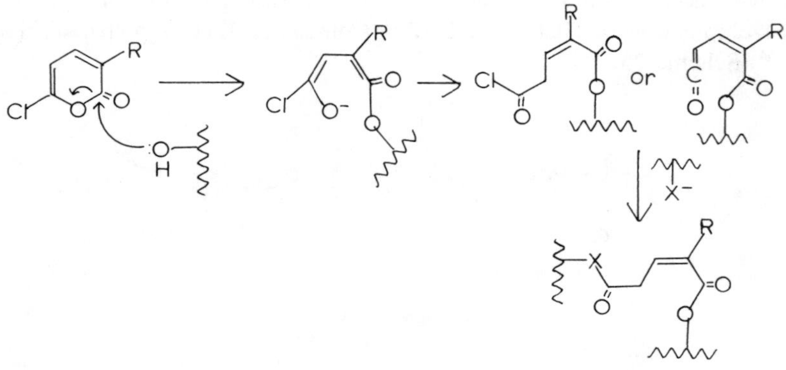

Scheme 27.

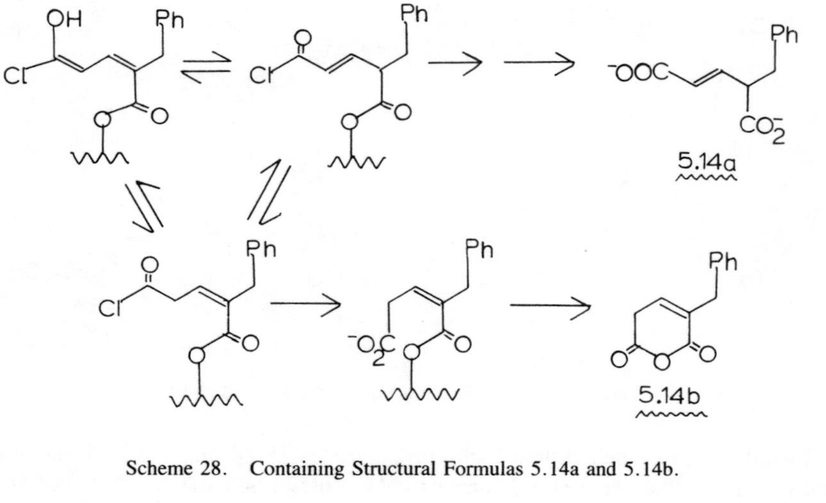

Scheme 28.   Containing Structural Formulas 5.14a and 5.14b.

The inactivation mechanism shown in Scheme 27 was modified by Gelb and Abeles.[24] The inactivator was synthesized with [^{13}C] labels at the 2-position, the 2,6-positions, and the 2,5-positions and ^{13}C NMR spectra were recorded. The inactivator is attached to the enzyme at only one point (the 2-position) through an ester linkage (presumably to serine-195); C-6 becomes a free carboxylate. This was confirmed by the incorporation of only one [^{18}O] atom from [^{18}O]H$_2$O upon hydrolysis and release from the enzyme. The revised mechanism is shown in Scheme 29. The *E*-isomer is believed to be involved in inactivation; the *Z*-isomer is rapidly deacylated. The 5-deuterio inactivator exhibits an isotope effect on reactivation, indicating that isomerization to the β,γ-position is important to reactivation. Reactivation occurs spontaneously with a $t_{1/2}$ = 23 hr. X-ray diffraction analysis at 1.9-Å resolution of the covalent adduct between chymotrypsin and 3-benzyl-6-chloro-2-pyrone shows that the oxygen of Ser-195 is attached to C-1 of (*Z*)-2-benzylpentenedioic acid.[24a] The benzyl group is in the hydrophobic specificity pocket of the enzyme, and the free carboxylate forms a salt bridge with His-57. It is suggested by Ringe et al.[24a] that the salt bridge prevents access of water and therefore slows hydrolysis of the acyl enzyme. Reactivation involves reorientation of the free carboxylate which breaks the salt bridge, causing rapid deacylation with release of (*E*)-4-benzyl-2-pentenedioic acid.

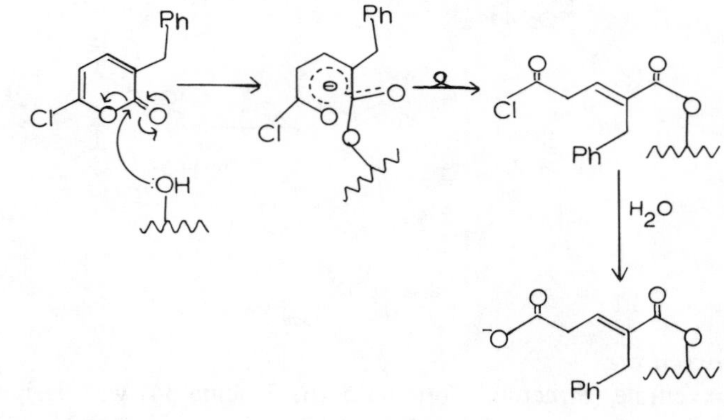

Scheme 29.

In the case of **5.15** (R = CH$_2$Ph) however, time-dependent inactivation gives a spectrum which suggests retention of the pyrone moiety.[23] This compound is not also a substrate. The mechanism proposed[23] is affinity labeling of the enzyme (Scheme 30). The carbon-13 NMR spectrum of chymotrypsin inactivated with 5-benzyl-6-chloro-2-pyrone-2,6-[^{13}C$_2$] contains two new resonances from the enzyme-inactivator adduct which are consistent with an intact pyrone ring and with the replacement of chlorine by an oxygen substituent.[25] X-ray analysis of the inactivated enzyme shows attachment of the serine-195 γ-oxygen to C-6 of the inactivator. The 5-benzyl substituent is ensconced in the hydrophobic pocket of the enzyme. These results corroborate the proposed mechanism as affinity labeling of the enzyme.[23] The position of the benzyl group, therefore, appears to determine the mode of serine attack: 3-benzyl gives acylation; 5-benzyl gives Michael addition.

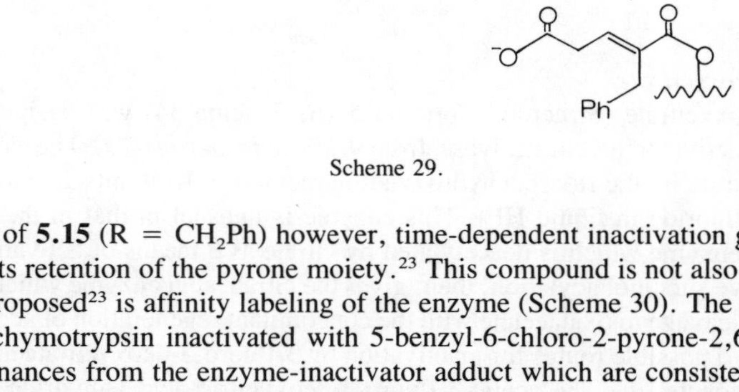

Scheme 30.

A variety of 2-pyrones substituted with alkyl, aryl, and arylalkyl groups at carbons 3, 4, and 5 and bearing a hydrogen, a halogen, or trifluoromethyl group at C-6 (Structural Formula 5.15a, Scheme 31) were synthesized by Boulanger and Katzenellenbogen.[25a] Only the 6-halo-2-pyrones are time-dependent inactivators of α-chymotrypsin. The 6-chloro-2-pyrones with 4-phenyl or 3-(2-napthylmethyl) substituents produce rapid enzyme inactivation; those with 3-benzyl or 3-(1-naphthylmethyl) substituents are poor inactivators, and those with 3-phenyl or 3-alkyl substituents are not inactivators. In general, the compounds were prepared by the route shown in Scheme 32.

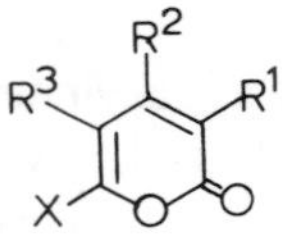

Scheme 31.   Structural Formula 5.15a.

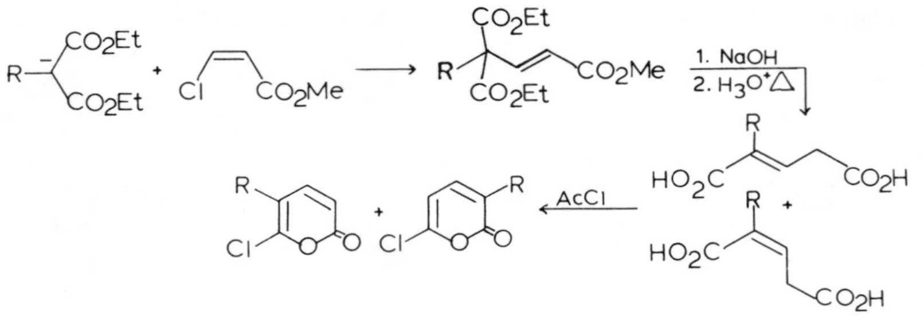

Scheme 32.

3. 3-Fluoro-3-deoxycitrate

3-Fluoro-3-deoxycitrate (Structural Formula 5.16, Scheme 34) was designed by Rokita et al.[26] as an inactivator of citrate lyase from *Klebsiella aerogenes*. The compound was prepared from citrate by the fluorodehydroxylation method of Kollonitsch et al.[27] (treatment with sulfur tetrafluoride in liquid HF). This enzyme is unusual in that in the resting state, it is an acetyl-S-enzyme which is deacetylated by citrate as a means of activating the citrate carboxylate; active site thiol acylation, then, gives the citrate acyl enzyme which decomposes by a retro-aldol cleavage to oxalacetate with the concomitant regeneration of acetyl-S-enzyme (Scheme 33). Two possible routes for inactivation by 3-fluoro-3-deoxycitrate are reasonable: initial deacetylation to give the acetic-3-fluoro-3-deoxycitric acid anhydride intermediate, which could diffuse from the active site and leave the enzyme in a deacetylated, inactive form; or, if the 3-fluoro-3-deoxycitrate acyl enzyme forms, it cannot undergo the retro-aldol reaction and it would be in an acylated inactive form[26] (Scheme 34). Inactivation kinetics are not pseudo first-order, but are time-dependent. With the use of 3-fluoro[1,5-^{14}C]-3-deoxycitrate, no products were detected by HPLC. The two proposed inactive enzyme species can be differentiated by treatment with acetic anhydride and acetate; the deacetylated enzyme is known to be reactivated by acetic anhydride; acetate should deacylate the acyl inactive enzyme and give active acetylated enzyme (Scheme 35). In fact, both acetic anhydride and sodium acetate lead to reactivation of 50% of the initial activity; however, acetate-reactivated enzyme, then loses activity with time, as expected for further reaction of the active enzyme with regenerated 3-fluoro-3-deoxycitrate. The acylated inactive enzyme hydrolyzes with a $t_{1/2} \approx 70$ min, which accounts for the reactivation by acetic anhydride. 3-Fluoro[1,5-^{14}C]-3-deoxycitrate was used to show that acylation of the enzyme is a viable pathway, and that the acylated enzyme is released upon treatment with acetate, but not acetic anhydride. As expected, the compound with hydrogen substituted for fluorine, i.e., tricarballylate, also is an inactivator, but with a much higher K_I value and much lower k_{inact} value. Inactive enzyme, however, is reactivated by acetic anhydride, not by acetate. Therefore, either acylation is too slow to detect or hydrolysis is too fast.

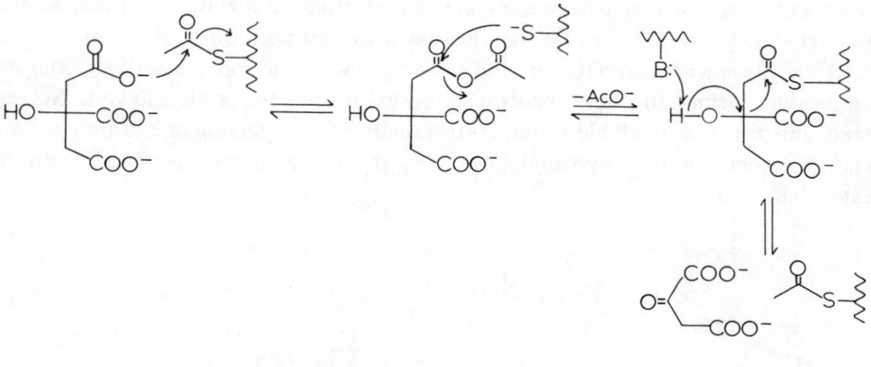

Scheme 33.

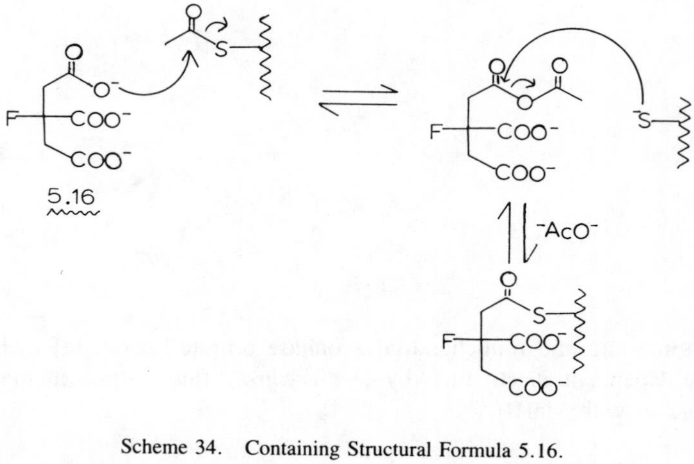

Scheme 34. Containing Structural Formula 5.16.

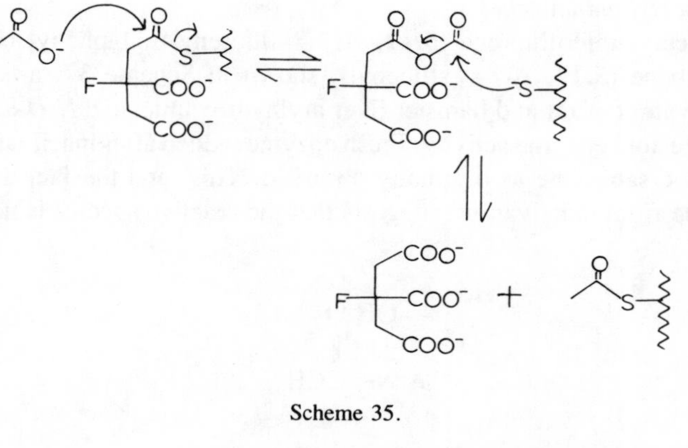

Scheme 35.

4. 2-Fluorocitrate

Bacterial citrate lyase is known to be unstable during citrate turnover; one out of 6500 to 12,000 turnovers leads to inactivation as a result of mistaken hydrolysis of the acetylated enzyme. When (−)-*erythro*-2-fluorocitrate is used as a substrate, a partition ratio of only 20 was observed by Rokita and Walsh.[28] Presumably, the 2-fluorocitryl enzyme hydrolyzes much more readily than does the citryl enzyme, leaving nonacetylated, inactive enzyme more frequently. Acetic anhydride reactivates the inactivated enzyme. (+)-*threo*- and (−)-*erythro*-2-Hydroxycitrate inactivate the enzyme with partition ratios of 55 and 15, respectively, in the presence of Mg(II) ion; the (+)-*erythro*-isomer only inactivates the enzyme without product formation. A mechanism suggested for this isomer is shown in Scheme 36. This mechanism is not available to the corresponding fluoro analogue and may account for the major difference in these two analogues; (+)-*erythro*-2-fluorocitrate has a partition ratio of greater than 100.

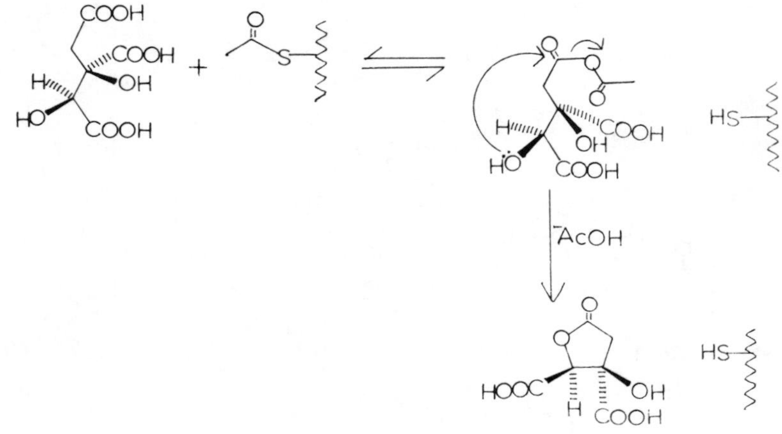

Scheme 36.

Pig liver cytoplasmic and mitochondrial aconitase (citrate [isocitrate] hydrolyase) also undergoes time-dependent inactivation by (−)-*erythro*-2-fluorocitrate in the presence of Mg(II) and more so with Mn(II).[29]

B. Acylation/Acylation/Addition

1. N-Hydroxy-2-acetylaminoarenes

N-Hydroxy-2-acetylamidofluorene (**5.17**, Ar = fluorenyl), -biphenyl (**5.17**, Ar = biphenyl), and -stilbene (**5.17**, Ar = stilbenyl), shown in Scheme 37, are time-dependent irreversible inactivators of rat and hamster liver arylhydroxamic acid *N,O*-acyltransferase.[30] Dialysis does not restore enzyme activity. Fresh enzyme, added after inactivation is complete, is inactivated at the same rate as originally observed. This, and the fact that thiols do not protect the enzyme from inactivation, suggests that the reactive species is not released prior to inactivation.

Scheme 37.   Structural Formula 5.17.

N-Hydroxy-2-acetamidofluorene was used by Yeh and Hanna[31] to inactivate N-arylhydroxamic acid N,O-acyltransferase in order to determine if this is the same enzyme in a partially purified hepatic preparation that activates N-hydroxy-2-propionamidofluorene and N-(2-fluorenyl)methoxyacetohydroxamic acid. Bioactivation is depressed for the latter two compounds after pretreatment with N-hydroxy-2-acetamidofluorene, suggesting that N-arylhydroxamic acid N,O-acyltransferase is responsible. A series of 7-substituted-N-hydroxy-2-acetamidofluorenes was synthesized (Scheme 38) and shown to be inactivators of hamster hepatic N-arylhydroxamic acid N,O-acyltransferase.[32] Dialysis does not cause return of activity. By inclusion of different low molecular nucleophiles in the inactivation buffer, **partial** protection (0 to 58% depending upon inactivator and nucleophile) of inactivation was observed, indicating that both mechanism-based inactivation and metabolically generated and released inactivation are involved. QSAR analysis of the compounds provides support for the proposal that a positively-charged species is involved in the inactivation process.

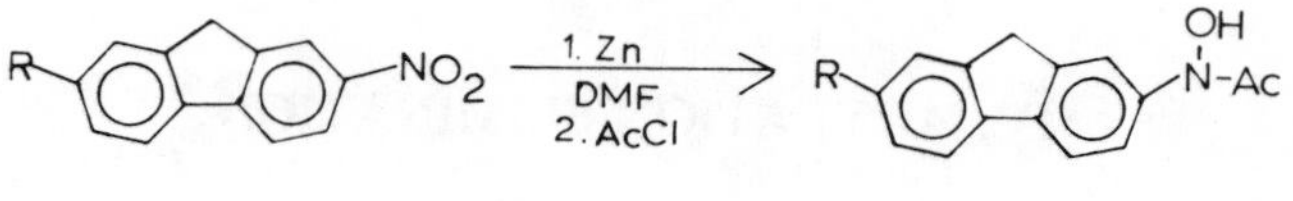

Scheme 38.

A series of N-hydroxyacetanilides (Structural Formula **5.17a**) was prepared by Mangold and Hanna[33] in which alkyl, alkenyl, and cycloalkyl substituent replaced the p-phenyl group of N-hydroxy-4-acetamidobiphenyl (Scheme 39). Of the 11 compounds tested, only N-hydroxy-2-acetamidobiphenyl and N-hydroxy-2-acetamido-4-cyclohexylbenzene are time-dependent inactivators of the enzyme. However, inactivation by the biphenyl analogue exhibits pseudo first-order kinetics and is not affected by the presence of β-mercaptoethanol in the buffer, whereas the rate of inactivation by the 4-cyclohexylphenyl analogue increases with time. When β-mercaptoethanol is added, the inactivation rate by the latter compound becomes pseudo first-order and is partially decreased. These results suggest that the biphenyl analogue is a true mechanism-based inactivator whereas the 4-cyclohexylphenyl analogue is partly mechanism-based and partly metabolically activated, with the reactive product being released into solution. The inactivation mechanism proposed[33] is shown in Scheme 40.

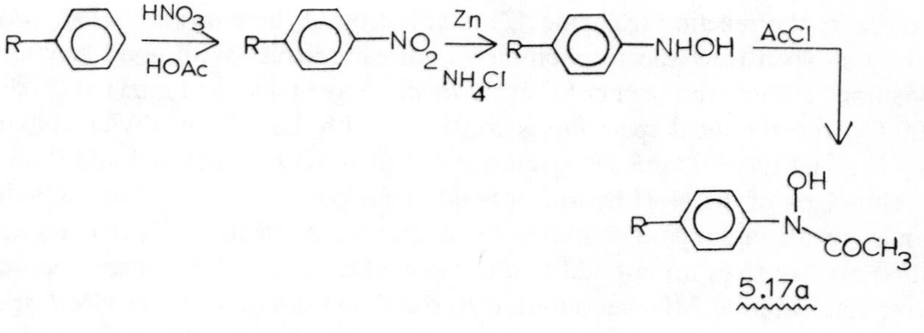

Scheme 39.   Containing Structural Formula 5.17a.

Compound **5.17** (Ar = fluorenyl) also is a mechanism-based inactivator of an acetyl CoA dependent N-acetyltransferase.[34] p-Aminobenzoic acid acetyl CoA-dependent N-acetyltransferase activity is protected from inactivation by cysteine, but the sulfamethazine acetyl CoA-dependent N-acetyltransferase activity is not; therefore, only the latter activity meets one of the criteria for mechanism-based inactivation, namely, that external nucleophiles do not protect.

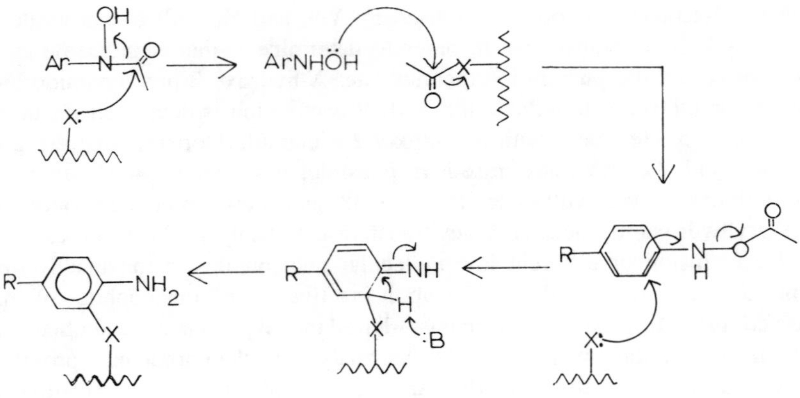

Scheme 40.

IV. ACYLATION/ELIMINATION

A. Acylation/Elimination

1. Cephalosporins

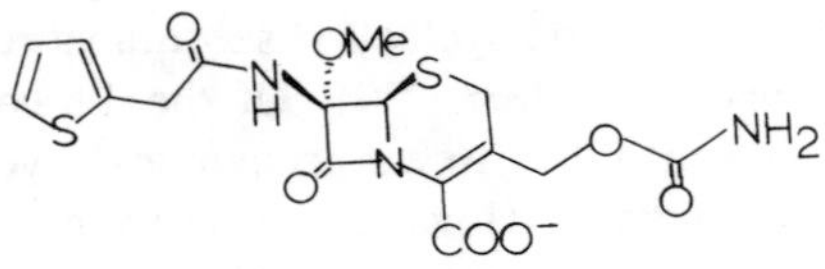

Scheme 41.   Structural Formula 5.18.

Cefoxitin (Structural Formula 5.18, Scheme 41), a poor substrate for RTEM β-lactamase, was used by Fisher et al.[35] to show that an acyl enzyme intermediate is involved in the reaction of β-lactams with β-lactamase. Fourier transform IR showed the formation of an ester, suggesting that a serine residue may be acylated. Cephalosporins with good leaving groups in the 3′-position transiently inhibit PC1 β-lactamase of *Staphylococcus aureus*.[36] The first step in the reaction (Scheme 42) is acylation of the enzyme (**5.19**), which can be observed by stopped-flow spectrophotometric measurements. With good leaving groups at the 3′-position, elimination occurs to give a much more stable acyl enzyme (**5.20**; $t_{1/2} \simeq 10$ min at 20°C) whose optical spectrum is consistent with that of a methylenedihydrothiazine structure. It is not clear why **5.20** is more stable than **5.19**, but Faraci and Pratt[36c] suggest that either the lack of the N–H hydrogen bond or the presence of a hydrogen bond acceptor in **5.20** may result in enzyme conformational change. At neutral pH, it is found in model studies that nucleophiles do not add to a compound related to **5.20**; therefore, stabilization by enzyme nucleophilic Michael addition to **5.20** was discounted. ^{1}H NMR spectroscopy of the hydrazinolysis of a variety of cephalosporins containing 3′ leaving groups and the β-lactamase-catalyzed hydrolysis of these compounds indicates that the acylation and elimination steps are not concerted,[36a] as was previously suggested by Page and Proctor[36b] from model studies. Pratt and Faraci[36a] also show that the elimination product (**5.20**) and **5.19** are in equilibrium. Further studies by Faraci and Pratt[36c] provided evidence for the effect of the 7α-methoxyl group and the 3′-leaving group of cephamycins on inactivation. The carbamoyloxy 3′ leaving group of **5.18** is eliminated and the compound produces a stable acylated enzyme with a 5-*exo*-methylene-1,3-thiazine structure (**5.20**, R′ = OMe). 3′-De(carbamoyloxy)cefoxitin (**5.18** with the carbamoyloxy group replaced by H) and 3′-

decarbamoylcefoxitin (**5.18** with carbamoyloxy replaced by OH) do not produce a stable acyl enzyme. The 7-methoxyl group weakens the noncovalent binding and slows down both acylation and deacylation rates, but mostly the acylation rate. The 7α-methoxyl group has a significant effect on the β-lactamase conformation in the acyl enzyme form, but not on that of the initial noncovalent complex. Boyd[36d] has postulated that the 3′ leaving group in cephalosporins affects the reaction with the active-site serine as a result of both an inductive effect and its leaving group ability and that a tetrahedral intermediate is involved.

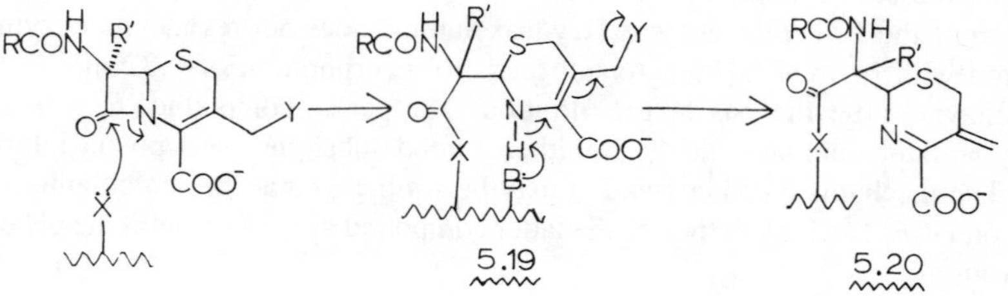

Scheme 42. Containing Structural Formulas 5.19 and 5.20.

Mobashery et al.[36e] substituted β-chloro-L-alanyl-β-chloro-L-alanine for the 3′-acetoxy group in cephalothin as an approach to drug delivery of that antibiotic dipeptide. The mechanism-based dipeptide is released by TEM β-lactamase catalyzed hydrolysis of the cepham. Therefore, this is an example of a multi enzyme-activated inactivator (see Chapter 1, Section III.D.).

B. Acylation/Elimination/Addition
1. Penam Sulfones

Knowles and co-workers[37,38] have suggested three criteria for mechanism-based inactivation of β-lactamases: (1) a β-lactam structure capable of forming a long-lived acyl enzyme, (2) an acidic 6α-proton, and (3) a good leaving group at C-5. By oxidizing the sulfide to a sulfone of various penams that are substrates of β-lactamase, criteria 2 and 3 would be satisfied; using poor β-lactamase substrates as the penam would satisfy criterion 1.

Structure-activity studies with various classes of β-lactam compounds as inactivators of a wide variety of β-lactamases were summarized by Reading and Cole.[38a]

a. Penicillin Derivatives

Four different penam sulfones of poor β-lactamase penam substrates were prepared by Fisher et al.[38] and shown to be mechanism-based inactivators. Quinacillin sulfone (**5.21**, R = 2-[3-carboxy]quinoxalinyl) labeled serine-70, which is believed to be involved in the normal hydrolytic mechanism. A mechanism was proposed[37,38] to account for the criteria mentioned above (Scheme 43). On the basis of the hypothesis that a good leaving group is required at C-5 in order to get inactivation, penicillin G sulfone was prepared;[39] however, it is a good substrate, but not an inactivator. The reason for this may be that the proton at C-6 is not acidic enough to permit isomerization.

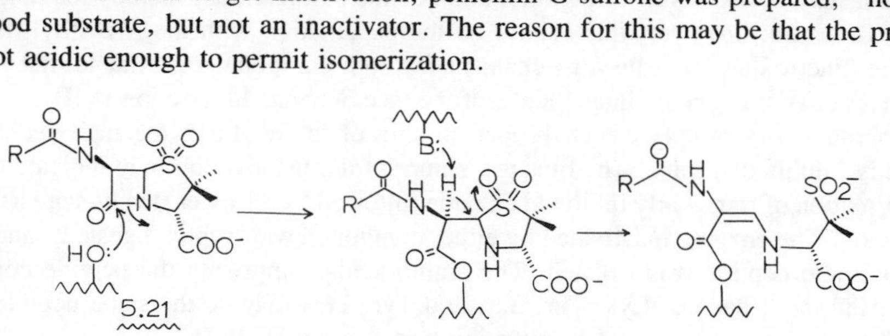

Scheme 43. Containing Structural Formula 5.21.

b. 6β-(Trifluoromethanesulfonyl)amidopenicillanic Acid Sulfone

6-Aminopenicillanic acid was converted by Mezes et al.[40,41] to 6β-(trifluoromethanesulfonyl)amidopenicillanic acid sulfone (Structural Formula 5.22, Scheme 44) in order to have an acidic proton at C-6 and a good leaving group at C-5. Inactivation of *Bacillus cereus* 569/H β-lactamase I and *Escherichia coli* RTEM β-lactamase by **5.22** is pseudo first-order and time dependent. The rate is inversely proportional to pH. Complete inactivation is achieved with a lower concentration of inactivator at lower pH values, e.g., 5- and 120-fold molar excesses of inactivator are needed at pH 4.8 and 6.8, respectively, for complete inactivation of the *B. cereus* enzyme. Hydroxylamine does not regenerate enzyme activity. Inactivation with [methyl-³H]-inactivator leads to incorporation of 0.82 mol of tritium per mole of enzyme after dialysis or gel filtration. The parent compound, 6β-(trifluoromethanesulfonyl)amidopenicillanic acid, is neither a good substrate nor a potent inhibitor of the enzyme. The mechanism of inactivation may be similar to that for penicillanic acid sulfone (see Section IV.B.1.c.), even though the latter compound does not exhibit the pH dependence of inactivation.

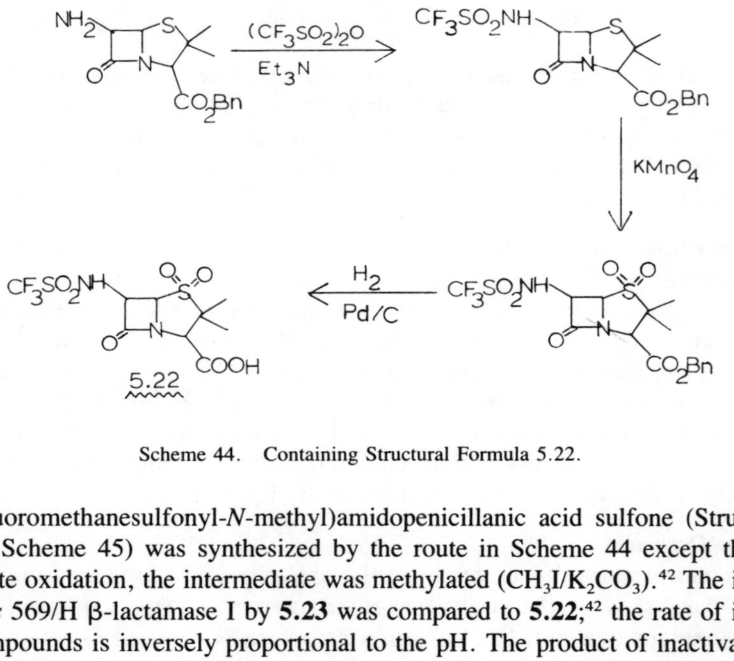

Scheme 44.   Containing Structural Formula 5.22.

6β-(Trifluoromethanesulfonyl-*N*-methyl)amidopenicillanic acid sulfone (Structural Formula 5.23, Scheme 45) was synthesized by the route in Scheme 44 except that prior to permanganate oxidation, the intermediate was methylated (CH_3I/K_2CO_3).[42] The inactivation of *B. cereus* 569/H β-lactamase I by **5.23** was compared to **5.22**;[42] the rate of inactivation by both compounds is inversely proportional to the pH. The product of inactivation at any pH is the same, as shown by varying the pH after inactivation and observing the same pH-dependent changes in the UV spectra. The pH dependence of inactivation for the parent compound may be the result of deprotonation of the sulfonamide, thereby leading to a less reactive β-lactam; however, this cannot be the explanation with the *N*-methyl derivative. Since the kinetic data indicate a mechanism for both compounds similar to that supported by Brenner et al.[43] for penicillanic acid sulfone (see Scheme 48, Section IV.B.1.c.), the N-Me compound may simply affect the partitioning of the acyl-enzyme intermediate to the transiently inhibited species. The biphasic nature of the inactivation may indicate the initial rapid formation of transiently inhibited enzyme followed by a slower steady-state irreversible inactivation. The enzyme inactivated by either compound was tryptic digested[42] and, in both cases, a single peptide was isolated. The amino acids comprising the peptide contained 2 Phe, 2 Ala, and 1 each of Lys, Thr, Ser, and Tyr. This may be the same peptide isolated from 6β-bromopenicillanic acid inactivation (see Section IV.B.7).

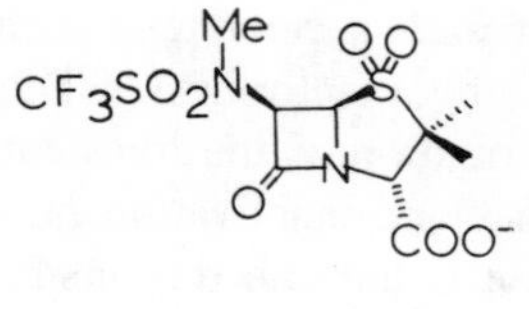

Scheme 45. Structural Formula 5.23.

A series of 6β-sulfonamidopenicillanic acid sulfones (Structural Formula 5.24, Scheme 46) was prepared by Dmitrienko et al.[44] as inactivators of *B. cereus* 569/H β-lactamase I. 6β-Phthalimidopenicillanic acid also was prepared. Partition ratios for **5.24**, where R = trifluoromethyl, methyl, tolyl, and anisyl, are 120, 18,000, 24,000, and 28,000, respectively. The phthalimido derivative has a partition ratio of 8000. A study was carried out with the trifluoromethyl analogue to determine if inactivation is accompanied by a conformational alteration of the acyl enzyme intermediate. This was suspected because it was found that the enzyme is completely inactivated by 780 equivalents of the phthalimido analogue in the presence of 10% acetonitrile, but by 8000 equivalents in the absence of a cosolvent. However, the rate of hydrolysis of benzyl penicillin by the enzyme is not affected by addition of the cosolvent. Therefore, the native and inactivated enzymes were studied by differential scanning calorimetry and optical rotatory dispersion (ORD) techniques.[44] Inactivated β-lactamase is less thermostable than the native enzyme; at least two different conformational variants were observed for inactivated enzyme, indicating an alteration in the secondary structure of inactivated enzyme. If inactivation leads to an intramolecular cross-link, the opposite would be true. ORD measurements show a helical content for native enzyme of 29%, but only 17% for inactivated enzyme. These results, therefore, support an alteration in structure.

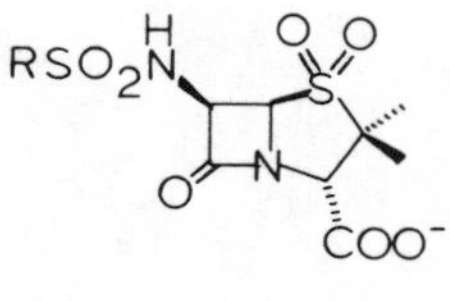

Scheme 46. Structural Formula 5.24.

c. Penicillanic Acid Sulfone

The most simple penam sulfone analogue is penicillanic acid sulfone ([2S-{2α, 5α}] 3,3-dimethyl-7-oxo-4-thia-1-azabicylo(3.2.0)heptane-2-carboxylic acid, 4,4-dioxide; sulbactam; CP-45,899) (Structural Formula 5.25, Scheme 47) a semi-synthetic antibiotic which is an irreversible inactivator of several penicillinases and cephalosporinases.[45] Several comparisons have been made between this inactivator and clavulanic acid (see Section IV.C.1). Although less potent than clavulanic acid against the *Staphylococcus aureus* enzyme, **5.25** is more potent against the *Shigella sonnei* enzyme; inactivation is time dependent.[46] Penicillanic acid sulfone also is more stable in aqueous solution than is clavulanic acid.[45] Compound **5.25** is both a substrate and inactivator of TEM-1, TEM-2, and Pitton's type 2 R-factor-mediated β-lactamases; the partition ratios are 525, 2280, and 1220, respectively.[47]

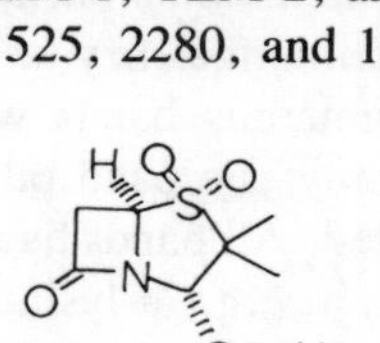

Scheme 47. Structural Formula 5.25.

The 6,6-dideuterio analogue of **5.25** accelerates the rate of the hydrolysis reaction and inactivation of RTEM β-lactamase by a factor of 3.[43] This inverse isotope effect apparently results from the slower rate of formation of the transiently inhibited enzyme, which leads to an acceleration in the hydrolysis and inactivation rates since more enzyme is available for these reactions. The 6β-proton is the one responsible for the formation of transiently inhibited enzyme (Structural Formula 5.25a, Scheme 48). Further support for this mechanistic scheme was gathered by Kemal and Knowles[48] in a study of this effect of pH on the enzyme reactions and model hydrolysis reactions at various pH values. The pH dependence for decomposition of the transiently inhibited enzyme is consistent with reversal to the imine followed by deacylation.

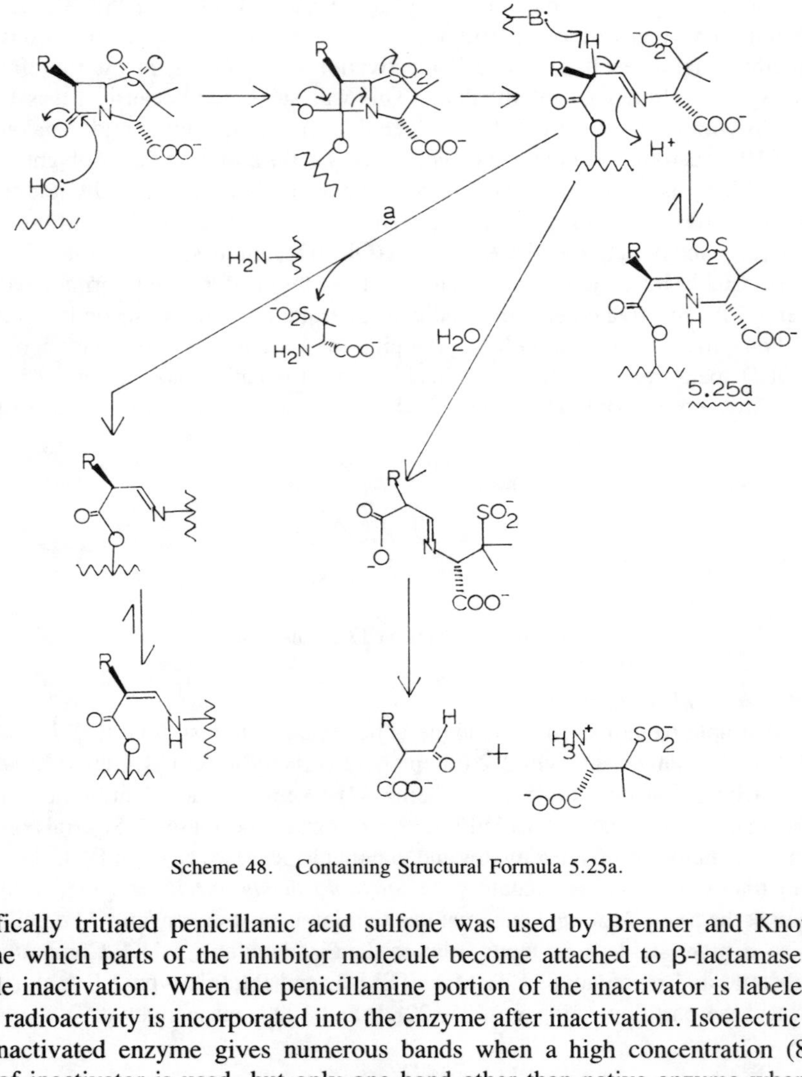

Scheme 48.   Containing Structural Formula 5.25a.

Specifically tritiated penicillanic acid sulfone was used by Brenner and Knowles[49] to determine which parts of the inhibitor molecule become attached to β-lactamase upon irreversible inactivation. When the penicillamine portion of the inactivator is labeled, essentially no radioactivity is incorporated into the enzyme after inactivation. Isoelectric focusing of the inactivated enzyme gives numerous bands when a high concentration (8000-fold excess) of inactivator is used, but only one band other than native enzyme when a lower concentration (800-fold excess) is used. All bands have pI values lower than native enzyme, indicating attachment to a basic group, e.g., a lysine. The same band was observed when β-lactamase was inactivated by clavulanic acid. The only fragment that is common to both of these inactivators is that derived from carbons 5, 6, and 7.

A model study also was carried out to support the proposed mechanism.[49] When penicillanic acid sulfone is treated with diethylamine in methanol at 25°C, the product obtained is methyl β-(diethylamino)acrylate, the expected product of solvolysis and transimination (Scheme 49). The same product was isolated when clavulanic acid was the starting β-lactam. This supports pathway a in Scheme 48.

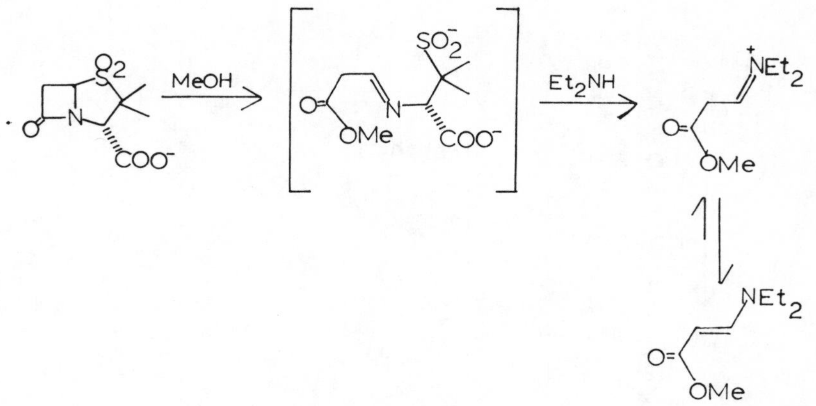

Scheme 49.

The rate constants for inactivation of *B. cereus* 569/H β-lactamase I by penicillanic acid sulfone and quinacillin sulfone are pH independent between pH 4.8 and 7.0; complete inactivation requires 20,000- and 1300-fold molar excess, respectively, of the two inactivators.[40]

d. 6-Hydroxybenzyl- and 6-Benzylidenepenicillanic Acid Sulfones

Analogues of penicillanic acid sulfone, namely (αR, 3S, 5R, 6R)- and (αS, 3S, 5R, 6R)-(α-hydroxybenzyl)penicillanic acids, their corresponding 1,1-dioxides (**5.26** and **5.27**, respectively) and (6E)- and (6Z)-benzylidenepenicillanic acids and their corresponding 1,1-dioxides (**5.28** and **5.29,** respectively) were prepared by Foulds et al.[50] as β-lactamase inhibitors (Schemes 50, 51, and 52). Compound **5.29** was prepared by the same route as for **5.28**, starting with the diphenylmethyl ester of **5.26**. All of the sulfone derivatives are irreversible inactivators of class C[51] β-lactamases, e.g., from *Pseudomonas aeruginosa*. There is no recovery of enzyme activity at pH 8.0, but slow recovery occurs at pH 5.5. Inactivation by all of these compounds occurs with an absorbance increase at 280 nm, which is thought to be the result of opening of the β-lactam ring (Scheme 52). These compounds are the first effective synthetic inactivators of class C β-lactamases. The partition ratio for inactivation of the β-lactamase from *P. aeruginosa* 18S by **5.26** is 10.[52] Similar results were obtained with the *amp* C β-lactamase from *E. coli*. At pH 8, no regain of enzyme activity was observed upon prolonged incubation; hydroxylamine, however, regenerates enzyme activity. Isoelectric focusing indicates that the protein is altered and one band predominates. The UV spectrum indicates the presence of an enamine.

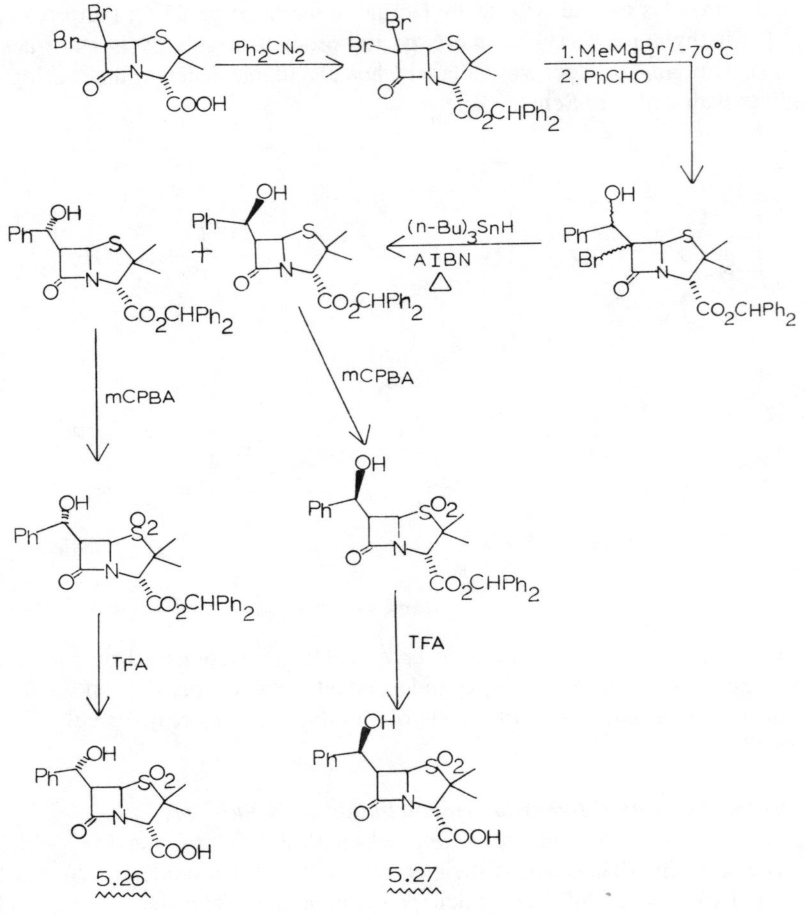

Scheme 50. Containing Structural Formulas 5.26 and 5.27.

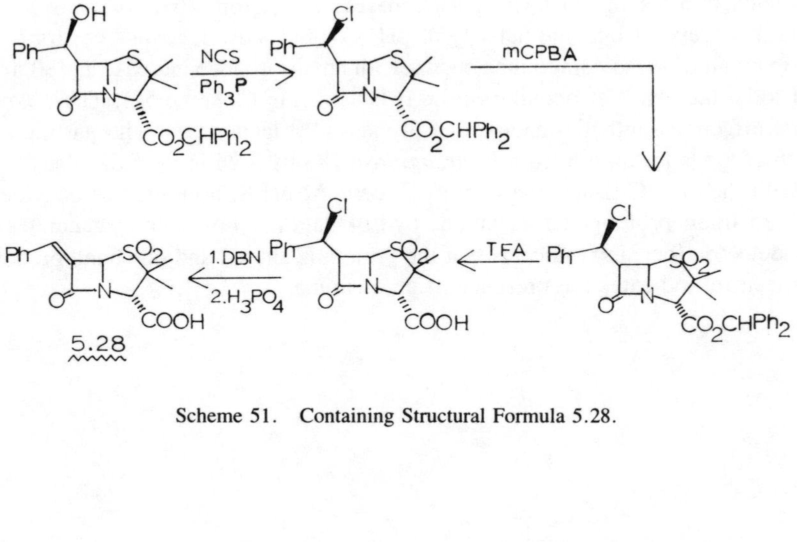

Scheme 51. Containing Structural Formula 5.28.

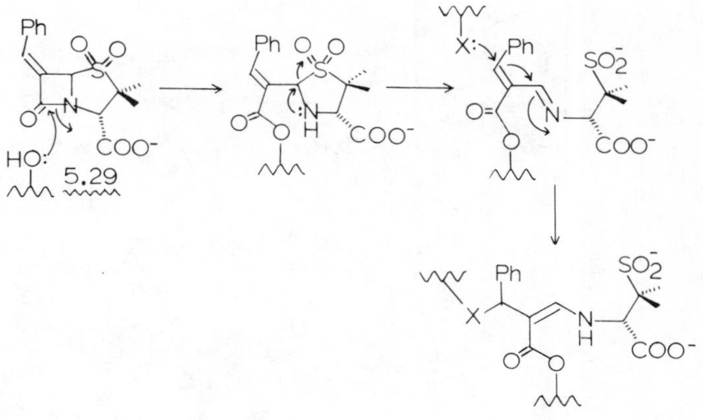

Scheme 52.   Containing Structural Formula 5.29.

e. 6-(2-Pyridyl)methylenepenicillanic Acid Sulfone

6-(2-pyridyl)methylenepenicillanic acid sulfone (Structural Formula 5.29a, Scheme 52A) inactivates a variety of β-lactamases.[52a] Compound **5.29a** was synthesized by Chen et al.[52a] by the route shown in Scheme 52A. A model study for the inactivation mechanism was carried out by treatment of the allyl ester of **5.29a** with NaOMe. The reaction shown in Scheme 52B (R = Me, R′ = CH₂CH=CH₂) was concluded after isolation of **5.29b** (R = Me, R′ = CH₂CH=CH₂). This suggests that inactivation of β-lactamase by **5.29a** may proceed by two mechanisms, one related to that for sublactam (Scheme 48) and the other as shown in Scheme 52B (R = Enz, R′ = H).

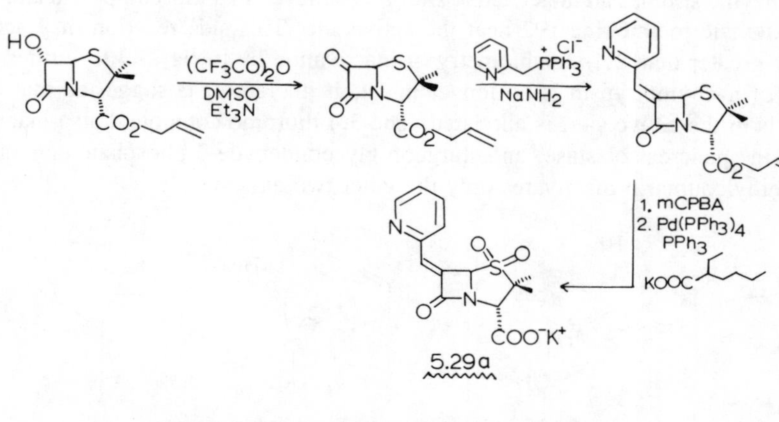

Scheme 52A.   Containing Structural Formula 5.29a.

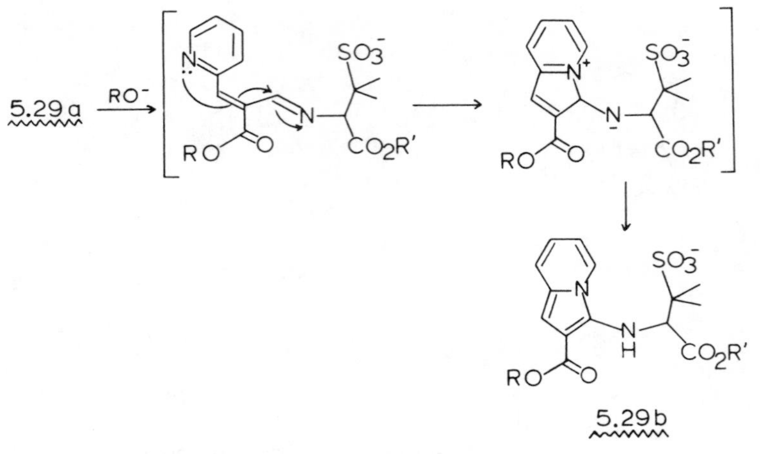

Scheme 52B. Containing Structural Formula 5.29b.

2. 6-Halomethylcoumarin Analogues

Inactivation of α-chymotrypsin by 3,4-dihydro-3,4-dibromo-6-bromomethylcoumarin (Structural Formula 5.30, Scheme 53) is time- and pH-dependent.[53] The enzyme is protected by acetylation of the active site serine-195 or by the presence of proflavin, a competitive inhibitor of the enzyme. The 6-bromomethyl substituent is required for irreversible inactivation since the 6-methyl analogue is only a substrate for the enzyme. With 3,4-dihydro-3,4-dibromo-6-[³H]-bromomethylcoumarin, 1.25 ± 0.1 equivalents of inactivator are incorporated with the loss of one histidine residue in the enzyme. The mechanism for inactivation shown in Scheme 53, in which **5.30** acts as a bifunctional reagent, was proposed by Béchet et al.[53] Evidence against direct affinity labeling of the enzyme is the observation that 6-bromomethylcoumarin, whose ester bond is stable, reacts much more slowly, and the modified enzyme still has an intact active site. It is believed that this compound and phenacyl bromides alkylate methionine-192 near the active site. This side reaction may account for the slightly greater than 1:1 stoichiometry of inactivation by [6-³H]-**5.30**. Furthermore, the active site of α-chymotrypsin is no longer intact. It, therefore, is suggested that histidine-57, which is in the active site, is alkylated. The 3,4-dibromo compound also inactivates β-trypsin, swine pancreas elastase, and sturgeon glyceraldehyde-3-phosphate dehydrogenase; 6-bromomethylcoumarin inactivates only the latter two enzymes.

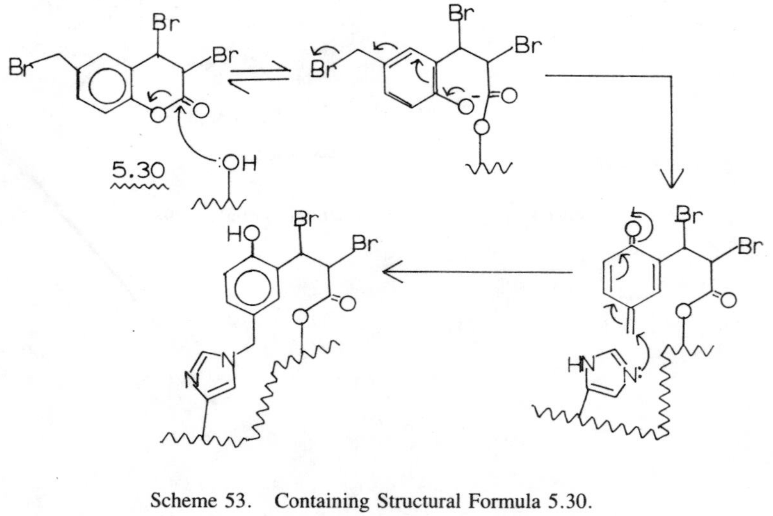

Scheme 53. Containing Structural Formula 5.30.

Pseudo first-order loss of enzyme activity was observed when **5.30** was incubated with the human urinary plasminogen activator, urokinase.[54] The hydrolysis products of the reagent do not inactivate urokinase. With the use of [6-^{3}H]-**5.30**, a stoichiometric (1.2 ± 0.1) amount of label is incorporated per active site. The incorporation of radioactivity into enzyme that is initially inactivated with the active site directed inactivator *p*-nitrophenyl-p′-guanidinobenzoate is blocked. Pepsin peptide maps show the presence of an acidic peptide containing the radioactivity, supporting a mechanism involving hydrolysis of an initial serine ester linkage. The labeled peptide replaces a Pauly-positive peptide in the native enzyme map, suggesting a histidine residue in or near the active site is modified, as appears to be the case with α-chymotrypsin (Scheme 53). Further purification and characterization of the peptides obtained from labeling of the high-molecular-weight human urokinase was carried out.[55] Three radioactively-labeled peptides from the pepsin digestion were isolated and shown to be residues 42 to 47, 44 to 47, and 41 to 47 located in the B-chain. The tritiated alkylated moiety is attached to the histidine residue (His-46).

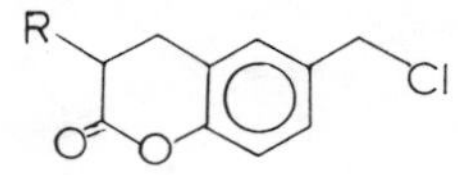

Scheme 54.   Structural Formula 5.31.

Because of the instability of **5.30** by dehydrohalogenation at C-3 and C-4, the simpler 3,4-dihydro-6-halomethylcoumarin (**5.31**, R = H), shown in Scheme 54, was prepared as an α-chymotrypsin inactivator.[56] In order to attain specificity of the inactivator for particular proteases and esterases, a 3-benzyl substituent (**5.31**, R = CH$_2$Ph) was attached. All of the compounds are mechanism-based inactivators of α-chymotrypsin. Compound **5.31** (R = H) is more stable than **5.30**, but has a half-life at pH 7 and 25°C of 65 min. All of the compounds form a slightly greater than 1:1 covalent complex to histidine-57, as evidenced by comparison of peptic digests of native α-chymotrypsin with those from radiolabeled α-chymotrypsin. Not all halomethyl derivatives are active; the lactam analogue of **5.31** (R = CH$_2$Ph), namely, 3-benzyl-6-chloromethyl-3,4-dihydroquinolin-2(1*H*)-one (Structural Formula 5.31a, Scheme 55) does not inactivate α-chymotrypsin.[57]

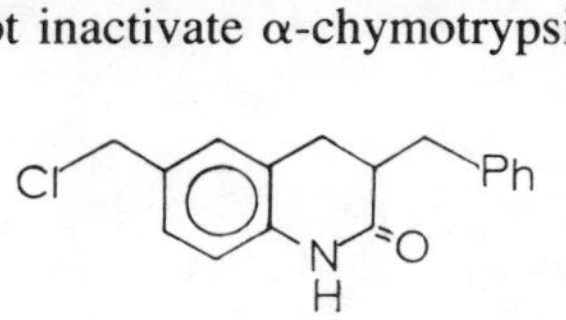

Scheme 55.   Structural Formula 5.31a.

The generality of inactivation of proteases and esterases by compounds **5.30** and **5.31** (R = H and R = CH$_2$Ph) was demonstrated by Béchet et al.[58] However, inactivation of α-chymotrypsin and subtilisin is rapid while that of papain is very slow. Elastase, kallikrein, and trypsin are inactivated at intermediate rates whereas chymotrypsinogen is not affected at all. Esterases, acetylcholinesterase and butyrylcholinesterase, are rapidly inactivated. Compound **5.31** (R = H) has no effect on aspartate aminotransferase, β-galactosidase, and adenosine triphosphatase, but **5.30** inactivates aspartate aminotransferase. A quantitative difference in inactivation rates is apparent for the benzyl-substituted coumarin; inactivation of all of the proteases except papain occurs at an increased rate. Papain is known to hydrolyze peptide bonds of bulky amino acids only very poorly and, therefore, is inactivated by the benzyl-substituted compound slowly. The benzyl-substituted analogue also is 130 times more reactive against the high-molecular-weight urokinase than is **5.30**.[55] Compound **5.31**

(R = H) is slightly less reactive towards the low-molecular-weight urokinase. 3,4-Dihydro-3-benzylcoumarin is not an inactivator, indicating the importance of the halomethyl group. These studies,[55] however, did not differentiate mechanism-based from affinity-labeling mechanisms.

As a model for the reactions of **5.30** and **5.31** with serine proteases, the reaction of these compounds with imidazole in water was carried out.[59] The product obtained from **5.30** is 3-bromo-6-imidazolylcoumarin (Structural Formula 5.32). None of this product is obtained when 3,6-dibromocoumarin is the starting material, supporting an acylation/elimination/addition mechanism, rather than an S_N2 mechanism (Scheme 56).

Compound **5.30** can be synthesized from 6-methylcoumarin[60] (Scheme 57) and **5.31** (R = H) from 3-(2-hydroxyphenyl)propionic acid[60] (Scheme 58). The 3-benzyl substituted analogue is prepared from 2-hydroxybenzyl alcohol[61] (Scheme 59).

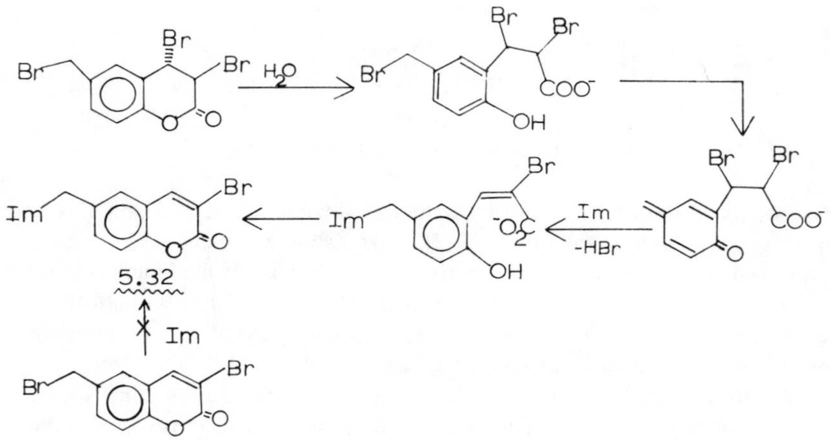

Scheme 56.   Containing Structural Formula 5.32.

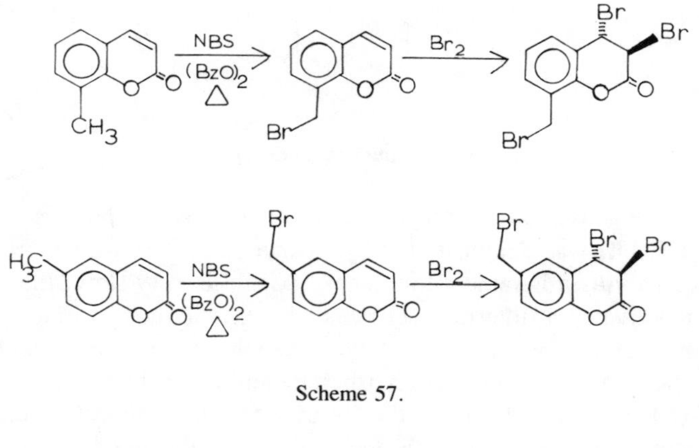

Scheme 57.

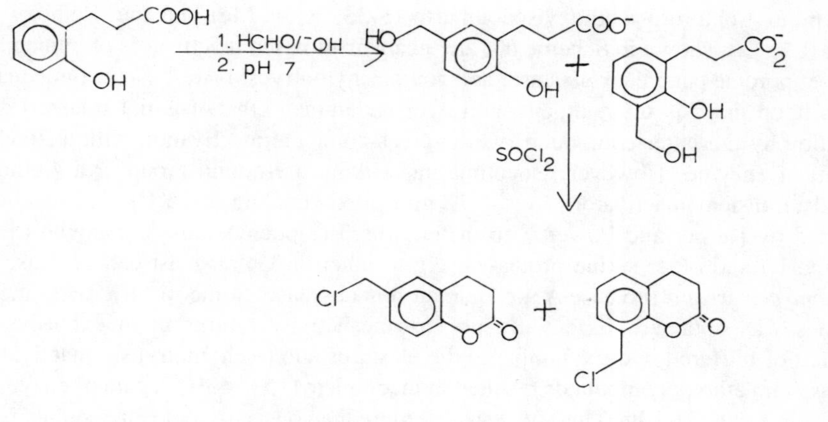

Scheme 58.

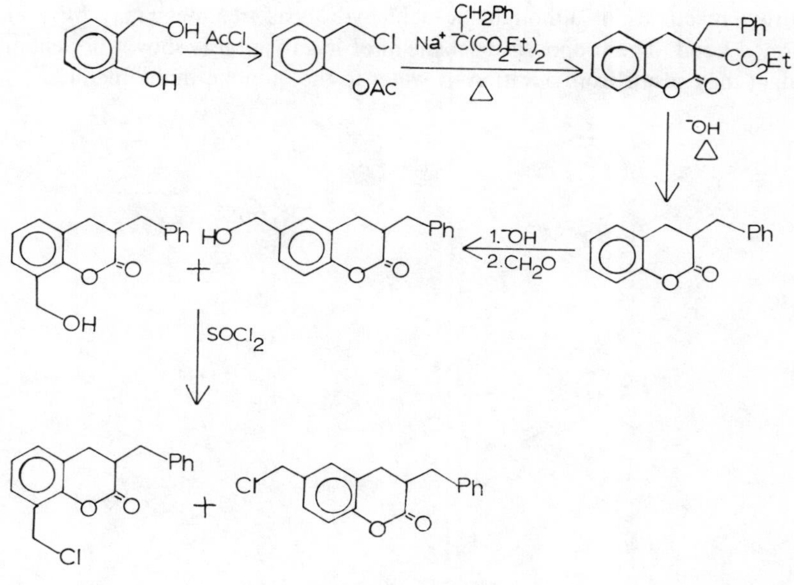

Scheme 59.

3. 3-Alkoxy-7-amino-4-chloroisocoumarins

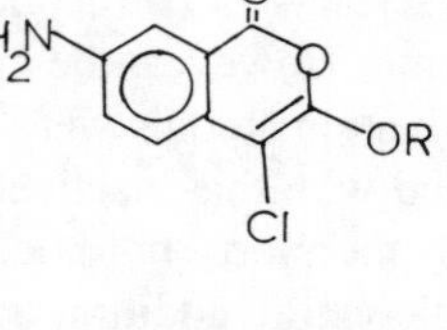

Scheme 60.   Structural Formula 5.33.

7-Amino-4-chloro-3-methoxyisocoumarin (**5.33**, R = Me) and the 3-ethoxy analogue (**5.33**, R = Et), shown in Scheme 60, are mechanism-based inactivators of human leukocyte elastase, porcine pancreatic elastase, and bovine chymotrypsin Aα.[62] No enzyme reactivation occurs upon dialysis or treatment with hydroxylamine. Ethanol is not released during inactivation by the latter compound, which gives complete inactivation with a stoichiometric amount of enzyme. However, isocoumarins without a 7-amino group or a 4-chloro group are only transient inactivators ($t_{1/2} < 72$ min upon standing at 25°C). These studies were extended by Harper and Powers[63] to include the 3-(2-phenylethoxy) analogue (**5.33**, R = PhCH$_2$CH$_2$) and other serine proteases, e.g., cathepsin G, rat mast cell protease I and II, *Streptomyces griseus* protease A, and human skin chymase. Some specificity for the different enzymes was obtained. Essentially no enzyme activity returns upon extensive dialysis. Addition of buffered hydroxylamine to the elastases and to chymotrypsin inactivated by the methoxy and ethoxy compounds resulted in incomplete (15 to 43%) regain of enzyme activity with $t_{1/2} = 6.7$ to 31.2 hr. This suggests that more than one active site nucleophile is involved or more than one adduct is formed (Structural Formula **5.34**). Acid hydrolysis of chymotrypsin Aα inactivated by the methoxy analogue indicates that neither histidine nor methionine is lost during inactivation, although the acid hydrolysis treatment may have cleaved the newly formed bond. The proposed mechanism of inactivation is shown in Scheme 61. The labile adduct may result from addition of water to the quinone imine methide.

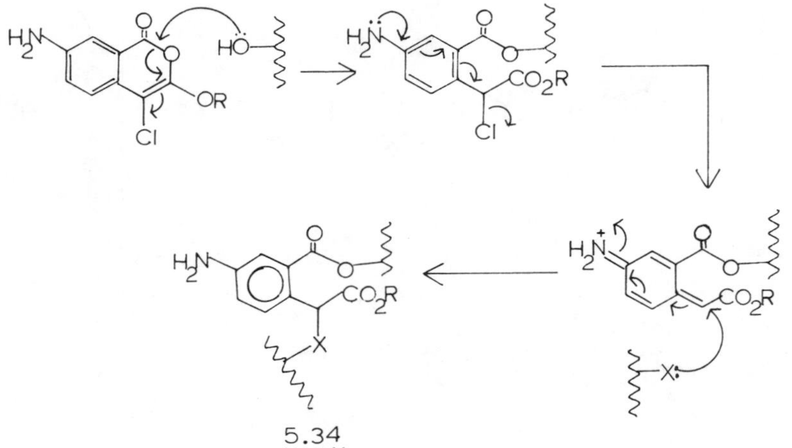

Scheme 61.   Containing Structural Formula 5.34.

A high resolution (1.8 Å) X-ray crystal structure of porcine pancreatic elastase inactivated by **5.33** (R = Me) was obtained by Meyer et al.[64] The inactivator is attached to the enzyme only through an ester linkage with serine-195 (Structural Formula 5.35, Scheme 62), suggesting affinity labeling of the enzyme. However, the crystals were grown in pH 5.0 buffer for solubility reasons, and the inactivation studies were carried out at pH 7.5. It is believed that the second covalent bond formed with this inactivator is with a histidine residue and at pH 5.0; it would be protonated and, therefore, not nucleophilic. The structure next will be investigated at pH 7.5 to see if nucleophilic addition occurs.

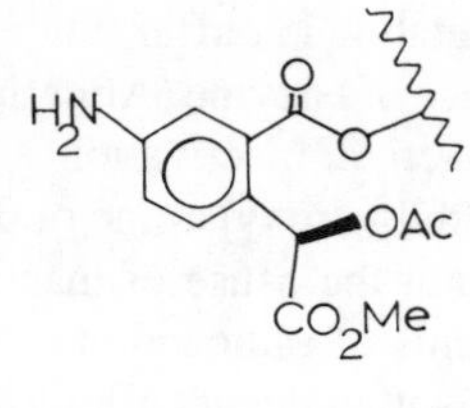

Scheme 62.   Structural Formula 5.35.

4. N-Nitrosoamides and Related

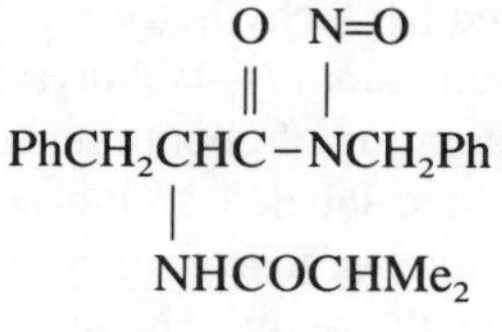

$$\underset{\displaystyle PhCH_2CHC-NCH_2Ph}{\overset{\displaystyle \overset{O}{\|} \quad \overset{N=O}{|}}{}}$$

NHCOCHMe₂

Scheme 63.   Structural Formula 5.36.

The *N*-nitroso-*N*-benzylamide of *N'*-isobutyrylphenylalanine (Structural Formula 5.36, Scheme 63) was selected by White et al.[65,66] as the initial target for inactivation of α-chymotrypsin by a mechanism involving enzyme-catalyzed carbocation generation. Although **5.36** decomposes in buffer, it does inactivate a small part of the enzyme. The low amount of inactivation was thought initially to be the result of the reactive species getting away from the active site. In order to prevent that problem, a cyclic nitrosoamide (Structural Formula 5.37) was prepared. In this case, initial acylation of the enzyme would generate the reactive species that is held in the active site as a result of the acylation; cross-linking of the enzyme would result (Scheme 64).

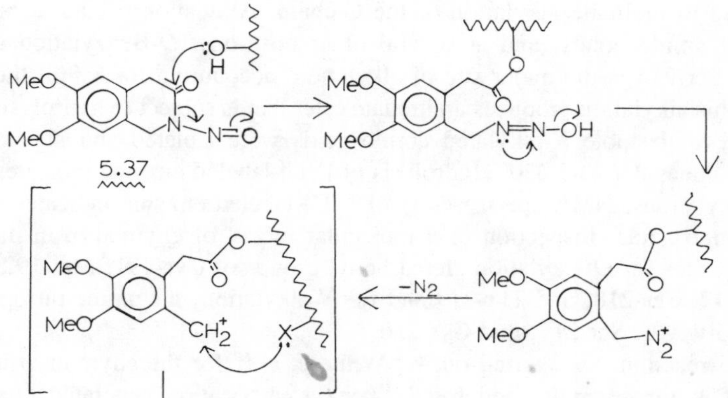

Scheme 64.   Containing Structural Formula 5.37.

The cyclic inactivator is still quite labile in buffer, but is hydrolyzed by the enzyme greater than 22 times faster than in the absence of enzyme. After the enzyme turns over 12 equivalents of the compound, it becomes 98% inactive. Inactivation is inhibited by competitive inhibitors. Treatment of inactivated enzyme with hydroxylamine or dialysis does not regenerate activity and, therefore, simple acylation is not the cause of inactivation. [^{14}C]-Labeled **5.36** results in the incorporation of 1.6 equivalents of radioactivity. Since the partition ratio is 11, it is not clear from these data if the inactivation occurs prior to release of the reactive intermediate from the active site.

The cause for the low levels of inactivation of α-chymotrypsin by **5.36** was tracked by White et al.[67] to contamination of the L-phenylalanine used in the synthesis by some D-phenylalanine. The two optically pure nitrosoamides were synthesized and it was found that both isomers are rapidly hydrolyzed by α-chymotrypsin (L is about 6 times faster), but only the D-isomer results in enzyme inactivation. N'-Isobutyryl-N-[^{14}C]-benzyl-N-nitroso-D-phenylalaninamide results in the incorporation of about one benzyl group per inactivated enzyme molecule; acid hydrolysis gives three labeled amino acids, indicating multiple alkylation sites. An explanation for the specificity of inactivation of the D-isomer is based on molecular models. The L-isomer can bind to the active site in a manner like typical L-amino acid substrates; however, in this case, the nitrosoamino group is positioned on the surface of the enzyme so that the generated benzyl carbocation would be captured by solvent water molecules. In the case of the D-isomer, the same binding arrangement is not possible. If the N-benzyl group is inserted into the hydrophobic cleft for phenylalanine, the binding interactions seem plausible. However, the nitrosoamino group is now embedded into the active site, so that its release and generation of the benzyl carbocation would be expected to result in trapping by an enzyme nucleophile. This explanation is supported by results obtained with N-ethyl analogues. Ethyl would not interact in the hydrophobic cleft as well as benzyl, so if this explanation is correct, the D-isomer of the N-ethyl analogue should hydrolyze slower, but not the L-isomer. This is what was observed.[67] The inactivation of α-chymotrypsin by the D-isomer of **5.36** and by the corresponding D-alanine derivative, D-N-nitroso-N-benzyl-N'-isobutyrylalaninamide, was investigated by Donadio et al.[68] in more detail. Both compounds lead to multisite alkylation of the C-chain. Alkylation occurs at both oxygen and nitrogen of amide bonds, and at several other positions. O-Benzylation of the carbonyl oxygen of Ser-214 is the major site of alkylation, accounting for 8% of the total inhibited enzyme. This alkylation produces an imidate ester that is subject to hydrolysis at that peptide linkage. Several stable N-alkylated amino acids were isolated and appear to be located between residues 216 and 230. Hydrolysis of [^{14}C]-labeled enzyme indicates a minimum of three N-alkylations; NMR spectroscopy of [^{13}C]-labeled enzyme indicates seven major alkylated amino acids. Inspection of a molecular model of chymotrypsin indicates that the most likely sites for O-alkylation are carbonyl groups of Cys-191, Met-192, Ser-214, Gly-216, Ser-217, Ser-218, and Thr-219 and for N-alkylation, the amide nitrogen of Met-192, Gly-193, Gly-216, Ser-217, and Cys-220.

A model reaction was carried out by White et al.[69] for the chymotrypsin-catalyzed hydrolysis of N-nitrosoamides and for the benzyl carbocation generated O-alkylated amide linkages to give imidate esters (Scheme 65). If this occurs at the active site of an enzyme, mild hydrolysis of the imidate ester would lead to a break in the peptide chain to give two unlabeled peptides. End group analysis of these two peptide chains would lead to the identification of the amino acid that was labeled. If the sequence of the enzyme is known, sequencing of the two peptides generated for just a few residues could pinpoint the position of the labeled amino acid. When α-chymotrypsin is inactivated by N-[^{14}C]benzyl-[**5.36**], one benzyl group is incorporated per enzyme molecule. Hydrolysis in 6 *M* HCl gives a mixture of amino acids which retain only about one-third of the benzyl label; the other two-thirds are labile and volatile. The model studies carried out indicate that the stable third of the label probably indicates attachment to N, S, and C atoms, and the labile two-thirds results largely from imidate ester formation. When hydrolysis of labeled chymotrypsin is carried out at pH 3 in urea, one-half of the benzyl groups are lost. Evidence for protein chain cleavage was obtained by gel filtration.[69]

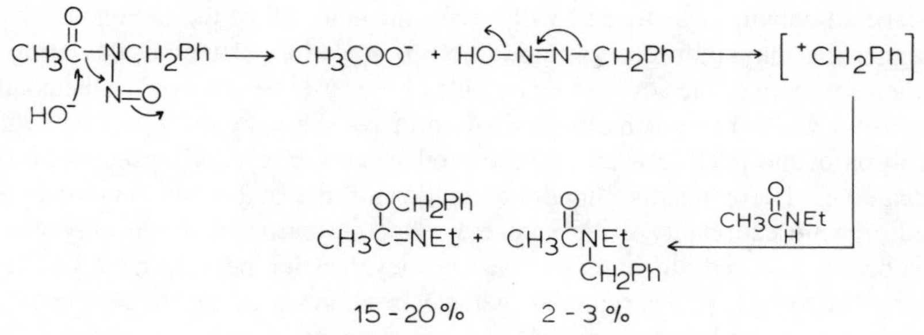

Scheme 65.

The N-nitroso analogues are synthesized from the parent amides by treatment with dinitrogen tetroxide and sodium acetate[70] (Scheme 66).

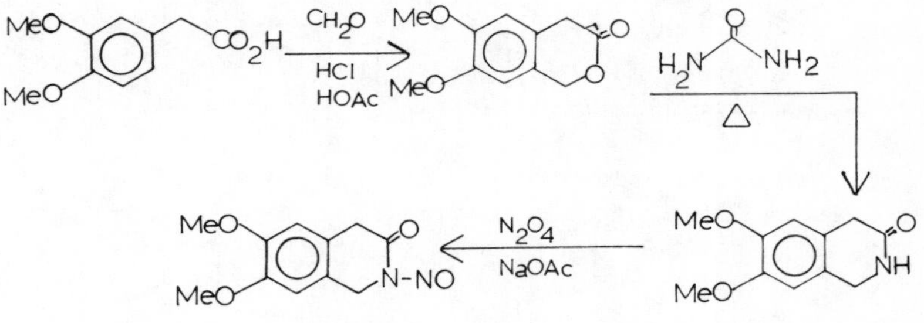

Scheme 66.

Hog liver esterase is irreversibly inactivated by *N*-substituted nitrosocarbamates with partition ratios on the order of 10.[71] The mechanism proposed is shown in Scheme 67. However, no experiment was done to determine if inactivation occurs prior to or subsequent to release of the α-hydroxynitrosoamines.

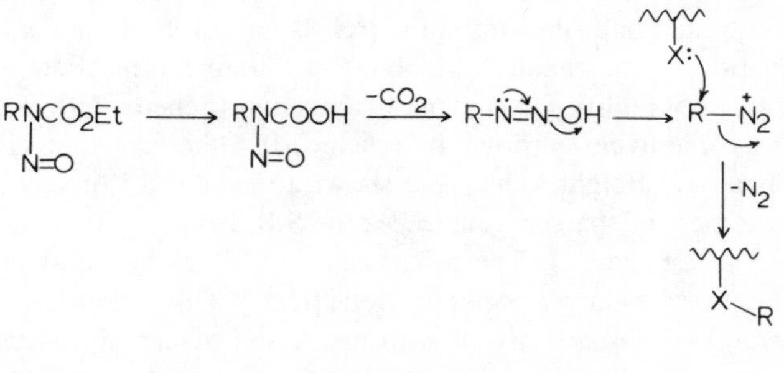

Scheme 67.

5. 3-Chloropropionyl CoA

Rat mammary gland fatty acid synthase is inactivated by 3-chloropropionyl CoA in a first-order process.[72] With the use of 3-chloro[1-^{14}C]propionyl CoA, 1 mol of radioactivity is incorporated per active site, but with 3-chloropropionyl[^{32}P]CoA, only 0.07 mol of radioactivity is incorporated. This supports a mechanism involving acylation of the enzyme. When the [^{14}C]-labeled enzyme was treated with performic acid, all of the radioactivity remained bound, suggesting alkylation occurs in addition to acylation. Potential alkylation sites are the pantetheine moiety of the acyl carrier peptide or the cysteine involved in the condensation reaction. Acetyl CoA, but not malonyl CoA, protects the enzyme from inactivation, and HCl hydrolysis of the [^{14}C]-labeled enzyme produces (carboxyethyl)cysteine not (carboxyethyl)cysteamine. These results support alkylation of the active site cysteine. Since the inactivated enzyme can catalyze the ketoacyl reductase partial reaction only when a cysteamine substrate is used, it suggests that the acyl carrier peptide may be affected by inactivation. NaDodSO$_4$-polyacrylamide gel electrophoresis of inactivated enzyme under mild conditions was used to demonstrate the formation of cross-linked enzyme. Treatment with hydroxylamine prior to electrophoresis eliminates the cross-linking. On the basis of these results, Miziorko et al.[72] suggest the mechanism shown in Scheme 68.

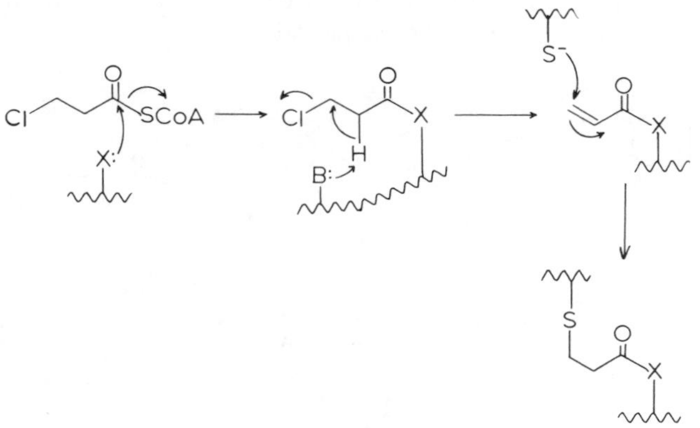

Scheme 68.

6. Sparsomycin and Related

An analogue of sparsomycin (Structural Formula 5.38) irreversibly inhibits *E. coli* polysomal peptidyl transferase.[73] Only the sulfoxide, not the sulfide or sulfone, gives greater inactivation with preincubation than without preincubation. With the [^{3}H]-inactivator, radioactivity is incorporated into ribosomes; the [^{3}H] is not released upon warming or density gradient centrifugation. A mechanism involving a Pummerer reaction was suggested by Flynn and Ash[73] as a possibility for enzyme inactivation (Scheme 69). The synthetic route to the analogue of sparsomycin is shown in Scheme 70. Other sparsomycin analogues were synthesized by Flynn and Beight[74] and were shown to exhibit a "preincubation" effect.

The four stereoisomers of sparsomycin (**5.39**; the S$_c$R$_s$ isomer is shown) were synthesized by Liskamp et al.[75] (Scheme 71). The S$_c$R$_s$ isomer (**5.39**) is the most potent competitive inhibitor and the only one to show a preincubation effect.[76] This is consistent with predictions made for the interaction of sparsomycin with the A site of peptidyl transferase, in which the sulfoxide oxygen occupies the site normally occupied by the participating nitrogen of aminoacyl tRNA. This supports the Pummerer rearrangement mechanism proposed in Scheme 69.

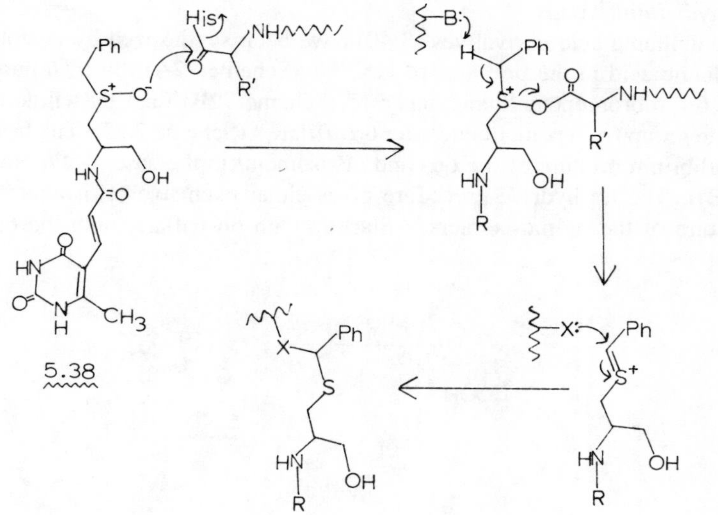

Scheme 69.   Containing Structural Formula 5.38.

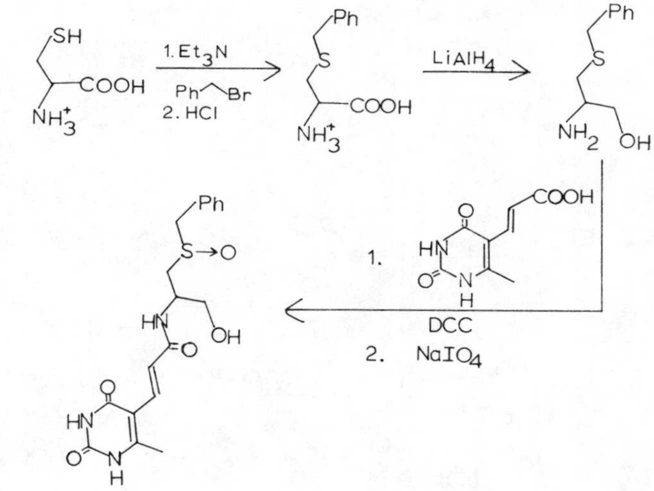

Scheme 70.

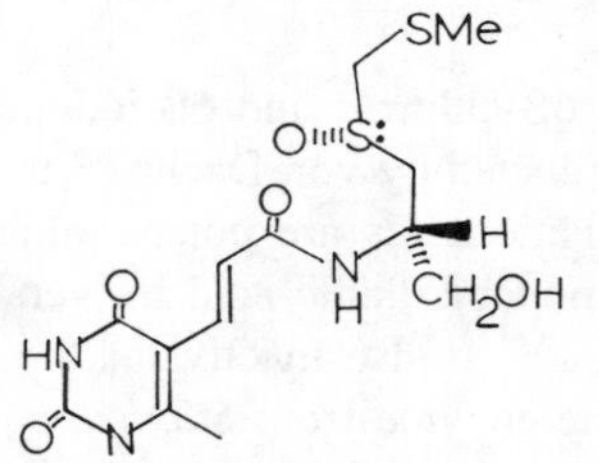

Scheme 71.   Structural Formula 5.39.

7. 6β-Halopenicillanic Acids

6β-Halopenicillanic acid derivatives (**5.40**) have been synthesized by diazotization of 6-aminopenicillanic acid in the presence of HX,[77,78] (Scheme 72A), by tri-*n*-butyltin hydride reduction of 6,6-dibromopenicillanic acid,[79,80] (Scheme 72B), and by halide displacement of a 6α-triflate group from penicillanic ester 6α-triflate[81] (Scheme 72C). The first method[77,78] gives an equilibrium mixture of the 6α- and 6β-isomers (in the case of the bromide, this is 88:12 (6α:6β)). The tin hydride procedure gives either exclusive 6β-isomer[79] or an 85:15 (6β:6α) mixture of the bromo-isomers.[80] Starting with 6α-triflate, only the 6β-halide was obtained.[81]

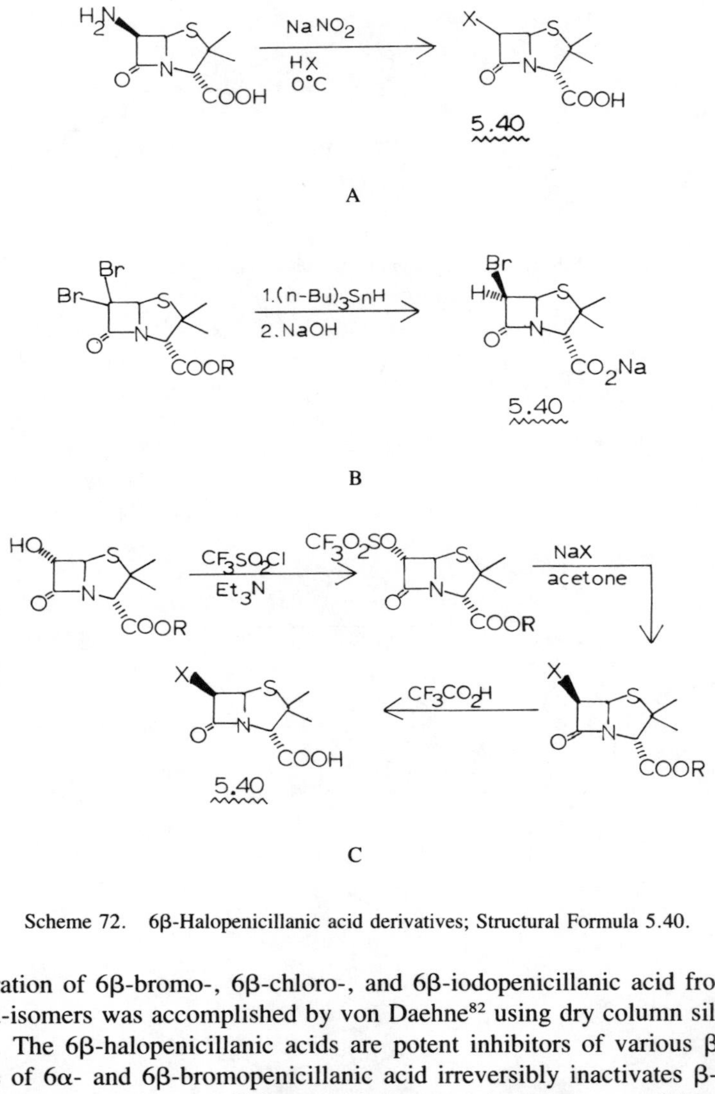

A

B

C

Scheme 72.   6β-Halopenicillanic acid derivatives; Structural Formula 5.40.

The separation of 6β-bromo-, 6β-chloro-, and 6β-iodopenicillanic acid from the corresponding 6α-isomers was accomplished by von Daehne[82] using dry column silica gel chromatogarphy. The 6β-halopenicillanic acids are potent inhibitors of various β-lactamases. The mixture of 6α- and 6β-bromopenicillanic acid irreversibly inactivates β-lactamases I from *Bacillus cereus* and *E. coli*.[77,83] It also inactivates the β-lactamase I from *B. licheniformis,* and, to a lesser extent, the enzyme from *Staphylococcus aureus,* but not *B. cereus* β-lactamase II.[83] 6α-Bromopenicillanic acid, freshly prepared, is not an inactivator of any of the β-lactamases tried; however, if the 6α-isomer is allowed to stand in buffer at pH 9 and 30°C, there is a time-dependent increase in the ability of the component in solution to inactivate the enzyme. An epimerization to the 6β-isomer was shown by Pratt and Loosemore[83] to be responsible for this activity. Similar conclusions were reached by Knott-Hunziker et al.[84] With the use of pure 6β-isomer, simple second-order kinetics were observed. The rate of inactivation of *B. cereus* 569 β-lactamase corresponds to the rate of appearance of a 326-nm chromophore, suggesting that the rate-determining step is cleavage of the β-lactam ring.[84] The same conclusion was reached by Loosemore et al.,[85] who also found that *B. cereus* β-lactamase I, which has been inactivated with clavulanic acid, is inert to 6β-bromopenicillinate, indicating a competition for the active site. Inactivation with [6-³H] 6β-bromopenicillate results in release of the tritium into solution. Spectral experiments indicate no turnover to product, only inactivation. All of the kinetic data suggest that there is a single rate-determining step leading to inactive enzyme with no evidence for accumulation of intermediate species; acylation of the enzyme is suggested as the rate-determining step.

Gel filtration of *B. cereus* β-lactamase I inactivated by 6β-bromopenicillanic acid does not regenerate enzyme activity.[86] 6α-Bromo[³H]penicillanate, which presumably epimerizes in the reaction buffer, reacts with the enzyme to incorporate 1 mol of [³H] per mole of enzyme. Sodium hydroxide releases a low-molecular-weight radioactive compound. These results suggest the possibility of an ester or ether linkage formed from reaction of a carboxyl or hydroxyl group on the enzyme with displacement of bromide.[86] The labeled enzyme was tryptic digested and the peptide containing the bound fragment was isolated by Orlek et al.[87,88] In order to identify the adduct structure, a model reaction of 6β-bromopenicillanic acid with sodium methoxide in methanol was carried out (Scheme 73.) The product obtained (5.41) was the same as that obtained by McMillan and Stoodley[89] using 6α-chloropenicillanic acid. A comparison of the spectra and general stability of 5.41 with the adduct-containing peptide indicated that the two species are related.[87,88] The structure of the labeled peptide is Phe–Ala–Phe–Ala–Ser–Thr–Tyr–Lys, comprising residues 40 to 47.[90] Further hydrolysis with a carboxypeptidase A and aminopeptidase M led to identification of the labeled amino acid as Ser-44. The same basic results were obtained, but a different conclusion was reached by Cohen and Pratt[91] who carried out a pronase digestion of *B. cereus* β-lactamase I inactivated by 6β-bromopenicillanic acid. A chromophoric tetrapeptide was isolated which, upon hydrolysis, was comprised of one each of Ala, Ser, Thr, and Tyr. N- and C-terminal analyses and knowledge of the amino acid sequence of the enzyme indicated that attachment was to residues 69 to 72 (Ala–Ser–Thr–Tyr), and that attachment of the chromophore was to either Ser or Thr. Conversion of the Ser to carboxymethylcysteine on treatment of the acylated peptide with thioglycollate was used to identify Ser-70 as the point of attachment. Evidence was presented to discount a migration of the adduct after initial acylation. Apparently, the sequence Ala–Ser–Thr–Tyr comprises residues 43 to 46 as well as 69 to 72 and that may be where the conflict arises. On the basis of the model study of McMillan and Stoodley,[89] the mechanism shown in Scheme 74 was proposed.[91]

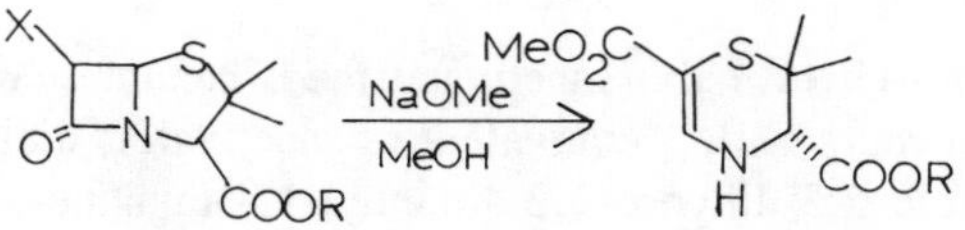

Scheme 73. Containing Structural Formula 5.41.

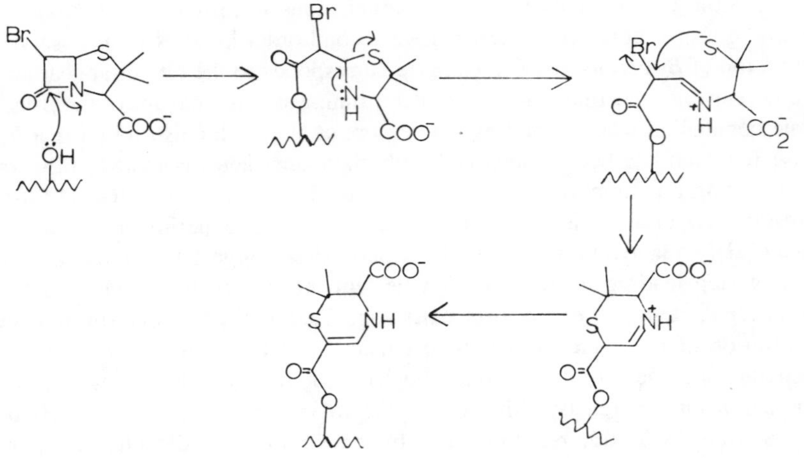

Scheme 74.

6β-Iodopenicillanic acid inactivates β lactamases from a variety of bacteria.[91a] The β-lactamases of *Streptomyces albus* G and *Actinomadura* R39 are inactivated by β-iodopenicillanate.[92] The partition ratio for the *Streptomyces* enzyme is 515 and for the *Actinomadura* enzyme is 80. The product was identified by Frère et al.[92] as 2,3-dihydro-2,2-dimethyl-1,4-thiazine-3,6-dicarboxylate (the diacid of **5.41**), the same product formed by saponification of β-bromopenicillanate. The same mechanism as shown in Scheme 74 was proposed.

6β-Iodopenicillanic acid also inactivates the Zn(II)-containing D-alanyl-D-alanine cleaving carboxypeptidase from *S. albus* G.[93] Irreversible inactivation was observed only at > 12,500 inactivator/enzyme ratios. X-ray diffraction studies of a crystal of the enzyme treated with β-iodopenicillanate indicate that the iodine atom is not present, and the site of interaction extends from His-190 just in front of the Zn(II) cofactor to an open cavity formed by the two segments (His-190)–(Gly-189)–(Pro-188) and (Asn-141)–Ser–Asn–(Val-144) –Gly–Gly–(Ala-147). This suggests alkylation of His-190 is responsible for inactivation, and is in contrast with the serine β-lactamases whose serine residue is acylated by β-bromopenicillanate.[91]

The P99 β-lactamase from *Enterobacter cloacae* also is inactivated by β-iodopenicillanate.[94,95] Both β-bromo- and β-iodopenicillanate titrate the purified enzyme in a pseudo first-order reaction to give a 1:1 adduct.[95] The new absorption band formed upon inactivation is very similar to that observed after inactivation of the *B. cereus* enzyme.[85] Tryptic digestion of the inactivated enzyme led to the identification of the amino acid which is labeled as serine-19.[94]

β-Iodopenicillanate also irreversibly inactivates the β-lactamase of *Streptomyces cacaoi*.[96] Complete inactivation occurs with 1 equivalent of inactivator; dialysis does not regenerate enzyme activity. No free 2,3-dihydro-2,2-dimethyl-1,4-thiazine-3,6-dicarboxylate is generated, consistent with the observation that 1 equivalent of inactivator produces complete inactivation. The UV spectrum changes to that observed when β-bromopenicillanate inactivates the β-lactamase of *B. cereus*.[85]

6β-Iodopenicillanic acid was shown by De Meester et al.[96a] to be useful as a probe for the classification of various β-lactamases. It is a substrate, not an inactivator, of class B Zn(II)-containing β-lactamases, but it inactivates β-lactamases from both classes A and C; however, it is much more efficient an inactivator of class A than class C enzymes. β-Lactamase from *Serratia marcescens* was inactivated by β-iodo[³H]-penicillanate and a 19-amino acid radioactive peptide was isolated by Joris et al.[96b] after trypsin digestion. The structure of the peptide establishes the enzyme as a class C β-lactamase. β-Iodo[³H]penicillanate inactivates β-lactamase K1 from *Klebsiella pneumoniae;* an active-site peptide was isolated by Joris et al.[96c] and a cysteine residue was identified as the residue immediately preceding the active-site serine. This is the only cysteine residue in the enzyme. In the *K. aerogenes* enzyme, there is an asparagine residue preceding the serine. β-Iodo[³H]penicillanate labels the active sites of the β-lactamases of *Streptomyces cacaoi* and *Streptomyces albus* G.[96d] The active site peptide sequence determined by De Meester et al.[96d] after tryptic digestion indicate that both enzymes are class A β-lactamases. Even though they are both class A enzymes, there is little amino acid homology.

6α-Chloropenicillanic acid was prepared and shown to be a poor substrate for the β-lactamase of *Staphylococcus aureus;* it is not an inactivator.[39] This was thought to be because the stereochemistry at C-6 is opposite that required and the leaving group is not good. Oxidation gave what was believed to be 6α-chloropenicillanic acid sulfone, which was as potent an inactivator as clavulanic acid (see Section IV.C.1.) but produced inactive enzyme which did not reactivate upon standing; the partition ratio was 100. However, it is likely that epimerization occurred either during oxidation or in solution prior to inactivation.

C. Acylation/Elimination/Isomerization
1. Clavulanic Acid and Analogues
X-ray crystallography of clavulanic acid isolated from *Streptomyces clavuligerus* proved its structure to be that of **5.42** (Scheme 75).[97] It is a potent, time-dependent, irreversible inhibitor of various β-lactamases.[97-100] Irreversible inactivation of β-lactamase I from *B. cereus* 569/H by clavulanate occurs even in the presence of benzylpenicillin and ampicillin; a complex mechanism involving at least two steps was suggested by Durkin and Viswanatha.[101] TEM-1, TEM-2, and Pitton's type 2 R-factor-mediated β-lactamases are inactivated by clavulanic acid with partition ratios of 160, 340, and 20, respectively.[47]

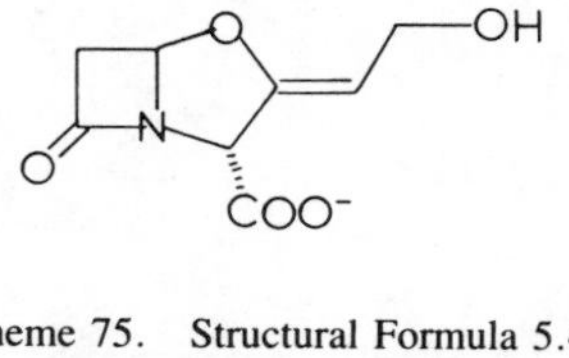

Scheme 75.   Structural Formula 5.42.

E. coli RTEM β-lactamase is destroyed by clavulanic acid and two inactive forms are produced.[102] One form is transiently stable and, upon standing, reverts to active enzyme, but the other form results in irreversible inactivation of the enzyme. Complete irreversible inactivation is accompanied by 115 turnovers to product. A kinetic analysis of the two inhibition processes led Fisher et al.[102] to postulate the minimal kinetic mechanism shown in Scheme 76.

Scheme 76.

In Scheme 76, E is enzyme, I is clavulanate, P is the normal turnover product, E-I is the irreversible inactivation product, E-T is the transiently inhibited enzyme, and P' is the product of decomposition of E-T. Chemical and spectroscopic evidence was presented by Charnas et al.[103] to support the kinetic mechanism shown in Scheme 76. Irreversibly inactivated enzyme results in three bands of equal intensity following polyacrylamide isoelectric focusing. Hydroxylamine treatment converts one of the bands to that of native enzyme and also restores one-third of the enzyme activity. Also, one-third of the absorbance of a 281-nm band formed during inactivation is diminished. The kinetic mechanism was, therefore, expanded to include three different E-I species (Scheme 77).

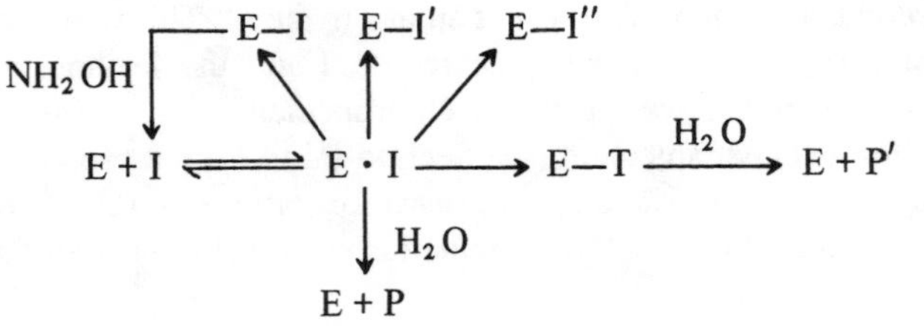

Scheme 77.

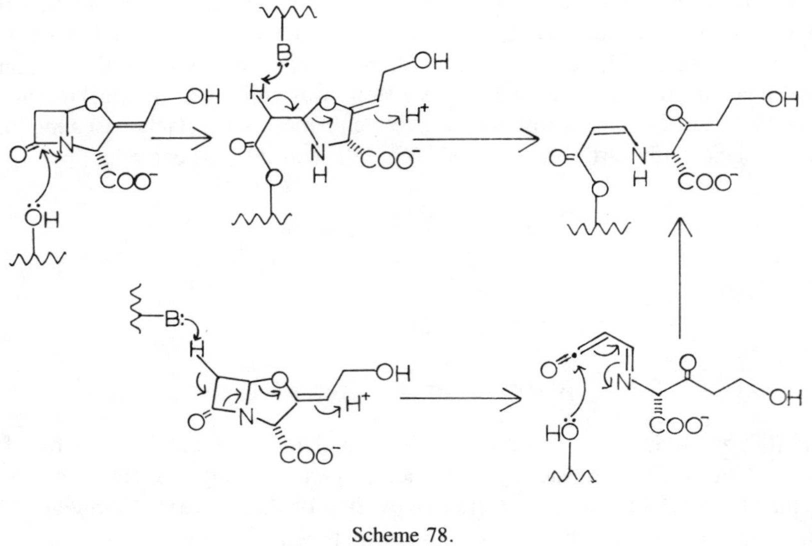

Scheme 78.

The mechanisms for inactivation shown in Scheme 78 were suggested[103] as possibilities. In contrast to this mechanism, Labia and Peduzzi[104] proposed an acylation/elimination/addition mechanism for the inactivation of three R-factor-mediated β-lactamases (TEM-1, TEM-2, and Pitton's type 2) (Scheme 79). However, Hunt et al.[105] have synthesized three analogues of clavulanic acid which do not contain an allylic hydroxyl group (Scheme 80) and all three are irreversible inactivators of a variety of β-lactamases. This is evidence against the mechanism in Scheme 79. Further evidence in support of the mechanism in Scheme 78 was provided by Charnas and Knowles.[106] 9-Deoxyclavulanic acid (**5.43**, Scheme 81) inactivates the RTEM β-lactamase from *E. coli;* very similar properties of inhibited enzyme were obtained as when clavulanate was used. When two radioactively labeled clavulanates (**5.44** and **5.45**, Scheme 81) were used, all of the carbon atoms remained bound to the enzyme in the transiently inhibited form, but some were lost in the irreversibly inactivated species. With the use of specifically radiolabeled clavulanate (**5.44**), it was shown that of the three irreversibly inactivated species formed, one retains only the carbons of the β-lactam ring and the other two retain all of the carbon atoms of the drug. Of the latter two, one reacts with hydroxylamine to regenerate active enzyme. A mechanism to account for these observations was proposed by Fisher et al.[37] (Scheme 82). However, an alternative mechanism is that related to the one proposed by Brenner and Knowles[49] for the inactivation of RTEM β-lactamase by penicillanic acid sulfone (see Scheme 83 for the mechanism following acylation).

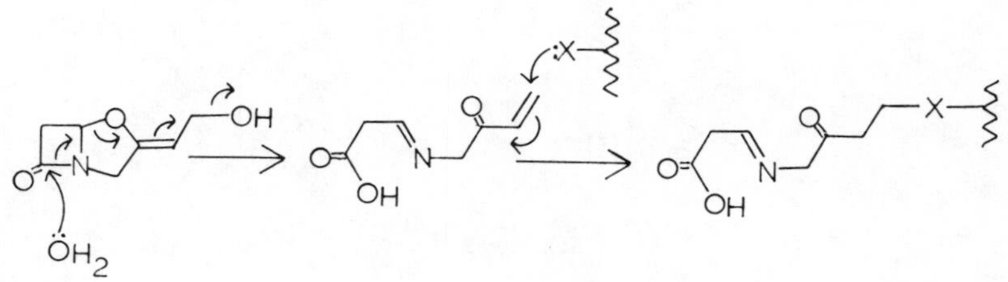

Scheme 79.

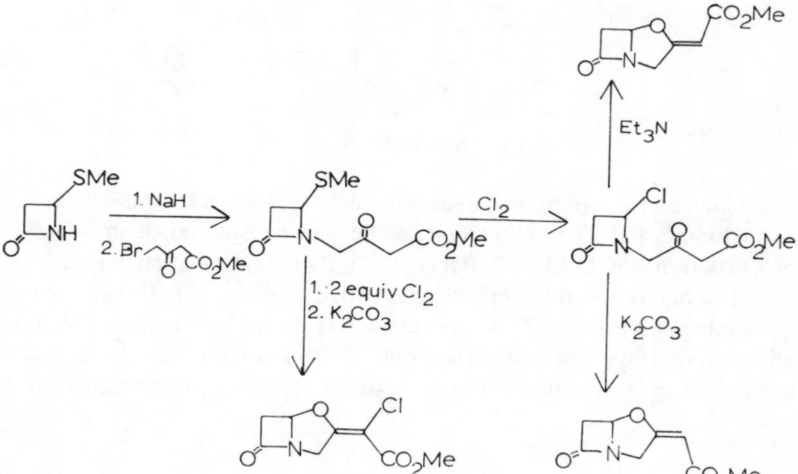

Scheme 80.

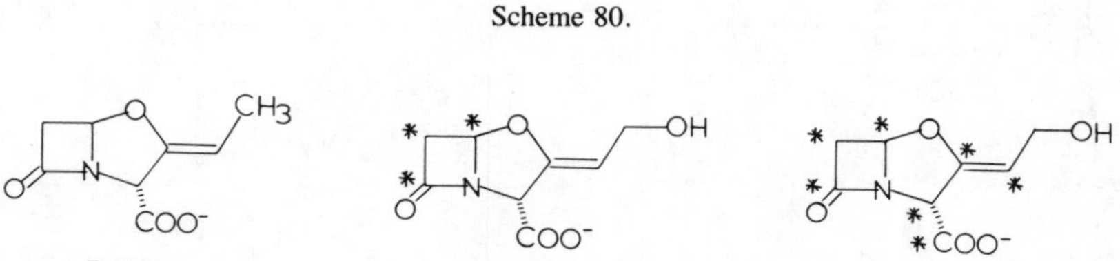

Scheme 81.   Structural Formulas 5.43 to 5.45.

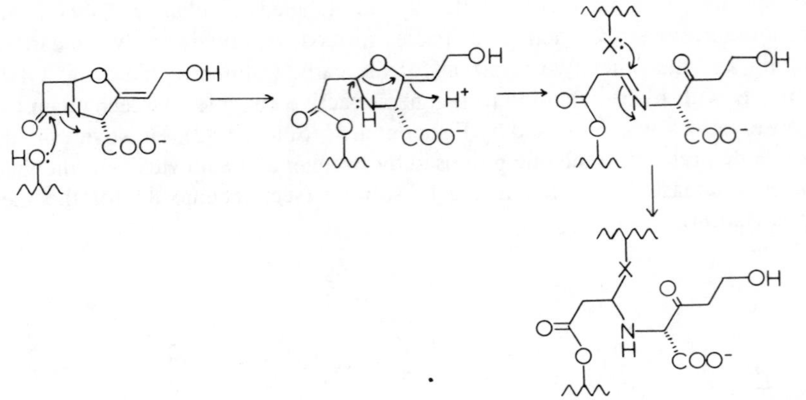

Scheme 82.

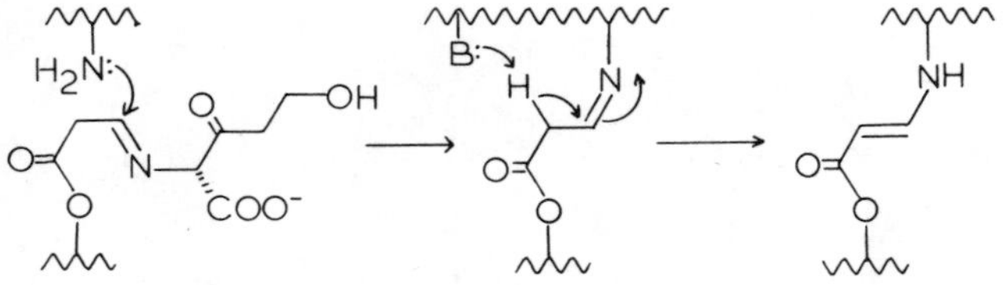

Scheme 83.

Sodium clavulanate also rapidly inactivates the β-lactamase from *Staphylococcus aureus* with a 1:1 stoichiometry.[39] A modification of the mechanism shown in Scheme 78 was suggested by Cartwright and Coulson[39] (Scheme 84). Partial return of enzyme activity occurs with time.[39,107] The rate of reactivation increases with decreasing pH.[107] Another mechanism proposed by Reading and Hepburn[107] for inactivation is shown in Scheme 85. A model study for the inactivation reaction also was carried out. Methanolysis of clavulanic acid gives the decarboxylated enamine (Structural Formula 5.46) in support of this mechanism (Scheme 86).

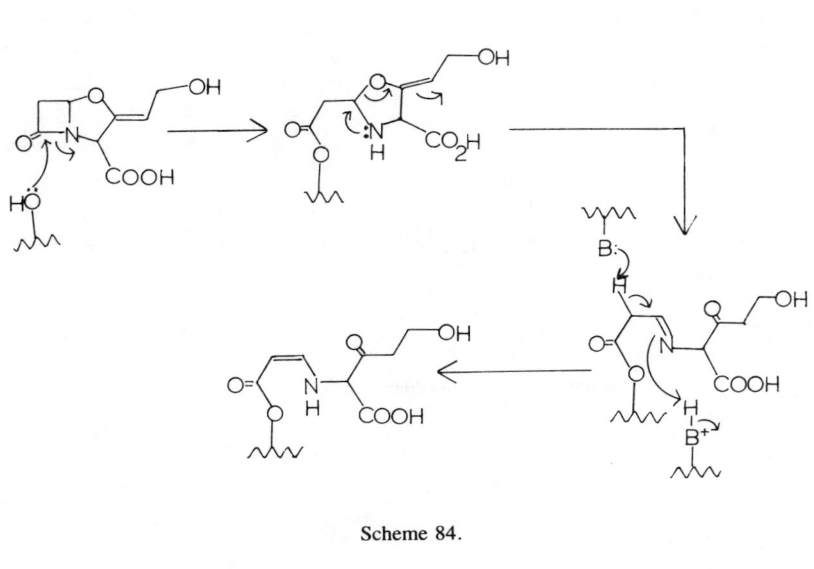

Scheme 84.

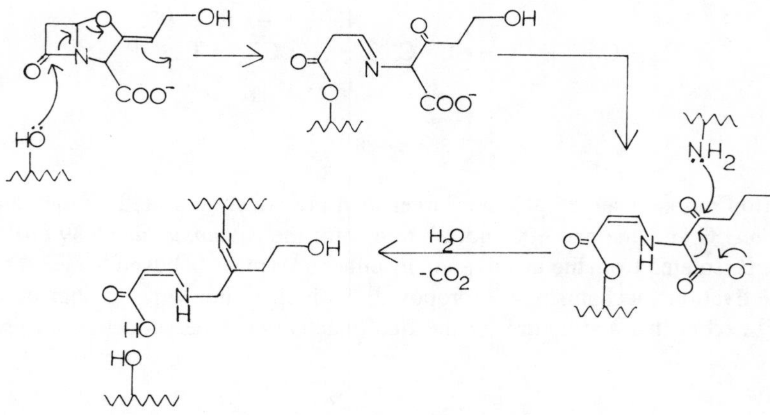

Scheme 85.

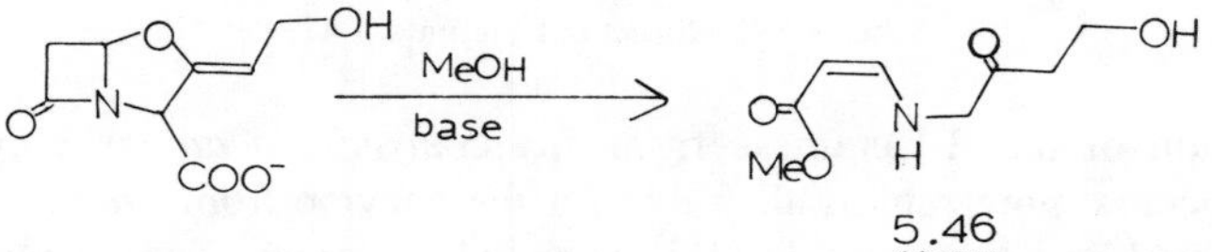

5.46

Scheme 86.   Containing Structural Formula 5.46.

The β-lactamases from *Klebsiella pneumoniae* E70 and TEM-2 behave similarly upon treatment with clavulanic acid.[108] Both enzymes form two types of complexes, a transiently stable species and an irreversibly inactivated enzyme; the TEM-2 enzyme partitions in a 3:1 ratio and the *Klebsiella* enzyme in a 1:1 ratio between transient and irreversible inhibition. A 115-molar excess of clavulanate is required to achieve complete inactivation of TEM-2 β-lactamase. The penicillinase from *Proteus mirabilis* C889 produces mostly a transiently inhibited complex with clavulanate.

The β-lactamases from *Actinomadura* R39 is inactivated by quinacillin sulfone (but not quinacillin), *N*-formamidoylthienamycin, and clavulanic acid, whereas the β-lactamase from *Streptomyces albus* G is inactivated only by clavulanic acid.[109] The inactivation of *Actinomadura* R39 β-lactamase by clavulanic acid resembles that between clavulanic acid and *E. coli* RTEM β-lactamase;[103] both reversible and irreversible interactions were observed. The partition ratio for inactivation is 27 to 42; with the RTEM β-lactamase, it is about 115.[102] A mechanism involving a branched pathway is suggested for inactivation of the *Actinomadura* R39 enzyme, but a simple linear pathway is suggested for inactivation of the *S. albus* G β-lactamase.

Further studies on the inactivation of the β-lactamases of *S. albus* G and *Actinomadura* R39 by clavulanic acid were carried out by Frère et al.[110] The model proposed for inactivation of *S. albus* G β-lactamase (Scheme 87) is not linear, as previously proposed.[109] The model proposed for inactivation of the *Actinomadura* R39 β-lactamase is even more complex (Scheme 88).

$$E + C \rightleftharpoons E \cdot C \rightarrow E \cdot C^* \nearrow^{E + P \rightarrow E + P^*}_{\searrow E-C}$$

Scheme 87.

$$E + C \rightleftharpoons E \cdot C \rightarrow E \cdot C^* \begin{array}{c} \nearrow E + P \rightarrow E + P^* \\ \rightarrow E \cdot C^{**} \rightarrow E + P' \\ \searrow E\text{--}C \end{array}$$

Scheme 88.

The partition ratios observed also are larger than previously reported;[109] they are 20,000 for the *S. albus* G enzyme and 400 (not 27 to 42) for the *Actinomadura* R39 enzyme. One of the products isolated from the inactivation by either enzyme is believed to be **5.47** (Scheme 89). An inactivation mechanism was proposed[110] which is the same as that described in Scheme 84[39] except that a structure for the final inactive enzyme adduct is not presented.

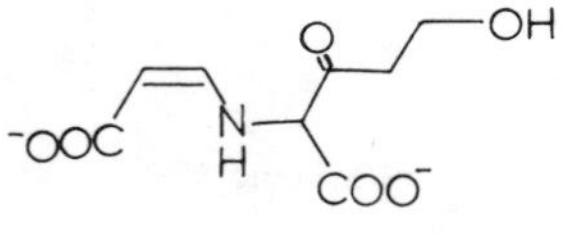

Scheme 89. Structural Formula 5.47.

The inactivation of the β-lactamase from *Streptomyces cacaoi* by clavulanate is time dependent and occurs more rapidly than that for the enzyme from *Bacillus cereus*.[111]

Streptomyces R61 and *Actinomadura* R39 D-alanyl-D-alanine serine peptidases are slowly inactivated by clavulanic acid and *N*-formamidoylthienamycin.[112] Quinacillin and its sulfone also inactivated the R39 enzyme, but slowly. None of the compounds inactivate the Zn(II)-containing G peptidase from *S. albus*.

A synthesis of clavulanic acid by Bentley et al.[113] is shown in Scheme 90.

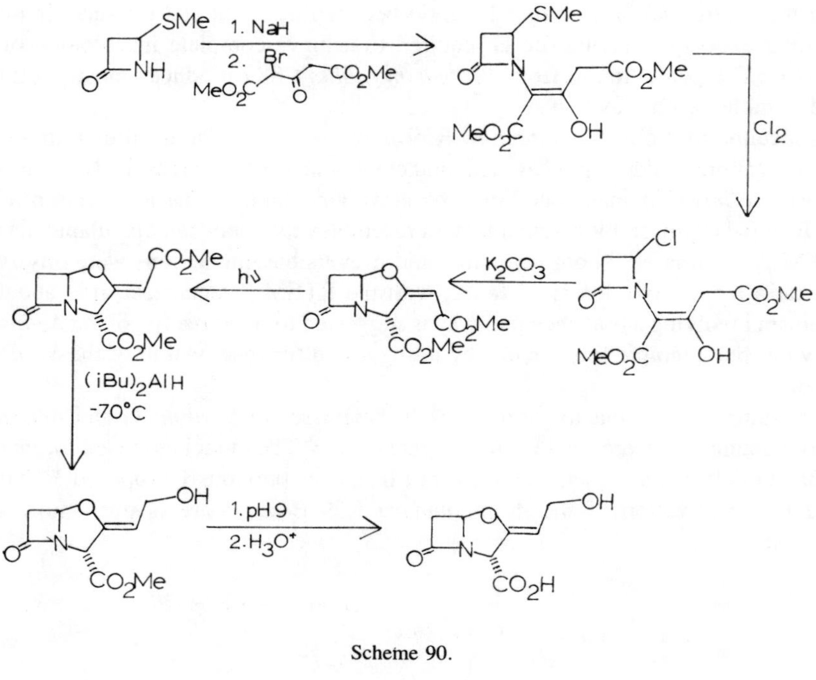

Scheme 90.

2-Carboxy-3-ethyl-7-oxo-4-oxa-1-azabicyclo[3.20]hept-2-ene (Structural Formula 5.48) was prepared by Cherry et al.[114] from a derivative of clavulanic acid, and was shown to be a potent inhibitor of β-lactamase from several organisms (Scheme 91).

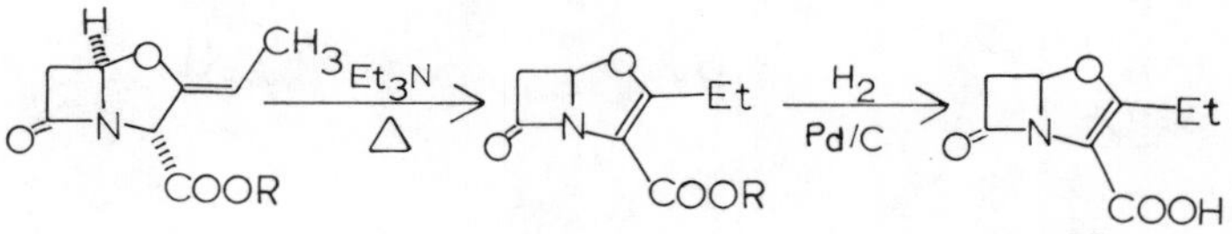

Scheme 91.   Containing Structural Formula 5.48.

V. ACYLATION/ISOMERIZATION

A. Acylation/Isomerization

1. Olivanic Acid and Analogues

Olivanic acids (**5.49**, R = SO_3^-, R' = −SCH=CHNHAc, −S(O)CH=CHNHAc, −SCH$_2$-CH$_2$NHAc), shown in Scheme 92, comprise a family of naturally-occurring carbapenam β-lactamase inhibitors produced by *Streptomyces olivaceus*,[115-117] *S. fulvoviridis*[118] (**5.49**, R = SO_3^-, R' = −S(O)CH=CHNHAc), *S. griseus* subsp. *cryophilus* nov. subsp.[119,120] (**5.49**, R = SO_3^-, R' = −S(O)CH=CHNHAc), and *S. pluracidomyceticus*[121] (**5.49**, R = SO_3^-, R' = −SCH=CHNHAc, −S(O)CH=CHNHAc, −SCH$_2$CH$_2$NHAc, −SO$_3$Na, −S(O)CH$_2$CO$_2$Na, −S(O)CH(OH)$_2$, −S(O)CHO, −S(O)CH$_2$OH, −SCH$_2$CO$_2$Na). An analogue of olivanic acid (**5.49**, R = R' = SO$_3$H) also was isolated from *S. sulfonofaciens*.[122]

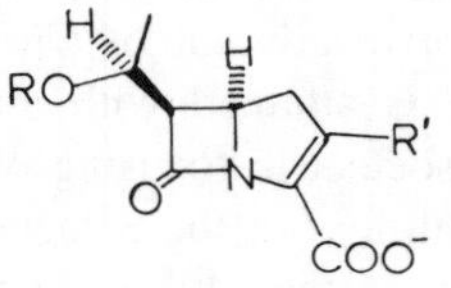

Scheme 92.   Structural Formula 5.49.

Twelve 5,6-*cis*-carbapenem antibiotics having two general structures (**5.49** R = H, SO_3^-; R' = −S(O)CH=CHNHAc and −SCH$_2$CH$_2$NHAc) were examined for β-lactamase inhibitory activity; the sulfonyloxyethyl carbapenems are the most active and are time-dependent inactivators.[123]

Two olivanic acid derivatives (**5.49**, R = H, R' = −SCH=CHNHAc and R = SO_3^-, R' = −SCH=CHNHAc) were tested by Charnas and Knowles[124] as inactivators of the RTEM β-lactamase from *E. coli*. The hydroxy compound is a good substrate and a poor inhibitor; the sulfate ester is a poor substrate and an excellent inactivator. The inactivation, however, is slowly reversible; after long incubations (ca. 5 hr) full enzyme activity returns. The initial hypothesis for inactivation by the sulfate ester was that a more stable α,β-unsaturated acyl enzyme was formed by elimination of sulfate (acylation/elimination; Scheme 93). This would explain why the hydroxy analogue is not an inactivator since hydroxyl is a much poorer leaving group than is sulfato. However, no release of sulfate was observed after inactivation, so this mechanism is not substantiated. The explanation offered instead was enzyme cationic stabilization of the sulfato substituent. Further studies by Easton and Knowles[125] revealed spectroscopic evidence for the formation of epimeric Δ^1-pyrrolines, and the modified acylation/isomerization mechanism shown in Scheme 94 was proposed. If Δ^1 pyrroline formation competes well with deacylation, then inactivation will result.

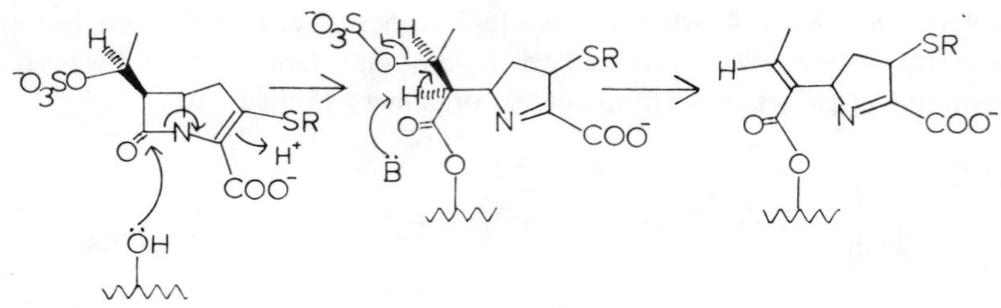

Scheme 93.

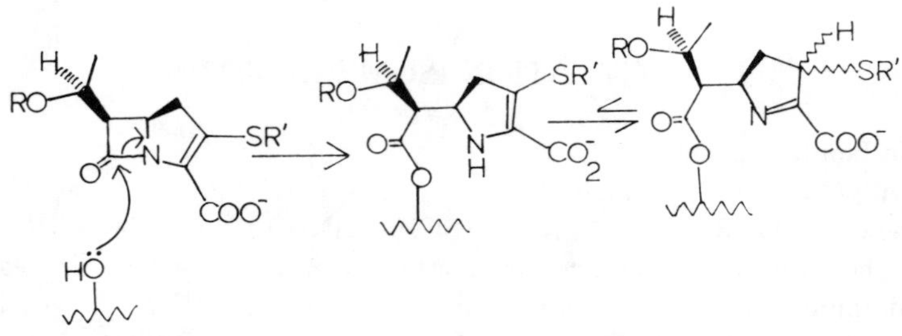

Scheme 94.

Other carbapenams related to olivanic acids are the asparenomycins A, B, and C (**5.50**, R = −S(O)CH=CHNHAc, −S(O)CH$_2$CH$_2$NHAc, and −SCH=CHNHAc, respectively), shown in Scheme 95, produced by *Streptomyces tokunonensis* and *S. argenteolus*.[126,127] Asparenomycin A also is a time-dependent inactivator of *Citrobacter freundii* 19 β-lactamase.[128] Complete irreversible inactivation is attained with an extrapolated value of 1.8 mol of inactivator per mole of enzyme. Isoelectric focusing of the inactivated enzyme shows one major and one minor new band with loss of the β-lactamase band. No mechanism is given other than that acylation is one step involved; however, it may be the only step, in which case it would be an affinity labeling agent.

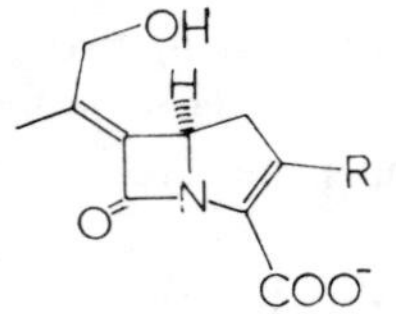

Scheme 95.   Structural Formula 5.50.

Another related carbapenam is PS-5 (Structural Formula 5.51, Scheme 96), isolated from *Streptomyces* sp. A271.[129] PS-5 is a time-dependent inactivator of the β-lactamase from *Proteus vulgaris;* enzyme activity slowly returns upon standing.[130] The time-dependent inactivation of the β-lactamase from *S. cacaoi* by PS-5 was compared with PS-5 inactivation of the β-lactamase from *B. cereus*.[111] PS-5 is a much more potent time-dependent inactivator of the *S. cacaoi* enzyme than of the *B. cereus* enzyme. Clavulanic acid is about 10 times more potent than is PS-5 as an inactivator. Compound **5.51** also is a time-dependent inactivator of the β-lactamase of *B. licheniformis* 749/C; at all concentrations of inactivator, there is always several percent of enzyme activity remaining, suggesting that a reversible inactivation may be occurring.[131]

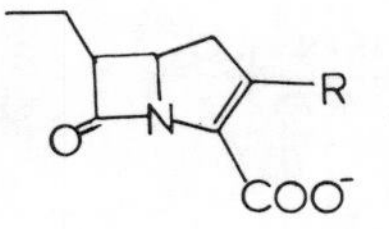

Scheme 96. Structural Formula 5.51.

B. Acylation/Isomerization/Addition

1. Ynenol Lactones

Ynenol lactones were incorrectly placed in the Acylation/Addition section (see Section II.E.). As is apparent from Scheme 17 these compounds undergo acylation/isomerization/addition reactions and therefore are more appropriately placed here. An extension of the work described in Section II.E. was described by Spencer et al.[131a] The syntheses of 5(*E* or *Z*)-(3-R'-2-propynylidene)-3-R-tetrahydro-2-furanones (Structural Formulas 5.52a to 5.52c, Scheme 96A) and 6(*E*)-3-R'-2-propynylidene)-3-R-tetrahydro-2-pyrones (Structural Formula 5.52d), where R = H, alkyl, or benzyl and R' = H, alkyl, or phenyl, were carried out (see Scheme 96B). Model reactions for the inactivation of serine proteases by ynenol lactones support the proposed inactivation mechanism (Scheme 17, Section II.E.). Hydrolysis leads to the corresponding α,β,γ-allenyl ketone which undergoes conjugate addition by nucleophiles. When R' is not H, an additional intermediate, the propargyl ketone intermediate shown in Scheme 17, is observed. Ynenol lactones (**5.52a to d**) that are both substituted α to the lactone carbonyl and unsubstituted at the acetylene terminus are potent mechanism-based inactivators of human leukocyte (and sputum) elastase.[131b] They inactivate porcine pancreatic elastase and trypsin weakly, and α-chymotrypsin very poorly. Compounds **5.52a** and **5.52d** (R = PhCH$_2$; R' = H) are the most potent inactivators. The partition ratio for **5.52a** (R = PhCH$_2$; R' = H) is 1.7 ± 0.5. The corresponding allenone carboxylic acid (i.e., hydrolyzed **5.52a**) does not inactivate the elastases, thus supporting the hypothesis that tethering of the inactivator via initial active-site serine acylation is vital to the inactivation process. Hydroxylamine does not reactivate inactivated enzyme, indicating that simple acylation is not responsible for inactivation.

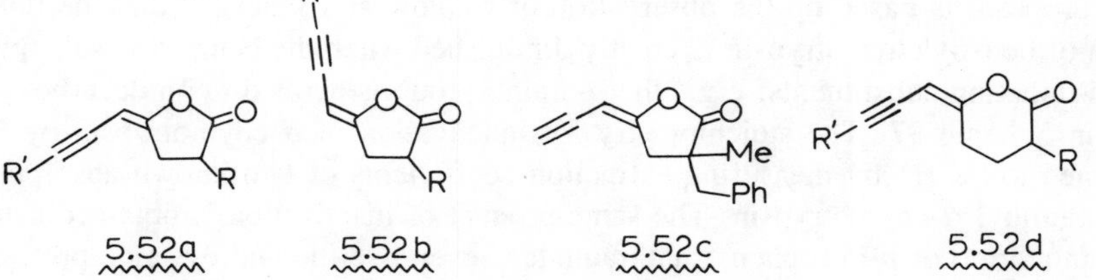

Scheme 96A. Structural Formulas 5.52a, b, c, and d.

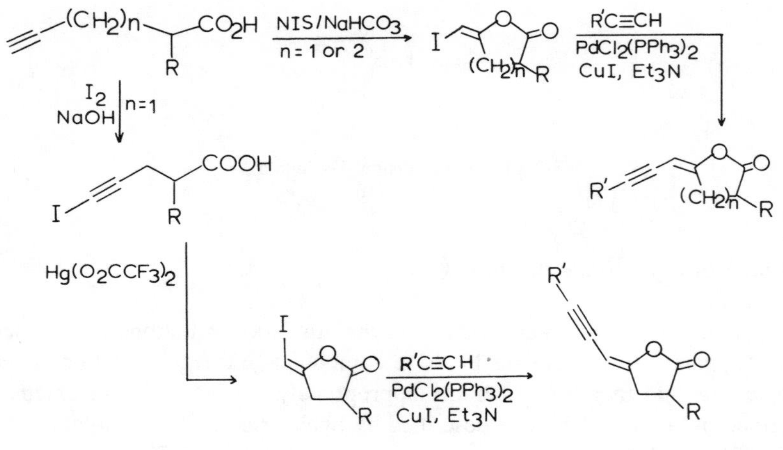

Scheme 96B.

VI. ACYLATION/DECARBOXYLATION

A. Isatoic Anhydride Analogues

An approach to mechanism-based inactivation which does not require "activation" of the inactivator, but rather formation of a stable adduct, is exemplified by isatoic anhydride (Structural Formula 5.52) inactivation of serine proteases.[132] Although anhydrides are somewhat reactive, acylation is not the cause for inactivation; rather, acylation followed by decarboxylation is (Scheme 97).

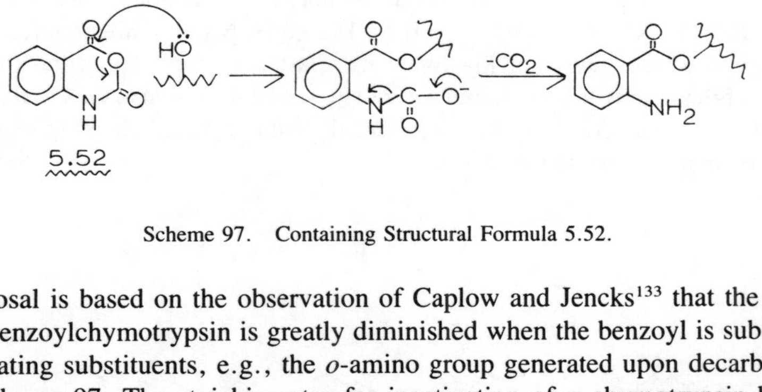

5.52

Scheme 97.   Containing Structural Formula 5.52.

This proposal is based on the observation of Caplow and Jencks[133] that the rate of deacylation of benzoylchymotrypsin is greatly diminished when the benzoyl is substituted with electron-donating substituents, e.g., the *o*-amino group generated upon decarboxylation as shown in Scheme 97. The stoichiometry for inactivation of α-chymotrypsin by **5.52** was determined to be 1:1 by measuring extinction coefficients of two known absorption peaks for anthraniloyl α-chymotrypsin. The same product of inactivation is obtained using either isatoic anhydride or *p*-nitrophenyl anthranilate. Several serine and cysteine proteases were tested; yeast aldehyde dehydrogenase, acetylcholinesterase, creatine kinase, carboxypeptidase A, yeast alcohol dehydrogenase, pig liver esterase and pancreatic elstase α-lytic protease, trypsin, papain, and α-chymotrypsin all are inactivated. Enzyme activity returns with pancreatic elastase, α-lytic protease, and trypsin. With α-chymotrypsin, inactivation is rapid, stoichiometric, and essentially irreversible. Analogues of isatoic anhydride, 3*H*-1,3-oxazine-2,6-diones (Structural Formula 5.53, Scheme 98) were prepared by Moorman and Abeles[132] in order to attempt to get specificity for particular proteases. With α-chymotrypsin and pancreatic elastase, inactivation by the 6-methyl derivative (**5.53**, R = CH₃, R' = H) is most rapid, then **5.53** (R = R' = H) (less by a factor of ~ 3); the 5-Me compound (**5.53**, R = H, R' = CH₃) is slower by two orders of magnitude. Therefore, specificity can be built into the inactivator. The mechanism for inactivation of chymotrypsin by **5.53** was modified after a more detailed investigation by Weidmann and Abeles[134] of the 5-*n*-butyl derivative, which was synthesized as shown in Scheme 99. Treatment of α-chymotrypsin with **5.53** (R = H, R' = C_4H_9) results in time-dependent inactivation, but not all of the enzyme activity is lost; the amount inactivated depends upon the concentration of the inactivator. Upon dilution, catalytic activity returns; the amount of activity returning depends on the dilution factor. The reversibility is not due to a simple equilibrium, and the data indicate that two complexes form, one of which can dissociate upon dilution. Inactivation with 5-butyl-3*H*-[2-^{14}C]oxazine-2,6-dione results in 1:1 stoichiometry after gel filtration at 4°C. This suggests that decarboxylation is not required for inactivation. When the temperature is raised to 25°C, there is an exponential loss of [^{14}C] from the enzyme, but a biphasic increase in catalytic activity; an initial rapid increase to about 30% enzyme activity is followed by a slow increase. Loss of [^{14}C] is more rapid than is recovery of enzyme activity. The initial adduct is not noncovalent because acid denaturation gives no intact inactivator back; everything remains bound to the enzyme. On the basis of these results, a revised mechanism for inactivation (Scheme 100) was proposed.[134] Both the carboxylated and decarboxylated adducts are believed to be the responsible species.

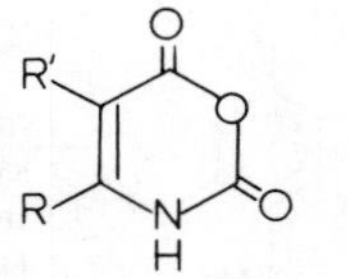

Scheme 98.   Structural Formula 5.53.

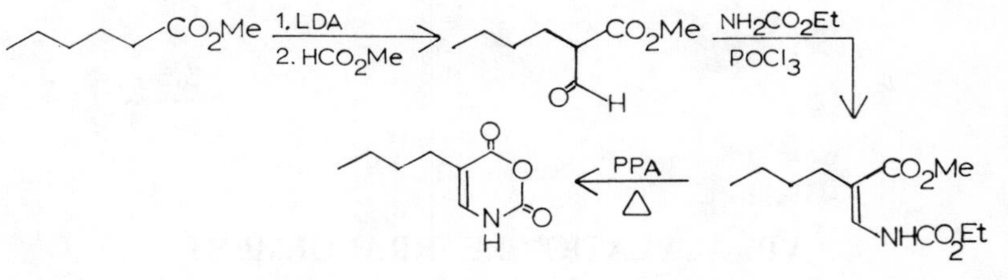

Scheme 99.

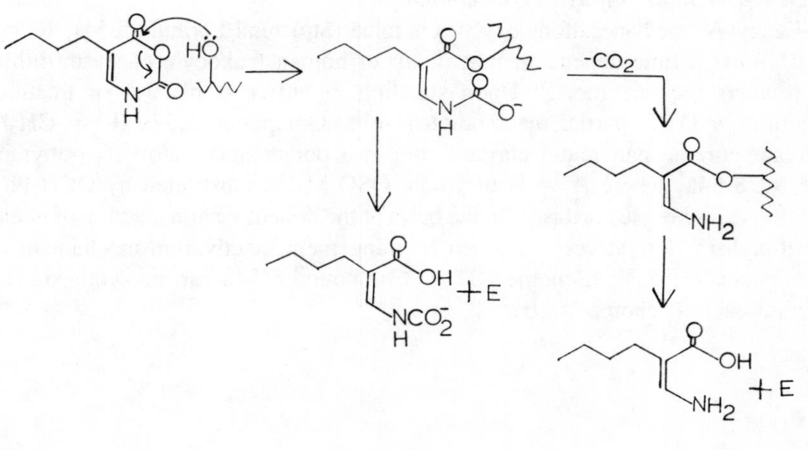

Scheme 100.

Substituted isatoic anhydrides with positively charged substituents, Scheme 101 (**5.54**, $R_1 = CH_2\overset{+}{N}H_3$, $R_2 = H$, CH_2Ph, or $CH_2[1\text{-Np}]$), were synthesized by Gelb and Abeles[135] in order to favor inactivation for trypsin-like proteases. Compound **5.54** ($R_1 = CH_2\overset{+}{N}H_3$, $R_2 = H$) preferentially inactivates bovine trypsin and bovine thrombin over bovine α-chymotrypsin and **5.54** ($R_1 = CH_2\overset{+}{N}H_3$, $R_2 = CH_2Ph$ or $CH_2[1\text{-Np}]$) are selective time-dependent inactivators of thrombin over chymoptrypsin and trypsin. The synthesis of **5.54** ($R_1 = CH_2\overset{+}{N}H_3$, $R_2 = CH_2Ph$ or $CH_2[1\text{-Np}]$) is shown in Scheme 102.

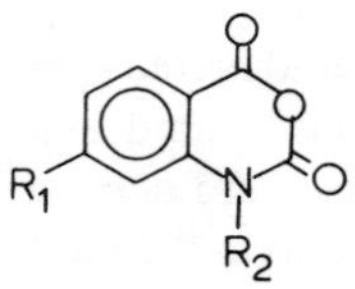

Scheme 101.   Structural Formula 5.54.

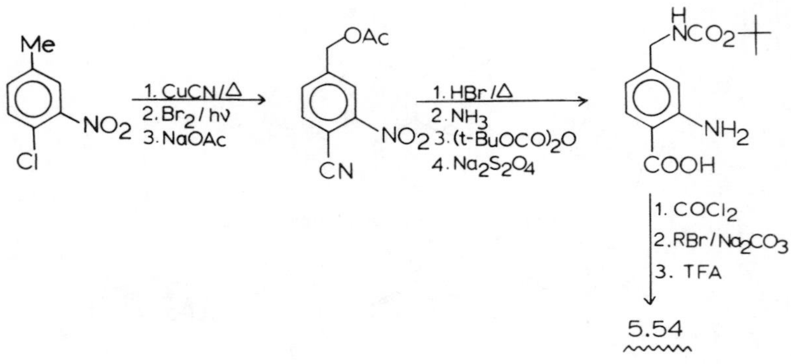

Scheme 102.

VII. ACYLATION/REARRANGEMENT

A. Acylation/Rearrangement/Acylation

1. 3-Benzyl-N-(methanesulfonyloxy)succinimide

(DL)-3-Benzyl-N-(methanesulfonyloxy)succinimide (Structural Formula 5.54a, R = CH_2Ph; Scheme 102A) is a time-dependent inactivator of human leukocyte elastase; dithiothreitol partially protects the enzyme.[135a] Upon standing in buffer at pH 7.5 or treatment with hydroxylamine or DTT, partial reactivation results. Compound **5.54a** (R = CH_2Ph) does not inactivate porcine pancreatic elastase, but is a potent inactivator of α-chymotrypsin. Analogues of **5.54a**, where R = H or where OSO_2Me is substituted by OCH_2Ph or OH, do not inactivate leukocyte elastase. On the basis of the nonenzymatic reaction of nucleophiles with N-sulfonyloxy derivatives, a Lossen rearrangement inactivation mechanism was proposed by Groutas et al.[135a] (Scheme 102A). Compound **5.54a** can be synthesized from 2-benzylsuccinic acid (Scheme 102B).

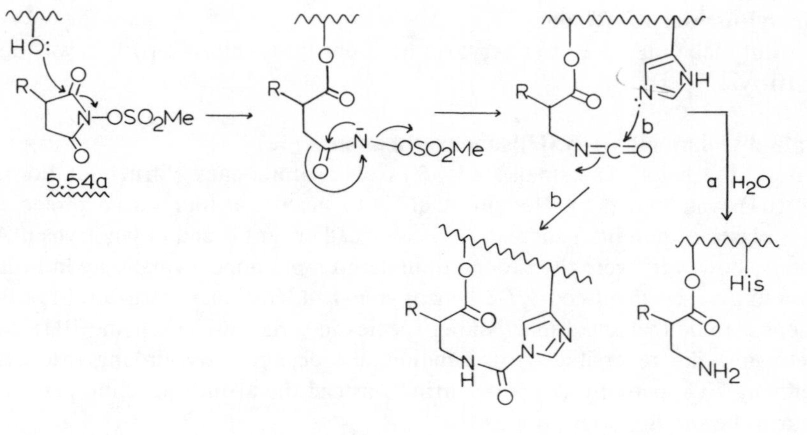

Scheme 102A.   Containing Structural Formula 5.54a.

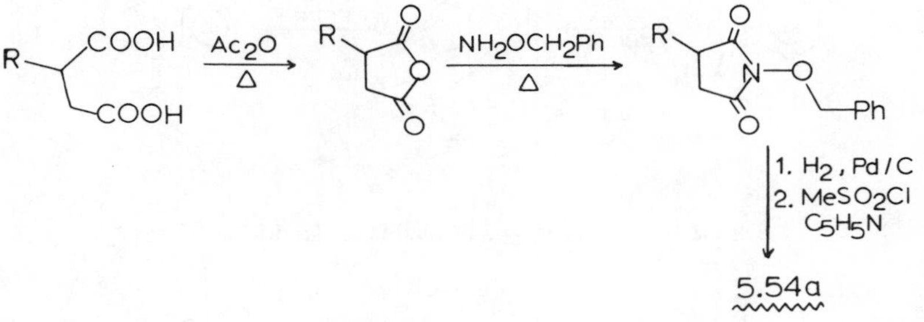

Scheme 102B.

VIII. NOT MECHANISM-BASED INACTIVATION

A. 2-Substituted-4*H*-3,1-benzoxazin-4-ones

The work of Abeles and co-workers[132,134] was extended by Hedstrom et al.[136] It was expected that 2-substituted-4*H*-3,1-benzoxazin-4-ones (**5.55**, R = OC_2H_5, CF_3) would inactivate serine protease by a mechanism similar to that which occurs with isatoic anhydride (see Scheme 97), *i.e.*, acylation of the enzyme followed by unmasking of the electron-releasing amino group. Chymotrypsin is inactivated by **5.55** (R = OC_2H_5 or CF_3), but enzyme activity slowly returns ($t_{1/2}$ = 11 and 0.8 hr for R = OEt and R = CF_3, respectively); the products formed during reactivation are 2-[*N*-(ethoxycarbonyl)amino]benzoic acid (**5.56**, R = OC_2H_5) and *N*-(trifluoroacetyl)anthranilic acid (**5.56**, R = CF_3), respectively. This suggests an affinity labeling mechanism (Scheme 103) in which the amino group is not unmasked. Trypsin and elastase also are inactivated by **5.55** (R = OC_2H_5). All of the enzymes are completely inactivated by a stoichiometric amount of inactivator.

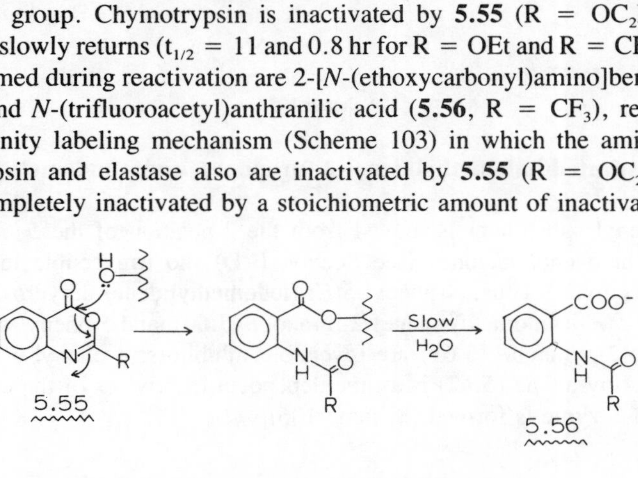

Scheme 103.   Containing Structural Formulas 5.55 and 5.56.

B. 5-Phenyl-6-chloro-2-pyrone

The affinity labeling of α-chymotrypsin by 5-phenyl-6-chloro-2-pyrone was discussed in Section III.A.2.

C. 3-Halo-3-(1-haloalkyl)-1(3*H*)isobenzofuranones

A series of 3-halo-3-(1-haloalkyl)-1(3*H*)-isobenzofuranones (Structural Formula 5.57, Scheme 104) was shown by Hemmi et al.[137] to inactivate four serine proteases, human leukocyte elastase, porcine pancreatic elastase, cathepsin G, and chymotrypsin Aα. These compounds, however, were then found to undergo rapid nonenzymatic hydrolysis in buffer at pH 7.5 to give 3-substituted-1*H*-2-benzopyran-1,4(3*H* diones (Structural Formula 5.58), which appear to be the actual inactivating species in some cases (Scheme 104). Since all of the inactivators are reversible upon standing, the desired cross-linking reaction (Scheme 105, pathway c) apparently is not occurring; instead the affinity labeling pathways (a and b) appear to be the important routes.

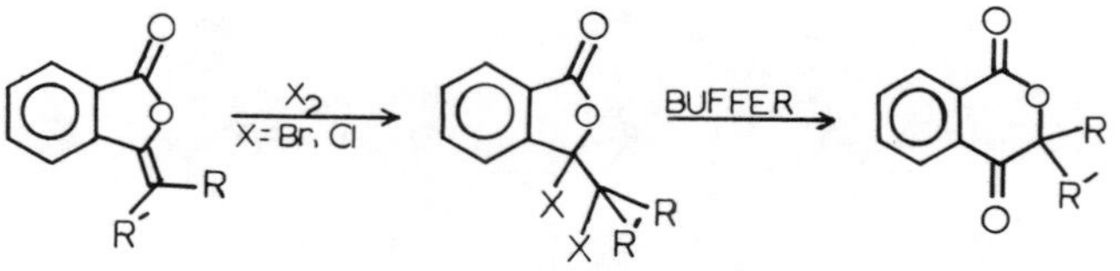

Scheme 104.   Containing Structural Formulas 5.57 and 5.58.

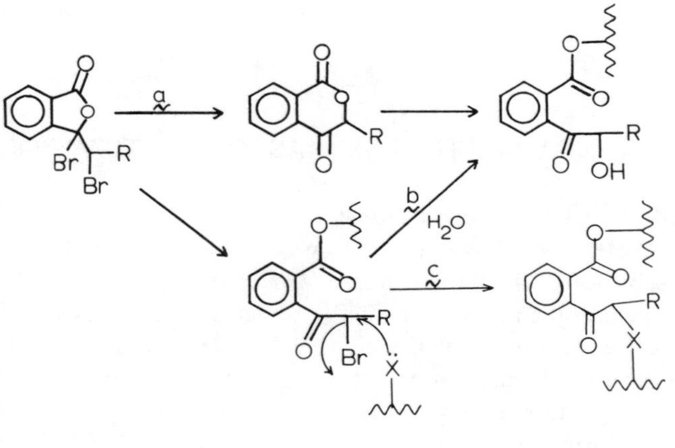

Scheme 105.

D. Phenyl-Substituted(halomethylidene)-2-furanones and -(halomethylidene)tetrahydropyranones

When the phenyl substituent is moved from the 3-position of the 5- and 6-membered lactone ring of halo enol lactones (see Section II.D), no irreversible inactivation of α-chymotrypsin occurs.[138] Thus, 4-phenyl-5(*E*)-(iodomethylidene)-dihydro-2(3*H*)-furnanone (**5.59**), 4-phenyl-5(*E*)-(iodomethylidene)-2-furanone (**5.60**), and 5-phenyl-6(*E*)-(iodomethylidene)tetrahydro-2-pyranone (**5.61**) are reversible inhibitors; 4-phenyl-6(*E*)-(iodomethylidene)tetrahydro-2-pyranone (**5.62**) is a time-dependent inactivator of the enzyme, but only a reversible acyl enzyme is formed (Scheme 106).

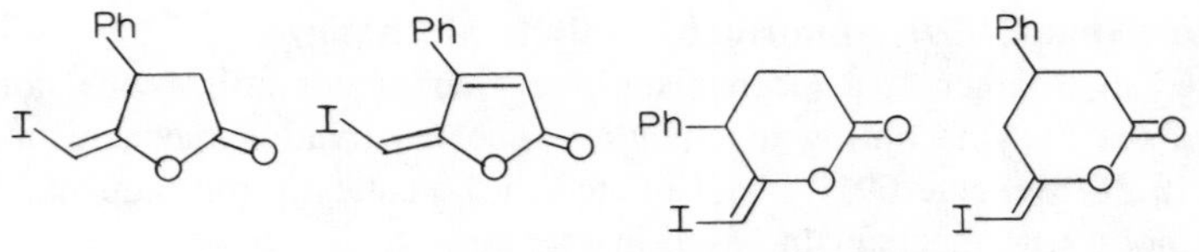

Scheme 106.   Structural Formulas 5.59 to 5.62.

E. Methicillin and Cloxacillin

Certain penicillin derivatives that have *O*-substituted aromatic or heterocyclic side chains, e.g., methicillin and cloxacillin, stoichiometrically inactivate β-lactamase I from *B. cereus*.[139,140] It was proposed by Waley and co-workers[139,140] that following enzyme acylation, an enzyme conformational change occurs. Since this is not observed with benzylpenicillin, it is believed that the unfolding of the enzyme results from specific interactions with the *O*-substituted side chain groups. Because the second step in this mechanism is a physical phenomenon, a process referred to[139,140] as substrate-induced deactivation, this is a type of affinity labeling. A similar process was observed for the inactivation of ox liver sulfatase A by aryl sulfates.[141,142]

F. 5-(Halomethyl)-2-pyranones

Compounds of the Structural Formula 5.63 (Scheme 107) were synthesized by Boulanger and Katzenellenbogen[143] as potential mechanism-based inactivators of α-chymotrypsin. The compounds where R = CF₃ are inactive, but when R = CH₂Cl they are time-dependent inactivators. However, no evidence for initial acylation was obtained and, therefore, these compounds may be affinity labeling agents.

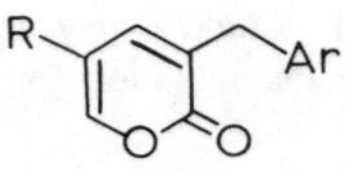

Scheme 107.   Structural Formula 5.63.

G. (Z)-2-(Alanylamino)-3-chloropropenoic Acid

(Z)-2-(Alanylamino)-3-chloropropenoic acid (Structural Formula 5.64, Scheme 108) was designed by Graham et al.[144] as a mechanism-based inactivator of porcine renal dipeptidase; however, it is a substrate, not a time-dependent inactivator.

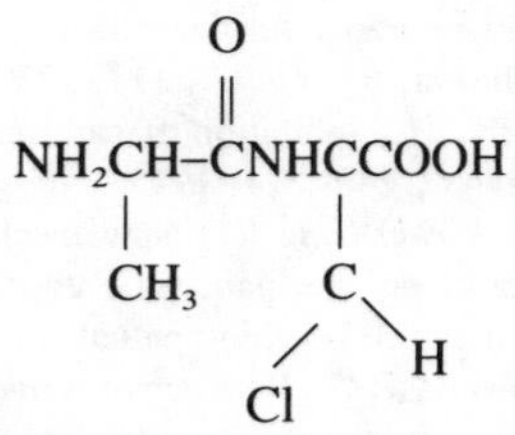

Scheme 108.   Structural Formula 5.64.

H. Inactivators that Utilize Abnormal Catalytic Mechanisms

As indicated in Chapter 1, a mechanism-based inactivator utilizes the normal catalytic mechanism of the enzyme during inactivation. The inactivation mechanism proposed for sparsomycin (**5.38**, Scheme 69)[73] is not related to the catalytic mechanism; therefore, this inactivator is not a true mechanism-based inactivator.

REFERENCES

1. **Arisawa, M. and Then, R. L.,** 6-Acetylmethylenepenicillanic acid (Ro 15-1903), a potent β-lactamase inhibitor. I. Inhibition of chromosomally and R-factor-mediated β-lactamases, *J. Antibiot.*, 35, 1578, 1982.
2. **Arisawa, M. and Then, R.,** Inactivation of TEM-1 β-lactamase by 6-acetylmethylenepenicillanic acid, *Biochem. J.*, 209, 609, 1983.
3. **Arisawa, M. and Adam, S.,** Mechanism of inactivation of TEM-1 β-lactamase by 6-acetylmethylene-penicillanic acid, *Biochem. J.*, 211, 447, 1983.
4. **Brenner, D. G. and Knowles, J. R.,** 6-(Methoxymethylene)penicillanic acid: inactivator of RTEM β-lactamase from *Escherichia coli*, *Biochemistry*, 23, 5839, 1984.
5. **Brenner, D. G.,** 6-(Methoxymethylene)penicillanic acid: a new β-lactamase inactivator, *J. Org. Chem.*, 50, 18, 1985.
6. **Béchet, J. J., Alazard, R., Dupaix, A., and Roucous, C.,** Intramolecular participation of the amide group in the hydrolysis of *p*-nitrophenyl *N*-(bromoacetyl)anthranilate, *Bioorg. Chem.*, 3, 55, 1974.
7. **Alazard, R., Béchet, J. J., Dupaix, A., and Yon, J.,** Inactivation of α-chymotrypsin by a bifunctional reagent, 2-bromomethyl-3,1-benzoxazin-4-one, *Biochim. Biophys. Acta*, 309, 379, 1973.
8. **Rando, R.,** Chemistry and enzymology of k$_{cat}$ inhibitors, *Science*, 185, 320, 1974.
9. **Krafft, G. A. and Katzenellenbogen, J. A.,** Synthesis of halo enol lactones. Mechanism-based inactivators of serine proteases, *J. Am. Chem. Soc.*, 103, 5459, 1981.
10. **Chakravarty, P. K., Krafft, G. A., and Katzenellenbogen, J. A.,** Haloenol lactones: enzyme-activated irreversible inactivators for serine proteases. Inactivation of α-chymotrypsin, *J. Biol. Chem.*, 257, 610, 1982.
11. **Daniels, S. B., Cooney, E., Sofia, M. J., Chakravarty, P. K., and Katzenellenbogen, J. A.,** Haloenol lactones, potent enzyme-activated irreversible inhibitors for α-chymotrypsin, *J. Biol. Chem.*, 258, 15046, 1983.
11a. **Daniels, S. B. and Katzenellenbogen, J. A.,** Halo enol lactones: studies on the mechanism of inactivation of α-chymotrypsin, *Biochemistry*, 25, 1436, 1986.
12. **Naruto, S., Motoc, I., Marshall, G. R., Daniels, S. B., Sofia, M. J., and Katzenellenbogen, J. A.,** Analysis of the interaction of haloenol lactone suicide substrates with α-chymotrypsin using computer graphics and molecular mechanics, *J. Am. Chem. Soc.*, 107, 5262, 1985.
13. **Sofia, M. J., Chakravarty, P. K., and Katzenellenbogen, J. A.,** Synthesis of five-membered halo enol lactone analogues of α-amino acids: potential protease suicide substrates, *J. Org. Chem.*, 48, 3318, 1983.
14. **Tam, T. F., Spencer, R. W., Thomas, E. M., Copp, L. J., and Krantz, A.,** Novel suicide inhibitors of serine proteinases. Inactivation of human leukocyte elastase by ynenol lactones, *J. Am. Chem. Soc.*, 106, 6849, 1984.
15. **Chase, J. F. A. and Tubbs, P. K.,** Conditions for the self-catalyzed inactivation of carnitine acetyltransferase, a novel form of enzyme inhibition, *Bichem. J.*, 111, 225, 1969.
16. **Zhong, J., Fong, J. C., and Schulz, H.,** Inhibition of carnitine acetyltransferase by metabolites of 4-pentenoic acid, *Arch. Biochem. Biophys.*, 240, 524, 1985.
17. **Harper, J. W., Hemmi, K., and Powers, J. C.,** New mechanism-based serine protease inhibitors: inhibition of human leukocyte elastase, porcine pancreatic elastase, human leukocyte cathepsin G, and chymotrypsin by 3-chloroisocoumarin and 3,3-dichlorophthalide, *J. Am. Chem. Soc.*, 105, 6518, 1983.
18. **Harper, J. W., Hemmi, K., and Powers, J. C.,** Reaction of serine proteases with substituted isocoumarins: discovery of 3,4-dichloroisocoumarin, a new general mechanism based serine protease inhibitor, *Biochemistry*, 24, 1831, 1985.
19. **Davies, W. and Poole, H. G.,** The action of phosphorus pentachloride on homophthalic acid, *J. Chem. Soc.*, 1616, 1928.
20. **Schnekenburger, J. and Kaiser, P.,** Acetylierungsprodukte des Homophthalsäureanhydrids 8. Mitt. über Acylderivate methylenaktiver Dicarbonylverbindungen, *Arch. Pharm.*, 304, 161, 1971.
21. **Milevskaya, V. B., Belinskaya, R. V., and Yagupol'skii, L. M.,** Reaction of homophthalic acid with phosphorus pentachloride, *J. Org. Chem. (USSR)*, 9, 2160, 1973.

22. **Ott, E.,** Symmetrical and unsymmetrical *o*-phthalyl chlorides, *Org. Synth. Coll. Vol.,* 2, 528, 1943.
23. **Westkaemper, R. B. and Abeles, R. H.,** Novel inactivators of serine proteases based on 6-chloro-2-pyrone, *Biochemistry,* 22, 3256, 1983.
24. **Gelb, M. H. and Abeles, R. H.,** Mechanism of inactivation of chymotrypsin by 3-benzyl-6-chloro-2-pyrone, *Biochemistry,* 23, 6596, 1984.
24a. **Ringe, D., Mottonen, J. M., Gelb, M. H., and Abeles, R. H.,** X-ray diffraction analysis of the inactivation of chymotrypsin by 3-benzyl-6-chloro-2-pyrone, *Biochemistry,* 25, 5633, 1986.
25. **Ringe, D., Seaton, B. A., Gelb, M. H., and Abeles, R. H.,** Inactivation of chymotrypsin by 5-benzyl-6-chloro-2-pyrone: ^{13}C NMR and X-ray diffraction analyses of the inactivator-enzyme complex, *Biochemistry,* 24, 64, 1985.
25a. **Boulanger, W. A. and Katzenellenbogen, J. A.,** Structure-activity study of 6-substituted 2-pyranones as inactivators of α-chymotrypsin, *J. Med. Chem.,* 29, 1159, 1986.
26. **Rokita, S. E., Srere, P. A., and Walsh, C. T.,** 3-Fluoro-3-deoxycitrate: a probe for mechanistic study of citrate-utilizing enzymes, *Biochemistry,* 21, 3765, 1982.
27. **Kollonitsch, J., Marburg, S., and Perkins, L. M.,** Fluorodehydroxylation, a novel method for synthesis of fluoroamines and fluoroamino acids, *J. Org. Chem.,* 44, 771, 1979.
28. **Rokita, S. E. and Walsh, C. T.,** Turnover and inactivation of bacterial citrate lyase with 2-fluorocitrate and 2-hydroxycitrate stereoisomers, *Biochemistry,* 22, 2821, 1983.
29. **Eanes, R. Z. and Kun, E.,** Inhibition of liver aconitase isozymes by (−)-*erythro*-fluorocitrate, *Mol. Pharmacol.,* 10, 130, 1974.
30. **Banks, R. B. and Hanna, P. E.,** Arylhydroxamic acid *N,O*-acyltransferase. Apparent suicide inactivation by carcinogenic *N*-arylhydroxamic acids, *Biochem. Biophys. Res. Commun.,* 91, 1423, 1979.
31. **Yeh, H. -M. and Hanna, P. E.,** Arylhydroxamic acid bioactivation via acyl group transfer. Structural requirements for transacylating and electrophile-generating activity of *N*-(2-fluorenyl)hydroxamic acids and related compounds, *J. Med. Chem.,* 25, 842, 1982.
32. **Marhevka, V. C., Ebner, N. A., Sehon, R. D., and Hanna, P. E.,** Mechanism-based inactivation of *N*-arylhydroxamic acid *N,O*-acyltransferase by 7-substituted-*N*-hydroxy-2-acetamidofluorenes, *J. Med. Chem.,* 28, 18, 1985.
33. **Mangold, B. L. K. and Hanna, P. E.,** Arylhydroxamic acid *N,O*-acyltransferase substrates. Acetyl transfer and electrophile generating activity of *N*-hydroxy-*N*-(4-alkyl-, 4-alkenyl-, and 4-cyclohexylphenyl)acetamides, *J. Med. Chem.,* 25, 630, 1982.
34. **Hanna, P. E., Banks, R. B., and Marhevka, V. C.,** Suicide inactivation of hamster hepatic arylhydroxamic acid *N,O*-acyltransferase, a selective probe of *N*-acetyltransferase multiplicity, *Mol. Pharmacol.,* 21, 159, 1982.
35. **Fisher, J., Belasco, J. G., Khosla, S., and Knowles, J. R.,** β-Lactamase proceeds via an acyl-enzyme with cefoxitin, *Biochemistry,* 19, 2895, 1980.
36. **Faraci, W. S. and Pratt, R. F.,** Mechanism of inhibition of the PC1 β-lactamase of *Staphylococcus aureus* by cephalosporins: importance of the 3'-leaving group., *Biochemistry,* 24, 903, 1985.
36a. **Pratt, R. F. and Faraci, W. S.,** Direct observation of ^{1}H NMR of cephalosporoate intermediates in aqueous solution during the hydrazinolysis and β-lactamase-catalyzed hydrolysis of cephalosporins with 3' leaving groups: kinetics and equilibria of the 3' elimination reaction, *J. Am. Chem. Soc.,* 108, 5328, 1986.
36b. **Page, M. I. and Proctor, P.,** Mechanism of β-lactam ring opening in cephalosporins, *J. Am. Chem. Soc.,* 106, 3820, 1984.
36c. **Faraci, W. S. and Pratt, R. F.,** Mechanism of inhibition of RTEM-2 β-lactamase by cephamycins: relative importance of the 7α-methoxy group and the 3' leaving group, *Biochemistry,* 25, 2934, 1986.
36d. **Boyd, D. B.,** Elucidating the leaving group effect in the β-lactam ring opening mechanism of cephalosporins, *J. Org. Chem.,* 50, 886, 1985.
36e. **Mobashery, S., Lerner, S. A., and Johnston, M.,** Conscripting β-lactamase for use in drug delivery. Synthesis and biological activity of a cephalosporin C_{10}-ester of an antibiotic dipeptide, *J. Am. Chem. Soc.,* 108, 1685, 1986.
37. **Fisher, J., Belasco, J. G., Charnas, R. L., Khosla, S., and Knowles, J. R.,** β-Lactamase inactivation by mechanism-based reagents, *Philos. Trans. R. Soc. London* Ser. B, 289, 309, 1980.
38. **Fisher, J., Charnas, R. L., Bradley, S. M., and Knowles, J. R.,** Inactivation of the RTEM β-lactamase from *Escherichia coli.* Interaction of penam sulfones with enzyme, *Biochemistry,* 20, 2726, 1981.
38a. **Reading, C. and Cole, M.,** Structure-activity relationships amongst β-lactamase inhibitors, *J. Enz. Inhib.,* 1, 83, 1986.
39. **Cartwright, S. J. and Coulson, A. F. W.,** A semi-synthetic penicillanase inactivator, *Nature (London),* 278, 360, 1979.
40. **Mezes, P. S. F., Clarke, A. J., Dmitrienko, G. I., and Viswanatha, T.,** 6-β-(Trifluoromethane sulfonyl)amidopenicillanic acid sulfone, a potent inhibitor for β-lactamases, *FEBS Lett.,* 143, 265, 1982.

41. **Mezes, P. S. F., Clarke, A. J., Dmitrienko, G. I., and Viswanatha, T.,** The inactivation of *Bacillus cereus* 569/H β-lactamase by 6β-(trifluoromethanesulfonyl)amidopenicillanic acid sulfone: pH dependence and stoichiometry, *J. Antibiot.*, 35, 918, 1982.

42. **Clarke, A. J., Mezes, P. S., Vice, S. F., Dmitrienko, G. I., and Viswanatha, T.,** Inactivation of *Bacillus cereus* 569/H β-lactamase I by 6β-(trifluoromethanesulfonyl)amidopenicillanic acid sulfone and its *N*-methyl derivative, *Biochim. Biophys. Acta*, 748, 389, 1983.

43. **Brenner, D. G., Knowles, J. R., and Rihs, G.,** Penicillanic acid sulfone: an unexpected isotope effects in the interaction of 6α- and 6β-monodeuterio and of 6,6-dideuterio derivatives with RTEM β-lactamase from *Escherichia coli*, *Biochemistry*, 20, 3680, 1981.

44. **Dmitrienko, G. I., Copeland, C. R., Arnold, L., Savard, M. E., Clarke, A. J., and Viswanatha, T.,** Inhibition of β-lactamase I by 6-β-sulfonamidopenicillanic acid sulfones: evidence for conformational change accompanying the inhibition process, *Bioorg. Chem.*, 13, 34, 1985.

45. **English, A. R., Retsema, J. A., Girard, A. E., Lynch, J. E., and Barth, W. E.,** CP-45,899, a beta-lactamase inhibitor that extends the antibacterial spectrum of beta-lactams: initial bacteriological characterization, *Antimicrob. Agents Chemother.*, 14, 414, 1978.

46. **Fu, K. P. and Neu, H. C.,** Comparative inhibition of β-lactamases by novel β-lactam compounds, *Antimicrob. Agents Chemother.*, 15, 171, 1979.

47. **Labia, R., Lelievre, V., and Peduzzi, J.,** Inhibition kinetics of three R-factor-mediated β-lactamases by a new β-lactam sulfone (CP 45899), *Biochim. Biophys. Acta*, 611, 351, 1980.

48. **Kemal, C. and Knowles, J. R.,** Penicillanic acid sulfone: Interaction with RTEM β-lactamase from *Escherichia coli* at different pH values, *Biochemistry*, 20, 3688, 1981.

49. **Brenner, D. G. and Knowles, J. R.,** Penicillanic acid sulfone: Nature of irreversible inactivation of RTEM β-lactamase from *Escherichia coli*, *Biochemistry*, 23, 5833, 1984.

50. **Foulds, C. D., Kosmirak, M., and Sammes, P. G.,** Substituted penicillanic acid 1,1-dioxides as β-lactam inhibitors: Studies on 6-benzylidene- and hydroxybenzylpenam sulphones, *J. Chem. Soc. Perkin Trans. I*, 963, 1985.

51. **Cartwright, S. J. and Waley, S. G.,** β-Lactamase inhibitors, *Med. Res. Rev.*, 3, 341, 1983.

52. **Knight, G. C. and Waley, S. C.,** Inhibition of class C β-lactamases by (1'R,6R)-6-(1'-hydroxy) benzylpenicillanic acid SS-dioxide, *Biochem. J.*, 225, 435, 1985.

52a. **Chen, Y. L., Chang, C.-W., and Hedberg, K.,** Synthesis of a potent β-lactamase inhibitor-1,1-dioxo-6-(2-pyridyl)methylenepenicillanic acid and its reaction with sodium methoxide, *Tetrahedron Lett.*, 27, 3449, 1986.

53. **Béchet, J-J., Dupaix, A., Yon, J., Wakselman, M., Robert, J. -C., and Vilkas, M.,** Inactivation of α-chymotrypsin by a bifunctional reagent, 3,4-dihydro-3,4-dibromo-6-bromomethylcoumarin, *Eur. J. Biochem.*, 35, 527, 1973.

54. **Reboud-Ravaux, M., Desvages, G., and Chapeville, F.,** Irreversible inhibition and peptide mapping of urinary plasminogen activator urokinase, *FEBS Lett.*, 140, 58, 1982.

55. **Reboud-Ravaux, M. and Desvages, G.,** Inactivation of human high and low molecular weight urokinase. Analysis of their active site, *Biochim. Biophys. Acta*, 791, 333, 1984.

56. **Béchet, J-J., Dupaix, A., and Blagoeva, I.,** Inactivation of α-chymotrypsin by new bifunctional reagents: halomethylated derivatives of dihydrocoumarins, *Biochimie*, 59, 231, 1977.

57. **Decodts, G. and Wakselman, M.,** Suicide inhibitors of proteases. Lack of activity of halomethyl derivatives of some aromatic lactams, *Eur. J. Med. Chim. Chim. Ther.*, 18, 107, 1983.

58. **Béchet, J-J., Dupaix, A., Roucous, C., and Bonamy, A-M.,** Inactivation of proteases and esterases by halomethylated derivatives of dihydrocoumarins, *Biochimie*, 59, 241, 1977.

59. **Hamon, J. F., Wakselman, M., and Vilkas, M.,** Lactones phénoliques halométhylées, inhibiteurs bifonctionnels de protéases. III. Réaction avec l'imidazole en milieu aqueux, *Bull. soc. Chim. Fr.*, 1387, 1975.

60. **Wakselman, M., Hamon, J. F., and Vilkas, M.,** Lactones phenoliques halomethylees, inhibiteurs bifonctionnels de proteases. II. Preparation de derives halomethyles de la dihydro-3,4 coumarine et de la benzofuranone-2, *Tetrahedron*, 30, 4069, 1974.

61. **Nicolle, J. P., Hamon, J. F., and Wakselman, M.,** Lactones phénoliques halométhylées, inhibiteurs bifonctionnels de protéases. IV. Préparation de dérivés chlorométhylés de la dihydro-3,4- benzyl-3 coumarine et de la tétrahydro-2,3,4,5 benzoxepinone-2, *Bull. Soc. Chim. Fr.*, 83, 1977.

62. **Harper, J. W. and Powers, J. C.,** 3-Alkoxy-7-amino-4-chloroisocoumarins, a new class of suicide substrates for serine protease, *J. Am. Chem. Soc.*, 106, 7618, 1984.

63. **Harper, J. W. and Powers, J. C.,** Reaction of serine protease with substituted 3-alkoxy-4-chloroisocoumarins and 3-alkoxy-7-amino-4-chloroisocoumarins: new reactive mechanism-based inhibitors, *Biochemistry*, 24, 7200, 1985.

64. **Meyer, E. F., Jr., Presta, L. G., and Radhakrishnan, R.,** Stereospecific reaction of 3-methoxy-4-chloro-7-aminoisocoumarin with crystalline porcine pancreatic elastase, *J. Am. Chem. Soc.*, 107, 4091, 1985.

65. **White E. H., Roswell, D. F., Politzer, I. R., and Branchini, B. R.,** Active site directed inhibition of enzymes utilizing deaminatively produced carbonium ions. Application of chymotrypsin, *J. Am. Chem. Soc.,* 97, 2290, 1975.
66. **White, E. H., Jelinski, L. W., Politzer, I. R., Branchini, B. R., and Roswell, D. F.,** Active-site-directed inhibition of α-chymotrypsin by deaminiatively produced carbonium ions: an example of suicide or enzyme-activated-substrate inhibition, *J. Am. Chem. Soc.,* 103, 4231, 1981.
67. **White, E. H., Jelinski, L. W., Perks, H. M., Burrows, E. P., and Roswell, D. F.,** Preferential inhibition of α-chymotrypsin by the D form of an amino acid derivative, *N'*-isobutyryl-*N*-benzyl-*N*-nitrosophenylalaninamide, *J. Am. Chem. Soc.,* 99, 3171, 1977.
68. **Donadio, S., Perks, H. M., Tsuchiya, K., and White, E. H.,** Alkylation of amide linkages and cleavage of the C chain in the enzyme-activated substrate inhibition of α-chymotrypsin with *N*-nitrosamides, *Biochemistry,* 24, 2447, 1985.
69. **White, E. H., Perks, H. M., and Roswell, D. F.,** Labeling of amide linkages in active site mapping: carbonium ion and extended photoaffinity labeling approaches, *J. Am. Chem. Soc.,* 100, 7421, 1978.
70. **White, E. H., Roswell, D. F., Politzer, I. R., and Branchini, B. R.,** Active site-directed inhibition with substrates producing carbonium ions: chymotrypsin, *Methods Enzymol.,* 46, 216, 1977.
71. **Gold, B. and Linder, W. B.,** α-Hydroxynitrosamines: transportable metabolite of dialkylnitrosamines, *J. Am. Chem. Soc.,* 101, 6772, 1979.
72. **Miziorko, H. M., Behnke, C. E., Ahmad, P. M., and Ahmad, F.,** Active site directed inactivation of rat mammary gland fatty acid synthase by 3-chloropropionyl coenzyme A, *Biochemistry,* 25, 468, 1986.
73. **Flynn, G. A. and Ash, R. J.,** Necessity of the sulfoxide moiety for the biochemical and biological properties of an analog of sparsomycin, *Biochem. Biophys. Res. Commun.,* 114, 1, 1983.
74. **Flynn, G. A. and Beight, D. W.,** A dehydroalanine route to an activated phenolic sparsomycin analog, *Tetrahedron Lett.,* 25, 2655, 1984.
75. **Liskamp, R. M. J., Colstee, J. H., Ottenheijm, H. C. J., Lelieveld, P., and Akkerman, W.,** Structure-activity relationships of sparsomycin and its analogues. Octylsparsomycin: the first analogue more active than sparsomycin, *J. Med. Chem.,* 27, 301, 1984.
76. **Ash, R. J., Flynn, G. A., Liskamp, R. M. J., and Ottenheijm, C. J.,** Sulfoxide configuration in sparsomycin determines time-dependent and competitive inhibition of peptidyl transferase, *Biochem. Biophys. Res. Commun.,* 125, 784, 1984.
77. **Loosemore, M. J. and Pratt, R. F.,** On the epimerization of 6α-bromopenicillanic acid and the preparation of 6β-bromopenicillanic acid, *J. Org. Chem.,* 43, 3611, 1978.
78. **Cignarella, G., Piferri, G., and Testa, E.,** 6-Chloro- and 6-bromopenicillanic acids, *J. Org. Chem.,* 27, 2668, 1962.
79. **Aimetti, J. A., Hamanaka, E. S., Johnson, D. A., and Kellogg, M. S.,** Stereoselective synthesis of 6β-substituted penicillanates, *Tetrahedron Lett.,* p. 4631, 1979.
80. **John, D. I., Thomas, E. J., and Tyrell, N. D.,** Reduction of 6α-alkyl-6β-isocyanopenicillanates by tri-*n*-butyltin hydride. A stereoselective synthesis of 6β-alkylpenicillanates, *J. Chem. Soc. Chem. Commun.,* p. 345, 1979.
81. **Kemp, J. E. G., Closier, M. D., Narayanaswami, S., and Stefaniak, M. H.,** Nucleophilic S$_N$2 displacements on penicillin-6 and cephalosporin-7-triflates; 6β-iodopenicillanic acid, a new β-lactamase inhibitor, *Tetrahedron Lett.,* 21, 2991, 1980.
82. **von Daehne, W.,** 6β-Halopenicillanic acids, a group of β-lactamase inhibitors, *J. Antibiot.,* 33, 451, 1980.
83. **Pratt, R. F. and Loosemore, M. J.,** 6β-Bromopenicillanic acid, a potent β-lactamase inhibitor, *Proc. Natl. Acad. Sci. U.S.A.,* 75, 4145, 1978.
84. **Knott-Hunziker, V., Orlek, B. S., Sammes, P. G., and Waley, S. G.,** Kinetics of inactivation of β-lactamase I by 6β-bromopenicillanic acid, *Biochem. J.,* 187, 797, 1980.
85. **Loosemore, M. J., Cohen, S. A., and Pratt, R. F.,** Inactivation of *Bacillus cereus* β-lactamase I by 6β-bromopenicillanic acid: kinetics, *Biochemistry,* 19, 3990, 1980.
86. **Knott-Hunziker, V., Orlek, B. S., Sammes, P. G., and Waley, S. G.,** 6-β-Bromopenicillanic acid inactivates β-lactamase I, *Biochem. J.,* 177, 365, 1979.
87. **Orlek, B. S., Sammes, P. G., Knott-Hunziker, V., and Waley, S. G.,** On the chemical binding of 6β-bromopenicillanic acid to β-lactamase I, *J. Chem. Soc. Chem. Commun.,* p. 962, 1979.
88. **Orlek, B. S., Sammes, P. G., Knott-Hunziker, V., and Waley, S. G.,** On the chemistry of β-lactamase inhibition by 6β-bromopenicillanic acid, *J. Chem. Soc. Perkin Trans. I,* p. 2322, 1980.
89. **McMillan, I. and Stoodley, R. J.,** Studies related to penicillins. I. 6α-Chloropenicillanic acid and its reaction with nucleophiles, *J. Chem. Soc. C,* p. 2533, 1968.
90. **Knott-Hunziker, V., Waley, S. G., Orlek, B. S., and Sammes, P. G.,** Penicillinase active sites: labelling of serine-44 in β-lactamase I by 6β-bromopenicillanic acid, *FEBS Lett.,* 99, 59, 1979.
91. **Cohen, S. A. and Pratt, R. F.,** Inactivation of *Bacillus cereus* β-lactamase I by 6β-bromopenicillanic acid: mechanism, *Biochemistry,* 19, 3996, 1980.

91a. **Moore, B. A. and Brammer, K. W.,** 6β-Iodopenicillanic acid (UK-38,006), a beta-lactamase inhibitor that extends the antibacterial spectrum of beta-lactam compounds: initial bacteriological characterization, *Antimicrob. Agents Chemother.,* 20, 327, 1981.

92. **Frère, J. -M., Dormans, C., Duyckaerts, C., and DeGraeve, J.,** Interaction of β-iodopenicillanate with the β-lactamases of *Streptomyces albus* G and *Actinomadura* R39, *Biochem. J.,* 207, 437, 1982.

93. **Charlier, P., Dideberg, O., Jamoulle, J. -C., Frère, J. -M., Ghuysen, J. -M., Dive, G., and Lamotte-Brasseur, J.,** Active-site-directed inactivators of the Zn^{2+}-containing D-alanyl-D-alanine-cleaving carboxypeptidase of *Streptomyces albus* G, *Biochem. J.,* 219, 763, 1984.

94. **Joris, B., Dusart, J., Frère, J. -M., Van Beeumen, J., Emanuel, E. L., Petursson, S., Gagnon, J., and Waley, S. G.,** The active site of the P99 β-lactamase from *Enterobacter cloacae, Biochem. J.,* 223, 271, 1984.

95. **Joris, B., DeMeester, F., Galleni, M., Reckinger, G., Coyette, J., Frère, J. -M., and Van Beeumen, J.,** The β-lactamase of *Enterobacter cloacae* P99. Chemical properties, N-terminal sequence and interaction with 6β-halopenicillanates, *Biochem. J.,* 228, 241, 1985.

96. **Lenzini, M. V. and Frère, J. M.,** The β-lactamase of *Streptomyces cacaoi:* interaction with cefoxitin and β-iodopenicillanate, *J. Enz. Inhib.,* 1, 25, 1985.

96a. **De Meester, F., Frère, J.-M., Waley, S. G., Cartwright, S. J., Virden, R., and Lindberg, F.,** 6-β-Iodopenicillanate as a probe for the classification of β-lactamases, *Biochem. J.,* 239, 575, 1986.

96b. **Joris, B., De Meester, F., Galleni, M., Masson, S., Dusart, J., Frère, J.-M., Van Beeumen, J., Bush, K., and Sykes, R.,** Properties of a class C β-lactamase from *Serratia marcescens, Biochem. J.,* 239, 581, 1986.

96c. **Joris, B., De Meester, F., Galleni, M., Frère, J.-M., and Van Beeumen, J.,** The K1 β-lactamase of *Klebsiella pneumoniae, Biochem. J.,* 243, 561, 1987.

96d. **De Meester, F., Joris, B., Lenzini, M. V., Dehottay, P., Erpicium, T., Dusart, J., Klein, D., Ghuysen, J.-M., Frère, J.-M., and Van Beeumen, J.,** The active sites of the β-lactamases of *Streptomyces cacaoi* and *Streptomyces albus* G, *Biochem. J.,* 244, 427, 1987.

97. **Howarth, T. I., Brown, A. G., and King, T. J.,** Clavulanic acid, a novel β-lactam isolated from *Streptomyces clavuligerus;* X-ray crystal structure analysis, *J. Chem. Soc. Chem. Commun.,* p. 266, 1976.

98. **Brown, A. G., Butterworth, D., Cole, M., Hanscomb, G., and Hood, J. D.,** Reading, C., and Rolinson, G. N., Naturally-occurring β-lactamase inhibitors with antibacterial activity, *J. Antibiot.,* 29, 668, 1976.

99. **Reading, C. and Cole, M.,** Clavulanic acid: a beta-lactamase-inhibiting beta-lactam from *Streptomyces clavuligerus, Antimicrob. Agents Chemother.,* 11, 852, 1977.

100. **Neu, H. C. and Fu, K. P.,** Clavulanic acid, a novel inhibitor of β-lactamases, *Antimicrob. Agents Chemother.,* 14, 650, 1978.

101. **Durkin, J. P. and Viswanatha, T.,** Clavulanic acid inhibition of β-lactamase I from *Bacillus cereus* 569/H, *J. Antibiot.,* 31, 1162, 1978.

102. **Fisher, J., Charnas, R. L., and Knowles, J. R.,** Kinetic studies on the inactivation of *Escherichia coli* RTEM β-lactamase by clavulanic acid, *Biochemistry,* 17, 2180, 1978.

103. **Charnas, R. L., Fisher, J., and Knowles, J. R.,** Chemical studies on the inactivation of *Escherichia coli* RTEM β-lactamase by clavulanic acid, *Biochemistry,* 17, 2185, 1978.

104. **Labia, R. and Peduzzi, J.,** Cinetique de l'inhibition de beta-lactamases par l'acide clavulanique, *Biochim. Biophys. Acta,* 526, 572, 1978.

105. **Hunt, E., Bentley, P. H., Brooks, G., and Gilpin, M. L.,** Total synthesis of β-lactamase inhibitors related to clavulanic acid, *J. Chem. Soc. Chem. Commun.,* 906, 1977.

106. **Charnas, R. L. and Knowles, J. R.,** Inactivation of RTEM β-lactamase from *Escherichia coli* by clavulanic acid and 9-deoxyclavulanic acid, *Biochemistry,* 20, 3214, 1981.

107. **Reading, C. and Hepburn, P.,** The inhibition of *Staphylococcal* β-lactamase by clavulanic acid, *Biochem. J.,* 179, 67, 1979.

108. **Reading, C. and Farmer, T.,** The inhibition of β-lactamases from gram-negative bacteria by clavulanic acid, *Biochem. J.,* 199, 779, 1981.

109. **Kelly, J. A., Frère, J. -M., Duez, C., and Ghuysen, J. -M.,** Interactions between non-classical β-lactam compounds and the β-lactamases of *Actinomadura* R39 and *Streptomyces albus* G, *Biochem. J.,* 199, 137, 1981.

110. **Frère, J. M., Dormans, C., Lenzini, V. M., and Duyckaerts, C.,** Interaction of clavulanate with β-lactamases of *Streptomyces albus* G and *Actinomadura* R39, *Biochem. J.,* 207, 429, 1982.

111. **Ogawara, H. and Mantoku, A.,** Interaction of β-lactamase of *Streptomyces cacaoi.* I. Clavulanic acid and PS-5, *J. Antibiot.,* 34, 1341, 1981.

112. **Kelly, J. A., Frère, J. -M., Klein, D., and Ghuysen, J. -M.,** Interaction between non-classical β-lactam compounds and the Zn^{2+}-containing G and serine R61 and R39 D-alanyl-D-alanine peptidases, *Biochem. J.,* 199, 129, 1981.

113. **Bentley, P. H., Berry, P. D., Brooks, G., Gilpin, M. L., Hunt, E., and Zomaya, I. I.,** Total synthesis of (±)-clavulanic acid, *J. Chem. Soc. Chem. Commun.,* p. 748, 1977.

114. **Cherry, P. C., Newall, C. E., and Watson, N. S.,** Preparation of the 7-oxo-4-oxa-1-azabicylo[3.2.0]hept-2-ene system and the reversible cleavage of its oxazoline ring, *J. Chem. Soc. Chem. Commun.*, p. 469, 1978.

115. **Brown, A. G., Corbett, D. F., Eglington, A. J., and Howarth, T. T.,** Structures of olivanic acid derivatives, MM 4550 and MM 13902; two new, fused β-lactams isolated from *Streptomyces olivaceus, J. Chem. Soc. Chem. Commun.*, p. 523, 1977.

116. **Corbett, D. F., Eglington, A. J., and Howarth, T. T.,** Structure elucidation of MM 17880, a new fused β-lactam antibiotic isolated from *Streptomyces olivaceus;* a mild β-lactam degradation reaction, *J. Chem. Soc. Chem. Commun.*, p. 953, 1977.

117. **Hood, J. D., Box, S. J., and Verrall, M. S.,** Olivanic acids, a family of β-lactam antibiotics with β-lactamase inhibitor properties produced by *Streptomyces* species. II. Isolation and characterisation of the olivanic acids MM 4550, MM 13902, and MM 17880 from *Streptomyces olivaceus, J. Antibiot.*, 32, 295, 1979.

118. **Maeda, K., Takahashi, S., Sezaki, M., Iinuma, K., Naganawa, H., Kondo, S., Ohno, M., and Umezawa, H.,** Isolation and structure of a β-lactamase inhibitor from *Streptomyces, J. Antibiot.*, 30, 770, 1977.

119. **Harada, S., Shinagawa, S., Nozaki, Y., Asai, M., and Kishi, T.,** C-19393 S_2 and H_2, new carbapenem antibiotics. II. Isolation and structures, *J. Antibiot.*, 33, 1425, 1980.

120. **Okonogi, K., Nozaki, Y., Imada, A., and Kuno, M.,** C-19393 S_2 and H_2 new carbapenem antibiotics. IV. Inhibitory activity against β-lactamases, *J. Antibiot.*, 34, 212, 1981.

121. **Tsuji, N., Nagashisma, K., Kobayashi, M., Terui, Y., Matsumoto, K., and Kondo, E.,** The structures of pluracidomycins, new carbapenem antibiotics, *J. Antibiot.*, 35, 536, 1982.

122. **Ito, T., Ezaki, N., Ohba, K., Amano, S., Kondo, Y., Miyadoh, S., Shomura, T., Sezaki, M., Niwa, T., Kojima, M., Inouye, S., Yamada, Y., and Niida, T.,** A novel β-lactamase inhibitor, SF-2103A produced by a *Streptomyces, J. Antibiot.*, 35, 533, 1982.

123. **Okonogi, K., Harada, S., Shinagawa, S., Imada, A., and Kuno, M.,** β-Lactamase inhibitory activities and synergistic effects of 5,6-*cis*-carbapenem antibiotics, *J. Antibiot.*, 35, 963, 1982.

124. **Charnas, R. L. and Knowles, J. R.,** Inhibition of the RTEM β-lactamase from *Escherichia coli.* Interaction of enzyme with derivatives of olivanic acid, *Biochemistry*, 20, 2732, 1981.

125. **Easton, C. J. and Knowles, J. R.,** Inhibition of the RTEM β-lactamase from *Escherichia coli.* Interaction of the enzyme with derivatives of olivanic acid, *Biochemistry*, 21, 2857, 1982.

126. **Tanaka, K., Shoji, J., Terui, Y., Tsuji, N., Kondo, E., Mayama, M., Kawamura, Y., Hattori, T., Matsumoto, K., and Yoshida, T.,** Asparenomycin A, a new carbapenem antibiotic, *J. Antibiot.*, 34, 909, 1981.

127. **Tsuji, N., Nagashima, K., Kobayashi, M., Shoji, J., Kato, T., Terui, Y., Nakai, H., and Shiro, M.,** Asparenomycins A, B, and C, new carbapenem antibiotics III. Structures, *J. Antibiot.*, 35, 24, 1982.

128. **Murakami, K., Doi, M., and Yoshida, T.,** Asparenomycins A, B, and C, new carbapenem antibiotics. V. Inhibition of β-lactamases, *J. Antibiot.*, 35, 39, 1982.

129. **Okamura, K., Hirata, S., Okumura, Y., Fukagawa, Y., Schimauchi, Y., Kouno, K., Ishikura, T., and Lein, J.,** PS-5, a new β-lactam antibiotic from *Streptomyces, J. Antibiot.*, 31, 480, 1978.

130. **Okamura, K., Sakamoto, M., and Ishikura, T.,** PS-5 inhibition of a β-lactamase from *Proteus vulgaris, J. Antibiot.*, 33, 293, 1980.

131. **Fukagawa, Y., Takei, T., and Ishikura, T.,** Inhibition of β-lactamase of *Bacillus licheniformis* 749/C by compound PS-5, a new β-lactam antibiotic, *Biochem. J.*, 185, 177, 1980.

131a. **Spencer, R. W., Tam, T. F., Thomas E., Robinson, V. J., and Krantz, A.,** Ynenol lactones: synthesis and investigation of reactions relevant to their inactivation of serine proteases, *J. Am. Chem. Soc.*, 108, 5589, 1986.

131b. **Copp, L. J., Krantz, A., and Spencer, R. W.,** Kinetics and mechanism of human leukocyte elastase inactivation by ynenol lactones, *Biochemistry*, 26, 169, 1987.

132. **Moorman, A. R. and Abeles, R. H.,** A new class of serine protease inactivators based on isatoic anhydride, *J. Am. Chem. Soc.*, 104, 6785, 1982.

133. **Caplow, M. and Jencks, W. P.,** The effect of substituents on the deacylation of benzoyl-chymotrypsins, *Biochemistry*, 1, 883, 1962.

134. **Weidman, B. and Abeles, R. H.,** Mechanism of inactivation of chymotrypsin by 5-butyl-3*H*-1,3-oxazine-2,6-dione, *Biochemistry*, 23, 2373, 1984.

135. **Gelb, M. H. and Abeles, R. H.,** Substituted isatoic anhydrides: selective inactivators of trypsin-like serine proteases, *J. Med. Chem.*, 29, 585, 1986.

135a. **Groutas, W. C., Giri, P. K., Crowley, J. P., Castrisos, J. C., and Brubaker, M. J.,** The Lossen rearrangement in biological systems. Inactivation of leukocyte elastase and alpha-chymotrypsin by (DL)-3-benzyl-N-(methanesulfonyloxy)succinimide, *Biochem. Biophys. Res. Commun.*, 141, 741, 1986.

136. **Hedstrom, L., Moorman, A. R., Dobbs, J., and Abeles, R. H.,** Suicide inactivation of chymotrypsin by benzoxazinones, *Biochemistry*, 23, 1753, 1984.

137. **Hemmi, K., Harper, J. W., and Powers, J. C.,** Inhibition of human leukocyte elastase, cathepsin G, chymotrypsin Aα, and porcine pancreatic elastase with substituted isobenzofuranones and benzopyrandiones, *Biochemistry,* 24, 1841, 1985.

138. **Sofia, M. J. and Katzenellenbogen, J. A.,** Enol lactone inhibitors of serine proteases. The effect of regiochemistry on the inactivation behavior of phenyl-substituted (halomethylene)tetra- and -dihydrofuranones and (halomethylene)tetrahydropyranones toward α-chymotrypsin: stable acyl enzyme intermediate, *J. Med. Chem.,* 29, 230, 1986.

139. **Kiener, P. A. and Waley, S. G.,** Substrate-induced deactivation of penicillanases. Studies of β-lactamase I by hydrogen exchange, *Biochem. J.,* 165, 279, 1977.

140. **Kiener, P. A., Knott-Hunziker, V., Petursson, S., and Waley, S. G.,** Mechanism of substrate-induced inactivation of β-lactamase I, *Eur. J. Biochem.,* 109, 575, 1980.

141. **Roy, A. B.,** The sulphatase of ox liver. XI. Kinetic studies of the substrate-induced inactivation of sulphatase A, *Biochim. Biophys. Acta,* 526, 489, 1978.

142. **Waheed, A. and Van Etten, R. L.,** Covalent modification as the cause of the anomalous kinetics of aryl sulfatase A, *Arch. Biochem. Biophys.,* 195, 248, 1979.

143. **Boulanger, W. A. and Katzenellenbogen, J. A.,** 5-(Halomethyl)-2-pyranones as irreversible inhibitors of α-chymotrypsin, *J. Med. Chem.,* 29, 1483, 1986.

144. **Graham, D. W., Ashton, W. T., Barash, L., Brown, J. E., Brown, R. D., Canning, L. F., Chen, A., Springer, J. P., and Rogers, E. F.,** Inhibition of the mammalian β-lactamase renal dipeptidase (dehydropeptidase-I) by (Z)-2-(acylamino)-3-substituted-propenoic acids, *J. Med. Chem.,* 30, 1074, 1987.

Chapter 6

ELIMINATION REACTIONS*

I. INTRODUCTION

Many of the inactivators whose mechanisms are initiated by elimination reactions are fluorine-containing compounds. Fluorine is unique among the halogens in two respects. First, its van der Waals radius (1.35 Å) is close in size to that of hydrogen (1.20 Å), and, therefore, steric factors are not important. Secondly, unlike the other halides, the strength of the C–F bond is such that S_N2 reactions generally proceed only under rigorous conditions;[1] elimination reactions involving fluorine, however, are quite favorable. For example, as a model for the enzymological activation of molecules containing C–F bonds, the hydrolysis of hydroxybenzotrifluorides and 5-difluoromethyluracils was carried out by Sakai and Santi.[2] The data show that S_N2 displacement is a highly unfavorable process, but when a carbanion is generated beta to the fluorine or through an extended π-system (Scheme 1), elimination of the fluoride ion is facile. These studies indicate the importance of fluorine in the design of mechanism-based enzyme inactivators that undergo elimination reactions.

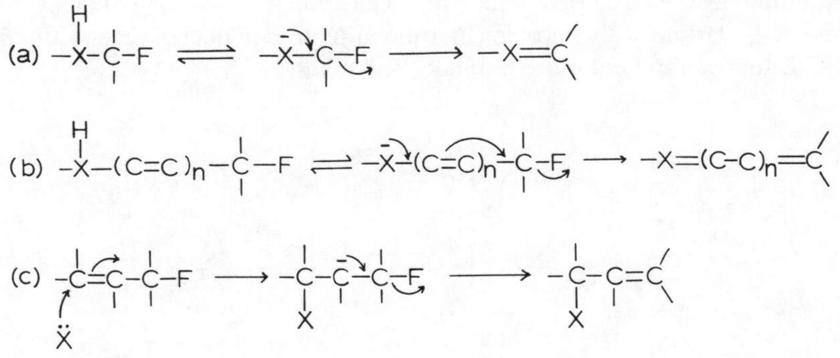

Scheme 1.

The syntheses of specific fluorine-containing inactivators are presented in the appropriate sections of this chapter and other chapters; however, several general methods for incorporation of fluorine into relevant molecules are shown here. Sulfur tetrafluoride in liquid hydrogen fluoride has been used by Kollonitsch et al.[3,4] to convert hydroxy amines and hydroxy amino acids to the corresponding fluoro amine and fluoro amino acids. Photofluorination with CF_3OF in HF of alkyl-substituted amino acids, e.g., alanine, has been described by Kollonitsch and co-workers.[5-7] A procedure for converting thiols to fluorides, termed fluorodesulfurization by Kollonitsch et al.,[8] involves treatment of a thiol-containing amino acid with CF_3OF in HF and either chlorine, *N*-chlorosuccinimide, or F_2. Diethylaminosulfur trifluoride is another useful reagent for the conversion of alcohols to fluorides,[9] ketones and aldehydes to *gem*-difluorides,[9] trimethylsilyl ethers to flourides,[9a] and carboxylic acids to acyl fluorides.[10] General syntheses of α-mono-, di-, and trifluoromethyl-substituted amino acids and amines were reported by Bey and co-workers.[11-14]

Before discussing the various elimination-initiated mechanism-based enzyme inactivation reactions, a comment is in order regarding the standard elimination/addition mechanism for

---

* A list of abbreviations and shorthand notations can be found prior to Chapter 1.

inactivation of PLP-dependent enzymes. This is a popular approach to inactivation of these enzymes and Scheme 2 (Pyr represents the pyridine ring of PLP) shows the previous universally accepted mechanism for this class of mechanism-based inactivators. Elimination of HX from the PLP-bound amino acid produces a PLP-bound aminoacrylate intermediate (Structural Formula 6.1), that can partition between hydrolysis to pyruvate (Scheme 2, pathway a) and Michael addition of an active site nucleophile (Scheme 2, pathway b). This was the widely-accepted mechanism for the inactivation of all PLP-dependent enzymes by amino acids containing a β-leaving group until Metzler and co-workers[15,16] proposed an alternative mechanism (the enamine mechanism, Scheme 3) on the basis of their results with glutamate decarboxylase and aspartate aminotransferase (see Section II.G.3.a). The important difference in the two mechanisms is that in the one shown in Scheme 2, an active site nucleophile is covalently bound to the inactivator directly (Structural Formula 6.2) whereas in Scheme 3, the nucleophile is attached to the PLP which is bound to the inactivator (Structural Formula 6.3). The enzyme-inactivator adduct in the former case would be stable to a variety of conditions; however, Metzler and co-workers[15,16] found that by increasing the pH to 11, a new species (Structural Formula 6.4) was released from their enzymes; an elimination mechanism was proposed (Scheme 4). Since this suggestion was presented, others have made similar observations,[17-19] and the mechanism shown in Scheme 3 may be much more prevalent than now appears. Because of the ubiquity of inactivators that undergo elimination/addition reactions, these three schemes will be referred to throughout the Elimination/Addition Section. The first section has been arranged according to the type of leaving group involved, starting with fluoride (first monofluoro compounds, then difluoro and trifluoro ones), followed by chloride, bromide, and other.

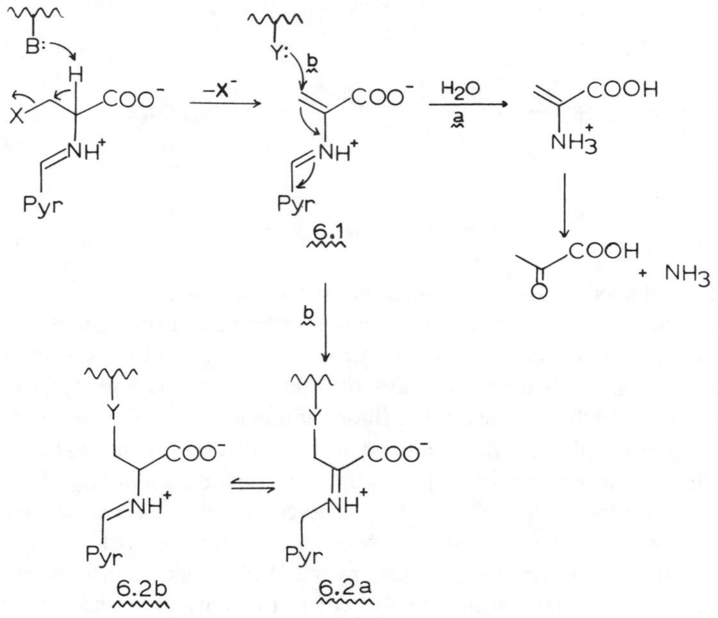

Scheme 2.   Containing Structural Formula 6.1 and 6.2a and b.

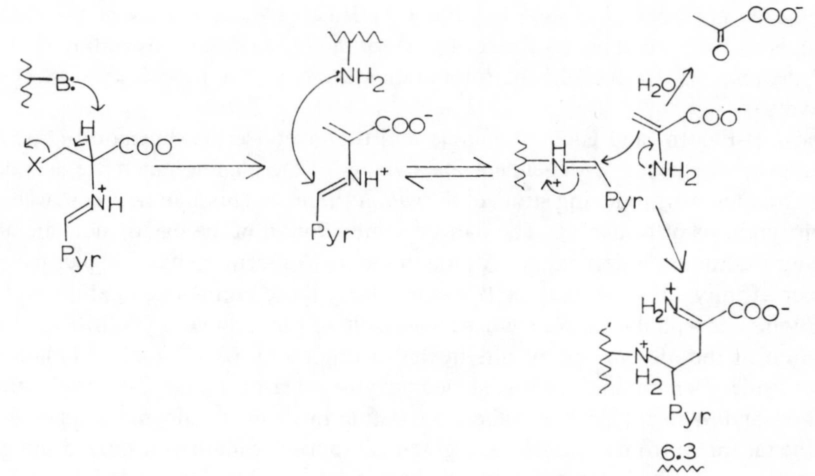

Scheme 3.   Containing Structural Formula 6.3.

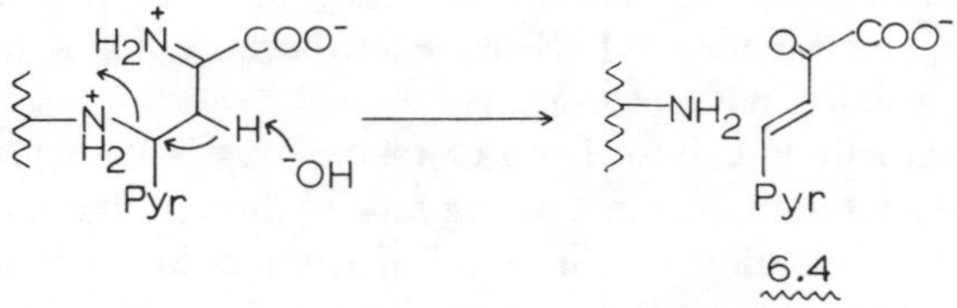

Scheme 4.   Containing Structural Formula 6.4.

II. ELIMINATION/ADDITION

A. Monofluorinated Inactivators

1. β-Fluoroalanine

β-Fluoroalanine was synthesized by photofluorination[6,7] and was reported by Kollonitsch and co-workers[7,20] to be an irreversible inactivator of bacterial alanine racemase. More detailed studies on the irreversible inactivation of *E. coli* B alanine racemase by both the D- and L-isomers of β-fluoroalanine and β-chloroalanine were carried out by Wang and Walsh.[21] All compounds partition between α,β-elimination to give pyruvate, ammonia, and halide ion (Scheme 2, pathway a) and inactivation; no racemization was observed. However, the partition ratios for all of the compounds are virtually identical (790 to 920). This suggests that a common intermediate, presumably the aminoacrylic acid-PLP complex (**6.1**, Scheme 2) is formed. To support this hypothesis, it was shown that *O*-carbamoyl-D-serine and *O*-acetyl-D-serine also had partition ratios similar to the halogen-substituted alanines (800 and 860). The L-isomers of the latter two compounds are not substrates or inactivators, but are competitive reversible inhibitors, suggesting steric interference of the bulky group that affects proton removal by the enzyme. The inactivation mechanism proposed by Wang and Walsh[21] is shown in Scheme 2 (pathway b).

The alanine racemase encoded by the *dadB* gene in *Salmonella typhimurium*, which functions in the catabolism of L-alanine, is inactivated by both D- and L-β-fluoroalanine.[17] The D-isomer has a K_m value 20 times lower than that for the L-isomer, yet the partition ratios are essentially the same (790 and 770, respectively). As with the *Escherichia coli* enzyme,[21] this suggests a common intermediate, namely, **6.1** (Scheme 2). However, with the use of β-chloroalanine, it was shown by Badet et al.[17] that the mechanism is that shown in Scheme 3, not Scheme 2 (see Section II.G.3.a. for more details of the mechanism). Similar results were obtained by Roise et al.[18] for the broad specificity amino acid racemase from *Pseudomonas striata*; the partition ratios were just a little higher (830 and 870, respectively).

D- and L-β-Fluoro- and β-chloroalanine and D-*O*-acetylserine, but not L-*O*-acetylserine, are pseudo first-order, time-dependent inactivators of the alanine racemase encoded by the *alr* gene from an overproducing strain of *S. typhimurium*.[21a] This alanine racemase is involved in the biosynthesis of D-alanine. The partition ratios for all of the inactivators are about 150, suggesting a common intermediate. Although the *dadB* racemase has a higher partition ratio and lower affinity for L-isomers of β-haloalanines, these compounds exhibit much higher inactivation rates with the *dadB* racemase than with the *alr* racemase. [^{14}C]D-β-Chloroalanine inactivation of the *alr* enzyme results in the incorporation of 1.2 mol of radioactivity per mol of enzyme. Denaturation of the labeled enzyme releases the radioactivity as **6.4**. Therefore, Esaki and Walsh[21a] proposed the inactivation mechanism shown in Scheme 3.

Alanine racemase from *P. putida*, and glutamate-pyruvate aminotransferase and glutamate-oxalacetate aminotransferase from pig heart are irreversibly inactivated by β-fluoro-D,L-alanine, but not by trifluoro-D,L-alanine or dichloro-D,L-alanine; gel filtration does not restore enzyme activity.[22] The enzymes in the pyridoxamine form are not affected by monohaloalanines. The aminotransferases and the racemase may be susceptible to inactivation as the result of formation of intermediate **6.1** (Scheme 2), because this is not an intermediate in the normal substrate reaction with enzyme.

Serine transhydroxymethylase from lamb or rabbit liver is known to catalyze transamination from D-alanine, but not L-alanine. Wang et al.[23] showed that this enzyme from either source also catalyzes elimination of fluoride ion from D- or L-fluoroalanine, producing pyruvate, fluoride, and ammonium ion, concomitant with inactivation of the enzyme. The rabbit enzyme produces 40 molecules of pyruvate or 44 molecules of fluoride ion for each inactivation; the lamb enzyme produces 67 and 60 molecules of pyruvate and fluoride ion, respectively, per inactivation event. Elimination of fluoride and inactivation are stimulated 500- to 700-fold by the addition of saturating tetrahydrofolate, a cosubstrate for the enzyme. D-Chloroalanine inactivates both enzymes at a slower rate than does D-fluoroalanine, but the partition ratio is the same as expected for the same intermediate. D-Difluoro- and D,L-trifluoroalanine both inactivate the enzyme at slower rates than does the monofluoro analogue, but have different partition ratios, 149 and 8, respectively. The rabbit enzyme is not inactivated by the L-isomers, but the lamb enzyme is. Inactivation of either enzyme by D-[^{14}C]-fluoroalanine results in incorporation of 0.9 equivalent of radioactivity per subunit. Titration of the enzyme cysteine groups with DTNB indicates the loss of one cysteine after inactivation. Acid hydrolysis of the labeled enzyme confirmed attachment to a cysteine residue.

β-Fluoro-D-alanine also inactivates the PLP-dependent 1-aminocyclopropane-1-carboxylate deaminase from *Pseudomonas* sp.[24] Other alanine derivatives with β-leaving groups also inactivate the enzyme (see the appropriate sections). Unlike inactivation of bacterial alanine racemase[21] and liver serine transhydroxymethylase,[23] where the partition ratios are the same, regardless of the leaving group, with 1-aminocyclopropane-1-carboxylate deaminase, the partition ratios are different (68, 190, and 300 for the β-chloro, β-fluoro, and β-acetoxy compounds, respectively). This suggests that the inactivating species is different for all three compounds and eliminates the aminoacrylate-PLP intermediate (**6.1**, Scheme 2) as that responsible for inactivation. The alternative mechanism (isomerization/substitution) proposed by Walsh et al.[24] is shown in Scheme 5. Because serine transhydroxymethylase can cleave the C_β-C_α bond of L-serine derivatives, and also abstract the α-proton of D-amino acids, leading to inactivation by β-fluoroalanine, this enzyme and 1-aminocyclopropane-1-carboxylate deaminase may be catalytically similar.

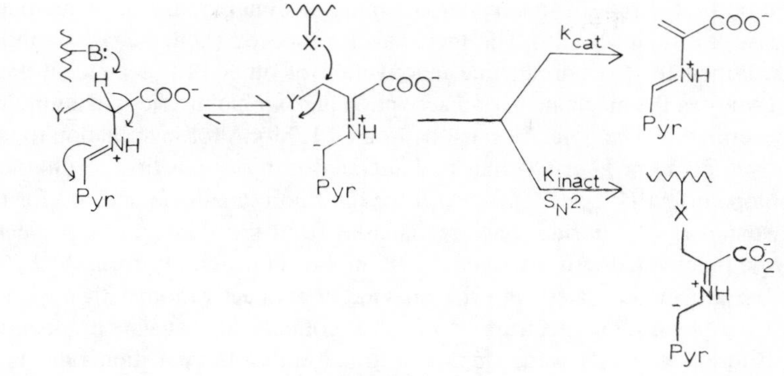

Scheme 5.

D-Amino acid aminotransferase from *B. sphaericus* shows a deuterium isotope effect of 2 to 3 for transamination of α-deuterio-D-alanine; however, no isotope effect was observed by Soper and Manning[25] for α-deuterio-β-fluoro-D,L-alanine. Since the rate of β-elimination, the partition ratio, and the K_I values are quite similar for β-fluoro-, β-chloro, and β-bromoalanine (the K_I values are much lower than the K_M for D-alanine), binding, α-proton removal, and halide elimination are not rate-determining. Solvolysis of the Schiff base intermediate is, therefore, the most likely rate-determining step.

2. 4-Amino-5-fluoropentanoic Acid (γ-Fluoromethyl GABA) and Related

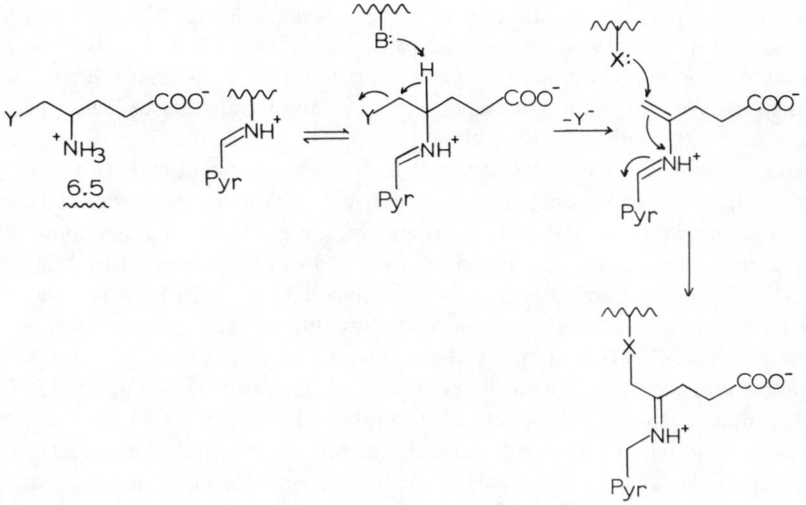

Scheme 6.   Containing Structural Formula 6.5.

Pig brain γ-aminobutyric acid aminotransferase, another PLP-dependent enzyme, was shown by Silverman and Levy[26] to be irreversibly inactivated by a series of (*S*)-4-amino-5-halopentanoic acids (Structural Formula 6.5, Y = F, Cl, and Br, Scheme 6). The bromo analogue undergoes nonenzymatic hydrolysis at a rate faster than enzyme inactivation at pH 8.5. The fluoro analogue has a K_I value 12 times lower than the K_m value for γ-aminobutyric acid (GABA). β-Mercaptoethanol has no effect on the inactivation rate; dialysis does not regenerate enzyme activity. (*S*)-[^{14}C]-4-Amino-5-chloropentanoic acid inactivation results in the incorporation of 1.7 mol of inactivator per mole of dimeric enzyme. Urea denaturation does not release the radioactivity. The mechanism proposed (Scheme 6) is similar to that previously described for β-fluoroalanine inactivation of other PLP-dependent enzymes. A more detailed study of the mechanism of inactivation of γ-aminobutyric acid aminotransferase by **6.5** was described.[27] The cofactor must be in the PLP form for inactivation to take place. Inactivation by **6.5** (Y = F or Cl) that is deuterated at the γ-position results in a kinetic deuterium isotope effect (k^H_{inact}/k^D_{inact}) of 5.5 for the fluoro analogue and 6.7 for the chloro analogue. Two moles of fluoride ions are released from the fluoro analogue per mole of dimeric enzyme inactivated. An average of 1.96 mol of radioactivity from the [^{14}C]-labeled chloro analogue are incorporated into the enzyme after inactivation with the concomitant conversion of the absorption spectrum of the PLP cofactor to that which appears to be of PMP. Four different methods were employed to show that the partition ratio is 0. All of these results are consistent with the proposed mechanism (Scheme 6), but they also would be consistent with a mechanism related to that described by Metzler and co-workers.[15,16] In order to differentiate these two mechanistic possibilities, inactivation of the enzyme reconstituted with [^{3}H]PLP was carried out by Silverman and Invergo.[19] The enzyme was inactivated by **6.5** (Y = F), then the pH was raised according to the procedure of Metzler and co-workers;[15,16] the expected adduct (Structural Formula 6.6, Scheme 7) resulting from a mechanism similar to that shown in Schemes 3 and 4 was obtained, thus supporting the enamine mechanism[15,16] for inactivation of an enzyme that catalyzes reactions on γ-amino acids. This mechanism is somewhat unexpected considering that the inactivator has a partition ratio of 0. Compound **6.5** (Y = F) exhibits no time-dependent inhibition of glutamate decarboxylase, alanine aminotransferase, or aspartate aminotransferase, but is a potent competitive reversible inhibitor of the former and a weak competitive inhibitor of the latter enzyme.[28] The series of compounds **6.5** (Y = F, Cl, and Br) was synthesized by Silverman and Levy[29] from L-glutamic by the route shown in Scheme 8. The fluoromethyl analogues of β-alanine (**6.7**, n = 1), γ-aminobutyrate (**6.5**) and δ-aminopentanoate (**6.7**, n = 3) were prepared (note Structural Formula 6.7, Scheme 9); all three compounds were shown by Bey et al.[30] to inactivate pig brain γ-aminobutyric acid aminotransferase. The potency of the inactivators decreases with increasing chain length. The best of the fluoro analogues are not as potent as γ-vinyl GABA (see Chapter 7 in Volume II, Section III.C.1.) or γ-ethynyl GABA (see Chapter 7 in Volume II, Section III.A.3.). None of the compounds inactivates mammalian glutamate decarboxylase, aspartate aminotransferase, alanine aminotransferase, or ornithine aminotransferase. Compound **6.7** (n = 1) was synthesized by Mathew et al.[31] (Scheme 10). A double bond was incorporated into **6.5** (Y = F) by Silverman et al.[32] to give (*S,E*)-4-amino-5-fluoropent-2-enoic acid (Structural Formula 6.8, Scheme 11). This compound inactivates GABA aminotransferase and no enzyme activity returns upon dialysis; β-mercaptoethanol does not protect the enzyme from inactivation. Compound **6.8** binds 50 times better than does the parent saturated compound (**6.5**, Y = F). No transamination occurs prior to inactivation, but five molecules of **6.8** are required to inactivate the enzyme (partition ratio of 4). The two most reasonable mechanisms for inactivation are those related to the ones shown in Schemes 2 and 3. In order to test the Michael addition pathway (Scheme 2), (*E*)-3-(1-aminocyclopropyl)-2-propenoic acid (Structural Formula 6.9) was synthesized as a cyclopropyl analogue of that intermediate which could, hypothetically, undergo the same type of inactivation mechanism (Scheme 12). However, this compound is a noncompetitive inhibitor and does not inactivate the enzyme. Therefore, no information was gained regarding the inactivation mechanism.

O=C–$CH_2CH_2COO^-$
|
CH
||
HC–Pyr

Scheme 7. Structural Formula 6.6.

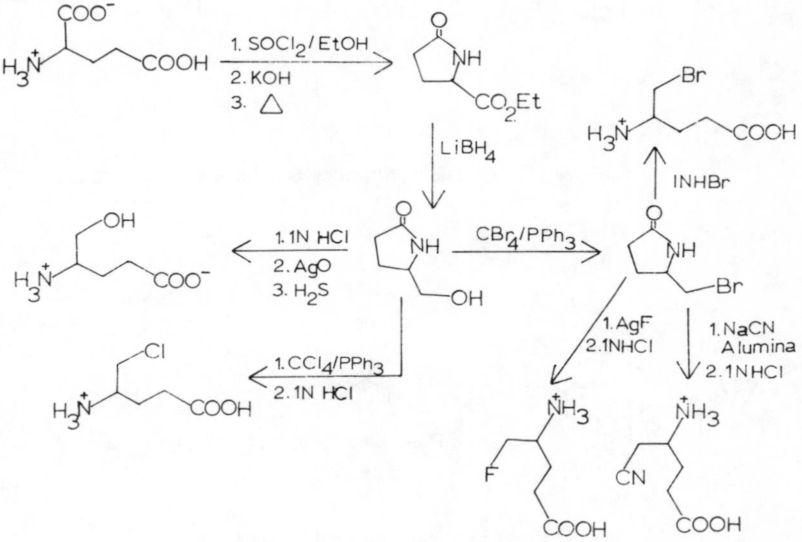

Scheme 8.

CH_2F
|
NH_3–$CH(CH_2)_nCOOH$

Scheme 9. Structural Formula 6.7.

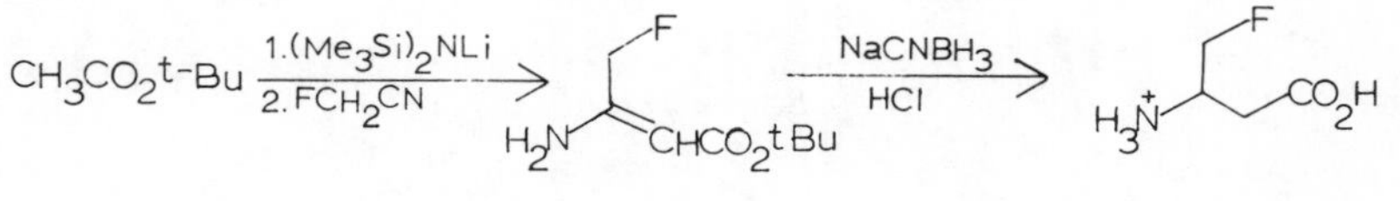

Scheme 10.

Compound **6.8** also was synthesized by Bey et al.[32a] (Scheme 12A), who found that it had a 17-fold lower K_I with pig brian GABA aminotransferase than does **6.5** (Y = F). Since this is a racemic mixture, the K_I for the (*S,E*) isomer would be comparable to that reported by Silverman et al.[32]

GABA aminotransferase from pig brain is irreversibly inactivated by 3-amino-4-chloro-4-fluorobutanoic acid (Structural Formula 6.9a; X = Cl, Y = H; Scheme 12B) and 3-amino-2,4-difluorobutanoic acid (**6.9a**; X = H, Y = F).[32b] The elimination/Michael addition mechanism (Scheme 2) or the elimination/enamine mechanism (Scheme 3) is suggested by Schirlin et al.[32b] A general synthetic scheme for the inactivators is shown in Scheme 12C.

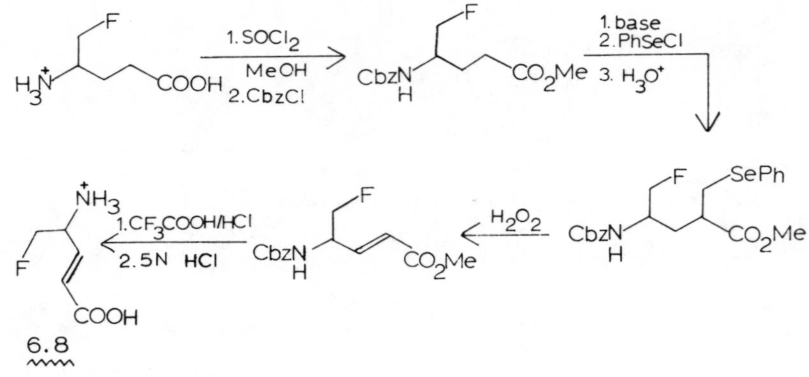

Scheme 11.   Containing Structural Formula 6.8.

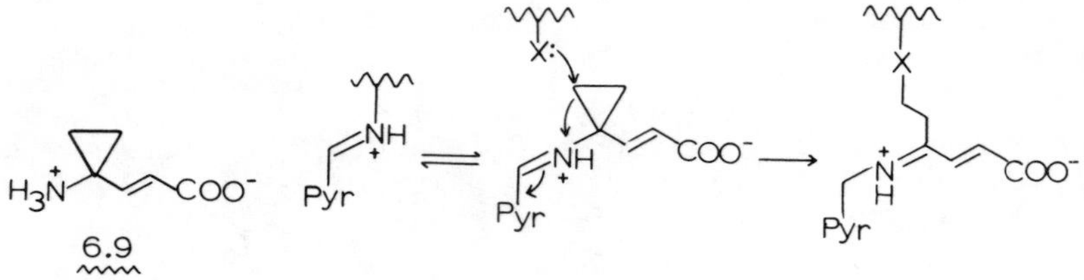

Scheme 12.   Containing Structural Formula 6.9.

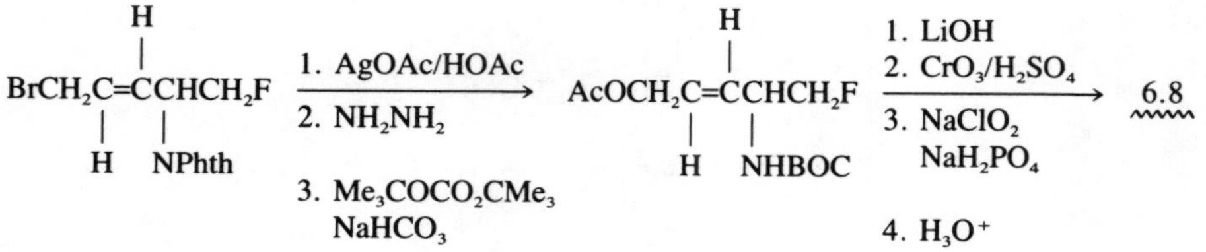

Scheme 12A.

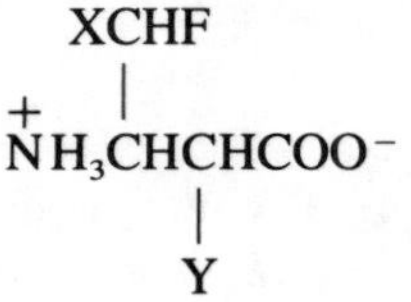

$$XCHF$$
$$\overset{+}{N}H_3CHCHCOO^-$$
$$Y$$

Scheme 12B.   Structural Formula 6.9a.

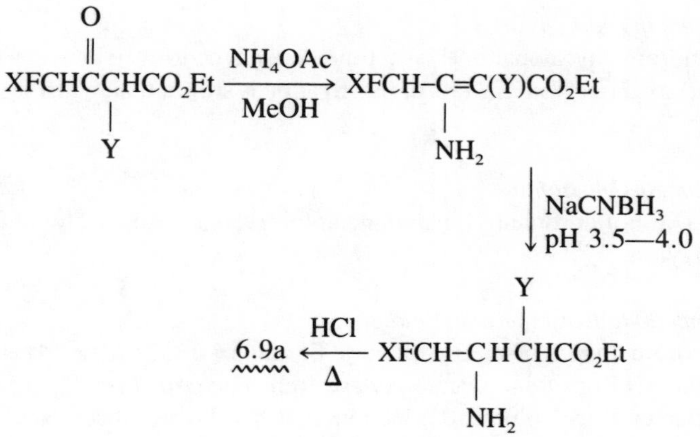

Scheme 12C.

In accord with the principle of microscopic reversibility, it should be possible to inactivate an aminotransferase in its PMP form with an appropriately substituted aldehyde or ketone. This approach was employed successfully by Lippert et al.[33] for the inactivation of GABA aminotransferase with 5-fluoro-4-oxopentanoic acid (Structural Formula 6.10, Scheme 13); only the PMP form of the enzyme is inactivated. Inactivation rate profiles parallel the GABA aminotransferase profiles. The presence of β-mercaptoethanol does not prevent inactivation. During inactivation, fluoride ion release was detected, but the amount was not stated. No inhibition of aspartate or alanine aminotransferase was observed, even at 10 mM concentration of **6.10**. The proposed mechanism (Scheme 13) proceeds via the same intermediate (Structural Formula 6.11) suggested by Silverman and Levy[26] for 4-amino-5-fluoropentanoic acid inactivation of this enzyme. However, since 4-amino-5-fluoropentanoic acid is now known[19] to inactivate GABA aminotransferase by the enamine mechanism[15,16] (Scheme 3), it is reasonable to suggest that **6.10** inactivates the PMP form of the enzyme from intermediate **6.11** by this same pathway. Compound **6.10** (also called 5-fluorolevulinic acid) can be synthesized by KF displacement of bromide from 5-bromolevulinate.[34]

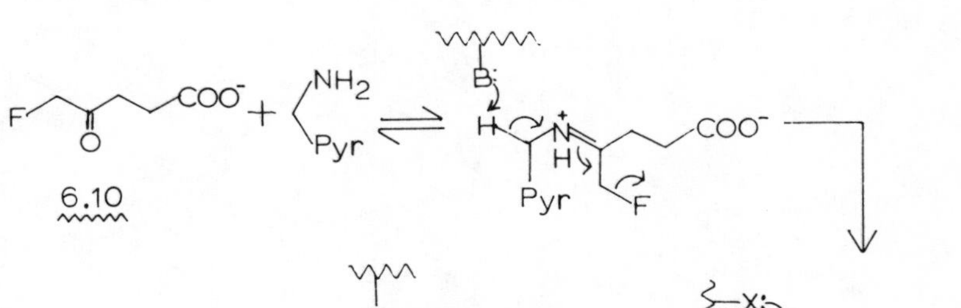

Scheme 13. Containing Structural Formulas 6.10 and 6.11.

3. α-(Monofluoromethyl)dopamine

(R)-α-(Monofluoromethyl)dopamine is a time-dependent irreversible inactivator of pig kidney dopa decarboxylase,[20] but the rate of inactivation is slower than with the corresponding dopa analogue.[34a]

4. α-(Monofluoromethyl)histamine

(R)- and (S)-α-(Monofluoromethyl)histamine are irreversible inactivators of fetal rat liver histidine decarboxylase.[20]

5. α-Monofluoromethylputrescine and Related

D,L-α-Monofluoromethylputrescine (Structural Formula 6.12, Scheme 14) is a potent time-dependent inactivator of ornithine decarboxylase from *Escherichia coli*; α-difluoromethyl-ornithine (see Chapter 8 in Volume II, Section II.B.5.) did not inactivate the enzyme at all.[35] Compound **6.12** has little effect on the ornithine decarboxylase from *Pseudomonas aeruginosa*. Therefore, α-difluoromethylornithine and D,L-monofluoroputrescine have complementary activities: the former inactivates ornithine decarboxylase from *P. aeruginosa*, but not from *E. coli*, and **6.12** has the reverse activities. Both compounds inactivate mammalian ornithine decarboxylase. Compound **6.12** also is a time-dependent irreversible inactivator of ornithine decarboxylase from *Trypanosoma brucei brucei*.[35a] This is another example of a mechanism-based inactivator designed with the principle of microscopic reversibility in mind. The product of decarboxylation of ornithine is putrescine; the final chemical step is believed to be protonation of the PLP-bound decarboxylated amino acid. In the reverse reaction, then, the first step is deprotonation of PLP-bound putrescine and that is presumed to be the step that initiates the elimination of fluoride from compound **6.12**. (Scheme 14). This compound also inactivates rat liver ornithine decarboxylase.[36] Because of the fluorine atom, the pKa of the β-amino group is low enough so that a reasonable population of only monoprotonated compound exists at physiological pH. Therefore, **6.12** is a substrate for monoamine oxidase (Scheme 15). Once it is converted to the amino aldehyde (**6.13**), aldehyde dehydrogenase oxidizes it to the corresponding carboxylic acid, i.e., 4-amino-5-fluoropentanoic acid (**6.14**), a known mechanism-based inactivator of γ-amino-butyric acid aminotransferase (see Section II.A.2.) Therefore, **6.12** is an example of a multi-enzyme activated inactivator (see Chapter 1, Section II.D.). A synthesis of **6.12** is shown in Scheme 16.[36,37]

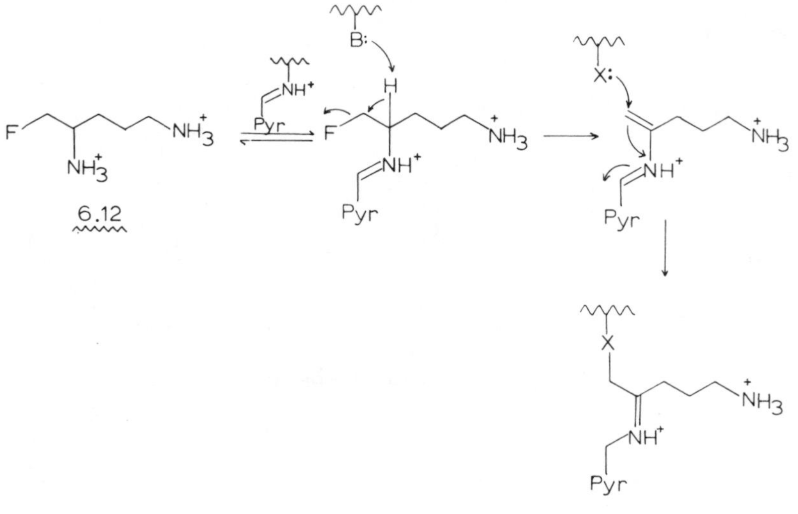

Scheme 14.   Containing Structural Formula 6.12.

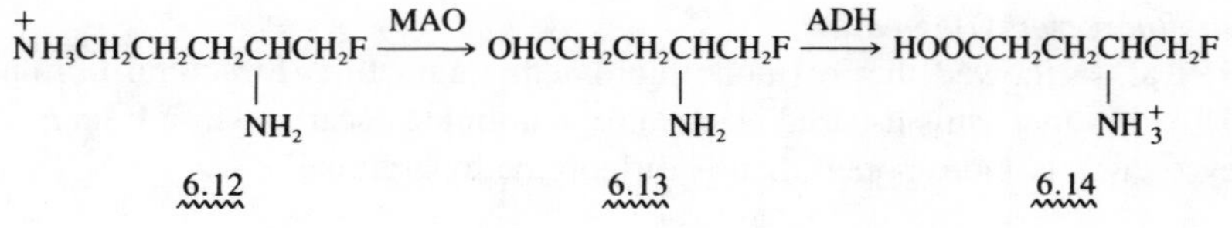

Scheme 15.   Containing Structural Formulas 6.12 to 6.14 in conversion.

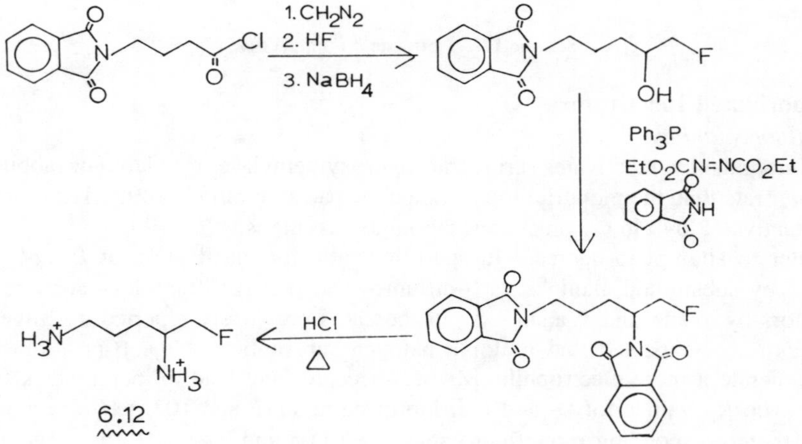

Scheme 16.

The unsaturated analogue of **6.12**, namely (*E*)-2,5-diamino-1-fluoropent-3-ene (Structural Formula **6.15**) was synthesized by Bey et al.[38] (Scheme 17). Compound **6.15** proved to be a mechanism-based inactivator of ornithine decarboxylase from livers of thioacetamide-treated rats; dialysis does not regenerate enzyme activity. It also inactivates *Trypanosoma brucei brucei* ornithine decarboxylase.[35a] In comparison with the saturated analogue (**6.12**), the incorporation of the double bond results in a sharp increase in the rate of inactivation with little effect on the K_1.[38]

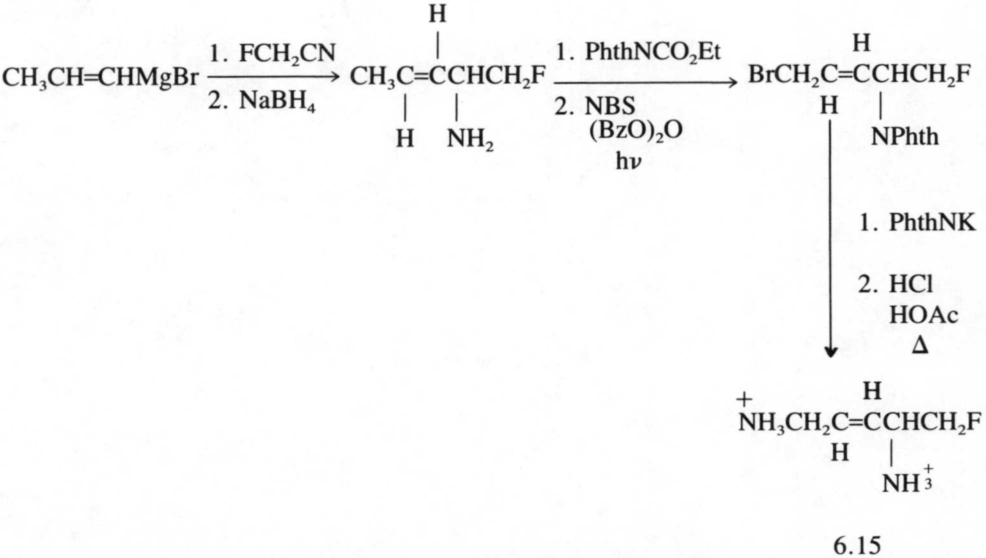

Scheme 17.   Containing Structural Formula 6.15.

6. α-(Monofluoromethyl)agmatine

Bitonti et al.[38a] showed that α-(monofluoromethyl)agmatine (Structural Formula 6.15a, Scheme 17A) is a mechanism-based inactivator of arginine decarboxylase from *E. coli,* oats, and barley, which is more potent than α-difluoromethylarginine.

$$
\begin{array}{ccc}
\text{FCH}_2 & & \text{NH} \\
| & & \| \\
\text{NH}_2\text{CHCH}_2\text{CH}_2\text{CH}_2\text{NHCNH}_2 &
\end{array}
$$

Scheme 17A.   Structural Formula 6.15a.

B. Difluorinated Inactivators

1. D-Difluoroalanine

D-Difluoroalanine inactivates serine transhydroxymethylase from lamb or rabbit liver, but at a slower rate than the monofluoro analogue; the partition ratio is 149.[23] The rabbit enzyme is not inactivated by the L-isomer, but the lamb enzyme is.

In order to attempt to decrease the partition ratio for inactivation of *E. coli* B alanine racemase by substituted alanines, β-β-difluoro- and β,β,β-trifluoroalanine were tested as inactivators by Wang and Walsh.[39] As in the cases previously reported by Silverman and Abeles,[22] it was expected that multiple halogen substitution at the β-position of alanine would generate a more electrophilic Michael acceptor and lead to increased rates of inactivation. The K_m values for D- and L-difluoroalanine (116 and 102 mM) are much greater than for the corresponding monofluoro analogue (0.05 and 2.4 mM); there also is a major difference in V_{max} and partition ratio values for all of these compounds. Dialysis of enzyme inactivated by difluoroalanine results in return of enzyme activity (45% overnight); a mechanism to account for these observations is shown in Scheme 18.[39]

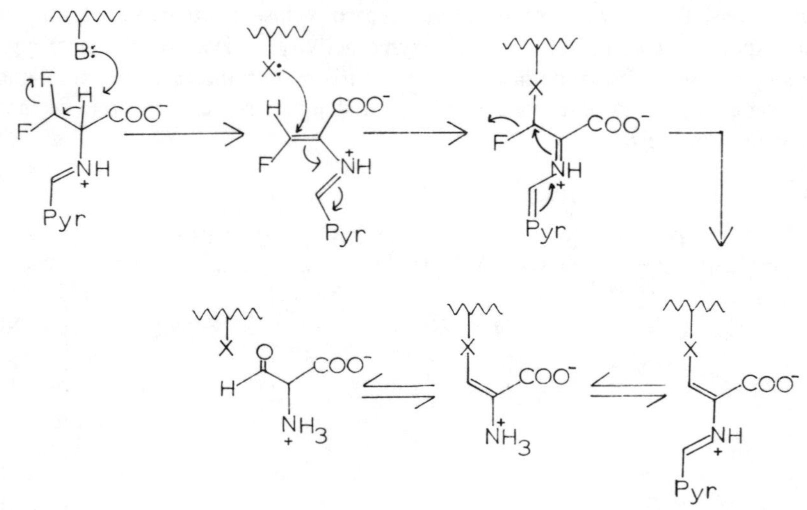

Scheme 18.

2. α-(Difluoromethyl)dopamine

α-(Difluoromethyl)dopamine inactivates pig kidney dopa decarboxylase.[12,20]

3. 4-Amino-5,5-difluoropentanoic Acid (γ-Difluoromethyl GABA) and Related

The difluoromethyl analogues of β-alanine (**6.16**, n = 1), γ-aminobutyrate (**6.16**, n = 2) and δ-aminopentanoate (**6.16**, n = 3) were prepared (note Structural Formula 6.16, Scheme 19) as potential inactivators of pig brain GABA aminotransferase.[30] All three of the analogues are time-dependent inactivators, but the latter compound does not exhibit saturation kinetics. The activity of the inactivator decreases with increasing chain length and with increasing fluorination. Only the (−)-isomer of **6.16** (n = 1) is an inactivator at a concentration of $1mM$.[32b] 3-Amino-3-deuterio-4,4-difluorobutanoic acid has no effect on the k_{inact}, but has an isotope effect of 4.5 on the K_I. A synthesis of this analogue is shown in Scheme 19A.

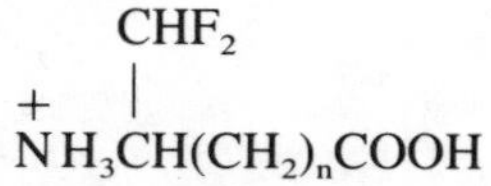

$$CHF_2$$
$$\overset{+}{N}H_3CH(CH_2)_nCOOH$$

Scheme 19.   Structural Formula 6.16.

$$CO_2t\text{-}Bu \qquad CHF_2$$

MeO$_2$CCH$_2$CH → MeO$_2$CCH$_2$CH

CO$_2$t-Bu

1. NaH
2. ClCHF$_2$
3. TFA
4. HOAc/Δ

CO$_2$H

1. SOCl$_2$
2. NaN$_3$
3. MeOH/Δ

$$CHF_2$$

HO$_2$CCH$_2$CHNH$_3^+$ ← MeO$_2$CCH$_2$CH

HCl/HOAc
Δ

NHCO$_2$Me

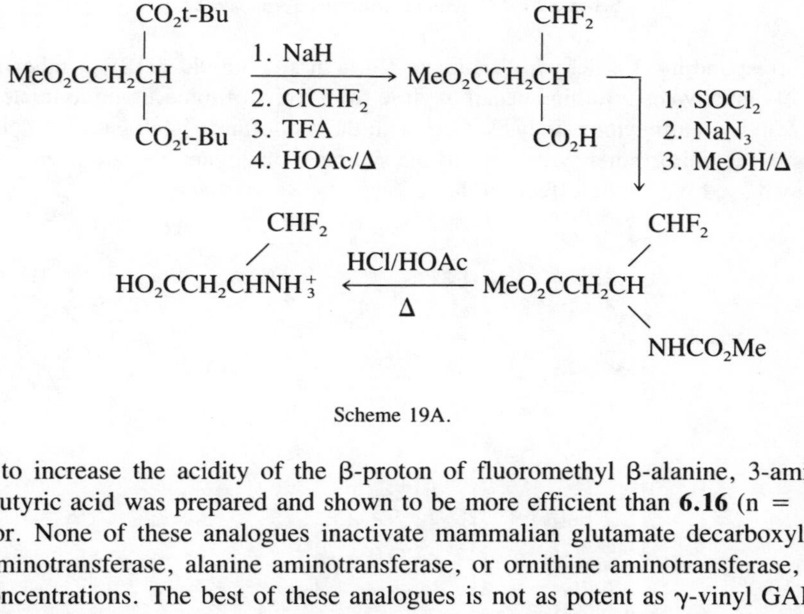

Scheme 19A.

In order to increase the acidity of the β-proton of fluoromethyl β-alanine, 3-amino-2,4-difluorobutyric acid was prepared and shown to be more efficient than **6.16** (n = 1) as an inactivator. None of these analogues inactivate mammalian glutamate decarboxylase, aspartate aminotransferase, alanine aminotransferase, or ornithine aminotransferase, even at 1 mM concentrations. The best of these analogues is not as potent as γ-vinyl GABA (see Chapter 7 in Volume II, Section III.C.1.) or γ-ethynyl GABA (see Chapter 7 in Volume II, Section III.A.3.).

4. α-Difluoromethylputrescine Related

α-Difluoromethylputrescine (Structural Formula 6.17), synthesized by Bey and Schirlin[12] (Scheme 20), was shown to produce time-dependent inactivation of rat liver ornithine decarboxylase.[36] No return of enzyme activity was observed upon dialysis; dithiothreitol in the buffer had no effect on the inactivation rate. The rate of inactivation by 4-deuterio-5,5-difluoropentane-1,4-diamine ([4-^{2}H]-α-difluoromethylputrescine) is the same as the 4-protio compound, indicating that there is no kinetic isotope effect on C−H bond breakage. However, the K_I increases by 3-fold when deuterium is incorporated into the inactivator. This is similar to the inactivation of ornithine decarboxylase by α-deuterio-α-ethynylputrescine and of glutamate decarboxylase by γ-deuterio-γ-ethynyl-γ-aminobutyric acid (see Chapter 7 in Volume II, Section III.A.3.) and arises from the effect of deuterium on the rate constant for hydrogen abstraction since the K_I is a function of this catalytic constant. Carbon-hydrogen bond breakage is not rate determining; possibly elimination of fluoride ion is. As with α-monofluoromethylputrescine, **6.17** is a multi-enzyme activated inactivator of γ-aminobutyric acid aminotransferase (see Section II.A.5.). Compound **6.17** also inactivates *Trypanosoma brucei brucei* ornithine decarboxylase.[35a]

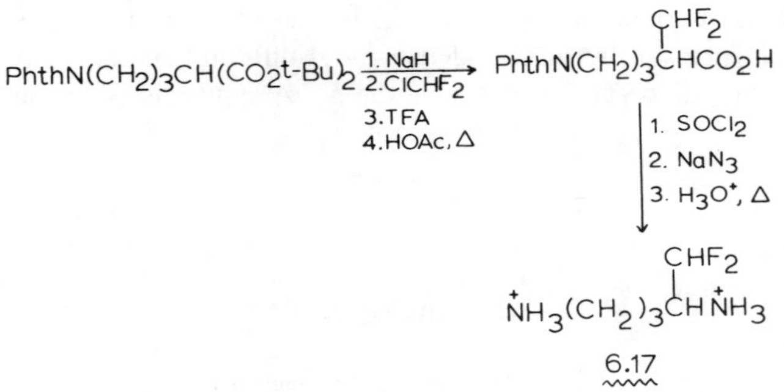

Scheme 20.   Containing Structural Formula 6.17.

The corresponding 3,4-dehydro analogue (Structural Formula 6.18), synthesized as in Scheme 21, inactivates ornithine decarboxylase from livers of thioacetamide-treated rats; no enzyme activity returns upon dialysis.[38] As with the monofluoro analogue, the double bond increases the inhibitor potency relative to the saturated analogue; the rate of inactivation is sharply increased with little effect on the K_I.

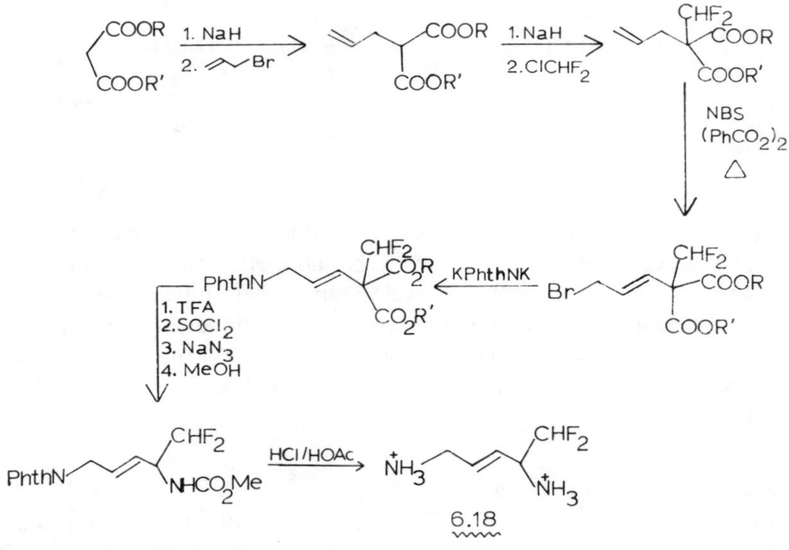

Scheme 21.   Containing Structural Formula 6.18.

C. Trifluorinated Inactivators
1. β,β,β-Trifluoroalanine

Scheme 22.   Structural Formula 6.19.

Several pyridoxal phosphate-dependent enzymes that catalyze β- and γ-elimination re-actions, namely, β-cystathionase (*E. coli*), γ-cystathionase (rat liver), tryptophanase (*E. coli*), tryptophan synthase (β₂ subunit and α₂β₂, both from *E. coli*), and threonine dehydrase (*E. coli*), undergo time-dependent inactivation by β,β,β-trifluoroalanine (Structural Formula 6.19, Scheme 22).[22] Enzyme activity is not regenerated by gel filtration. The apoenzyme of tryptophanase was prepared, and shown to be unaffected by the inactivators. Inactivation of γ-cystathionase by [1-^{14}C]trifluoroalanine leads to incorporation of 2 mol of label per mole of tetrameric enzyme, suggesting half-of-the-sites reactivity. No radioactive nonamines are generated, which indicates that the partition ratio is 0. Acid denaturation of the labeled enzyme results in release of radioactivity as $^{14}CO_2$. A mechanism was proposed by Silverman and Abeles[22] to account for these results and is shown in Scheme 23. None of these enzymes is inactivated by monohaloalanines; instead, these enzymes catalyze elimination of the halide to the PLP-bound 2-aminoacrylate, the same intermediate generated by these enzymes with their normal substrates. The inactivation of these enzymes by polyhaloalanines, however, can be rationalized in two ways. First, Michael acceptors are activated by substitution at the γ-position by electron-withdrawing substituents such as halogens. This may be enough activation to allow the addition to occur. Alternatively, Michael addition of an active site nucleophile to the PLP-bound 2-aminoacrylate intermediate generated from substrates may occur, but is reversible. When a halogen is substituted, however, addition leads to halide elimination, an irreversible reaction. In contrast with these results, enzymes which catalyze transamination reactions (glutamate-pyruvate aminotransferase and glutamate-oxaloacetate aminotransferase, both from pig heart) **are** inactivated by monohaloalanines (see Section II.A.1.), but not by the polyhaloalanines; the polyhaloalanines either do not bind or depro-tonation is too slow. The mechanism of the inactivation of rat liver γ-cystathionase by β,β,β-trifluoroalanine was investigated by Silverman and Abeles[40] in detail. No radioactivity from [1-^{14}C] inactivator is incorporated into enzyme which does not have intact PLP. The rate of incorporation of radioactivity into the enzyme, the rate of loss of enzyme activity, and the rate of fluoride ion release from the inactivator are all equal. Complete inactivation corre-sponds to release of about 3 fluoride ions, a result consistent with a partition ratio of 0. It was noted above[22] that denaturation leads to decarboxylation. When denaturation is carried out in $^{3}H_2O$, tritium is incorporated into the enzyme, as predicted for a decarboxylation reaction. Acid hydrolysis of the tritiated enzyme produces tritiated glycine as expected from Scheme 23. The tritiated protein was tryptic digested and tritiated peptides were subjected to different reagents in order to identify the type of carboxylic acid derivative linkage (amide, ester, anhydride, or thioester) that is involved. From these reactions, it was suggested[40] that the linkage is that of an amide, probably involving a lysine residue. Cyanogen bromide cleavage of the labeled enzyme, which was carried out by Fearon et al.,[41] gave a heptapeptide with the sequence **6.20** (Scheme 24).

This sequence is the same as the sequence obtained by reduction and cyanogen bromide cleavage of native enzyme (except with reduced pyridoxal in place of the Gly). Therefore β,β,β-trifluoroalanine labels the lysine residue that is bound to the active site pyridoxal phosphate. It also was proposed that this lysine residue functions as a proton transfer agent in the enzyme.

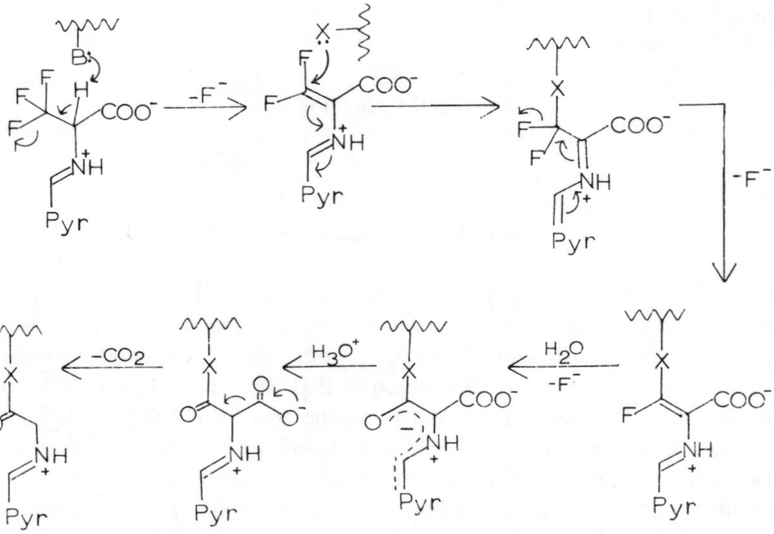

Scheme 23.

Cys–Ser–Ala–Thr–Lys–Tyr–Met
|
Gly

Scheme 24.   Sequence 6.20.

It was previously shown by Abeles and Walsh[42] that [^{14}C]propargylglycine also inactivates rat liver γ-cystathionase with incorporation of 2 mol of inactivator per enzyme (see Chapter 7 in Volume II, Section VII.B.1.). Inactivation by trifluoroalanine prevents incorporation of propargylglycine.[40] However, if the enzyme is first inactivated by propargylglycine, [1-^{14}C]trifluoroalanine is incorporated to the same extent as with native enzyme, but at one-third the rate. Three fluoride ions also are released during incorporation of the inactivator. Concomitant with inactivation by propargylglycine followed by trifluoroalanine is a large increase in absorbance at 519 nm. This is attributed by Silverman and Abeles[40] to a carb-anionic adduct. They suggest that both inactivators occupy the same two subunits and that propargylglycine, which becomes attached to an amino acid residue at its γ-position, can move out of the way in order for trifluoroalanine to undergo turnover and attachment at its β-position (Scheme 25).

β,β,β-Trifluoro-D,L-alanine also inactivates the alanine racemase from *E. coli* B, but the K_m could not be determined; the partition ratio is 10 ± 2.[39] No enzyme activity returns after dialysis, suggesting a mechanism similar to that shown is Scheme 23.

Liver serine transhydroxymethylase also is inactivated by β,β,β-trifluoro-D,L-alanine, but at a slower rate than the monofluoro analogue; the partition ratio is 8.[23]

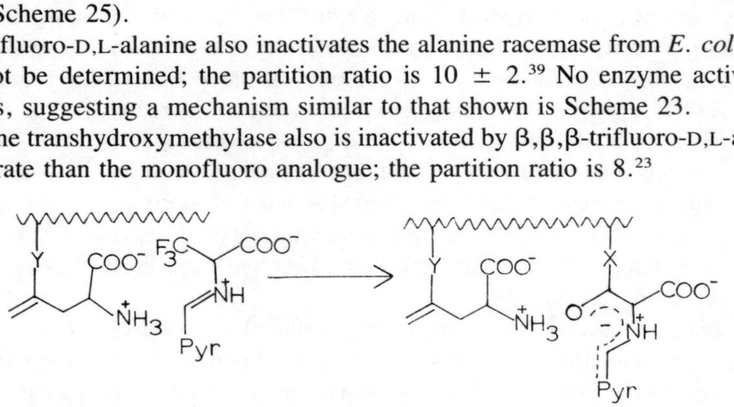

Scheme 25.

2. α-(Trifluoromethyl)dopamine

α-(Trifluoromethyl)dopamine inactivates pig kidney dopa decarboxylase.[20]

3. α-(Trifluoromethyl)histamine

Another example of inactivation based on the principle of microscopic reversibility is the inactivation of hamster placental histidine decarboxylase by α-trifluoromethylhistamine (Structural Formula 6.21), which exhibits nonpseudo first-order kinetics.[43] The mechanism proposed by Metcalf et al.[43] is shown in Scheme 26. The synthesis of **6.21**[43] is outlined in Scheme 27.

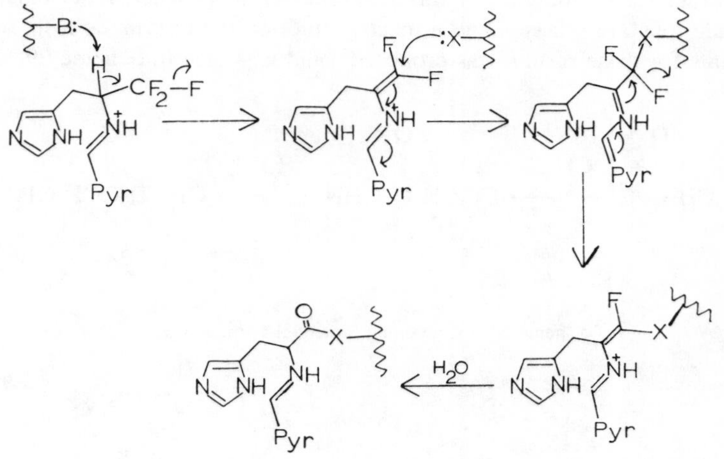

Scheme 26.

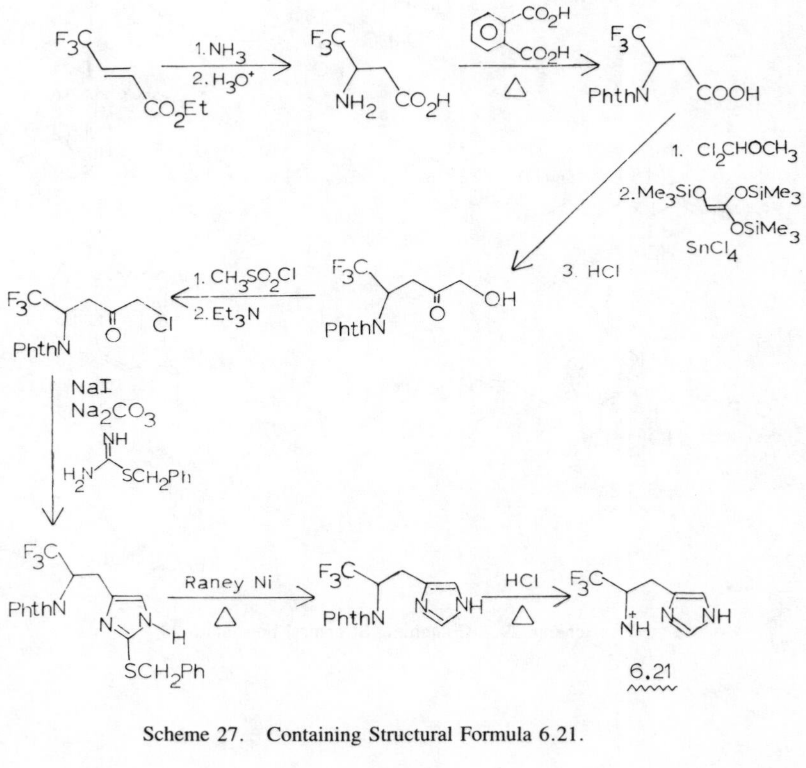

Scheme 27.   Containing Structural Formula 6.21.

4. 2-Keto-4,4,4-trifluorobutyl Phosphate

2-Keto-4,4,4-trifluorobutyl phosphate (Structural Formula 6.22) was synthesized by Magnien et al.[44] (Scheme 28) as an inactivator of fructose-1,6-bisphosphate aldolase from rabbit muscle. This compound produces time-dependent inactivation that is pseudo first-order; the presence of thiols in the buffer does not affect the rate of inactivation. Both dihydroxyacetone phosphate and glyceraldehyde 3-phosphate protect the enzyme from inactivation. When glyceraldehyde 3-phosphate is in the inactivation buffer, a compound that chromatographs on Dowex® 1 like fructose 1,6-bisphosphate was isolated. It contains fluorine, but its structure was not determined; the condensation product of the inactivator with glyceraldehyde 3-phosphate (Structural Formula 6.23) was suggested as the product. After inactivation, one to two fluoride ions are released and a reactive thiol on the enzyme is lost. A mechanism which accounts for these results was proposed[44] and is shown in Scheme 29.

$$CF_3CH_2\overset{\overset{O}{\|}}{C}{-}Cl \xrightarrow{CH_2N_2} CF_3CH_2\overset{\overset{O}{\|}}{C}{-}CHN_2 \xrightarrow{H_3PO_4} \underset{\underset{\textstyle 6.22}{}}{CF_3CH_2\overset{\overset{O}{\|}}{C}CH_2OPO_3H}$$

Scheme 28. Containing Structural Formula 6.22.

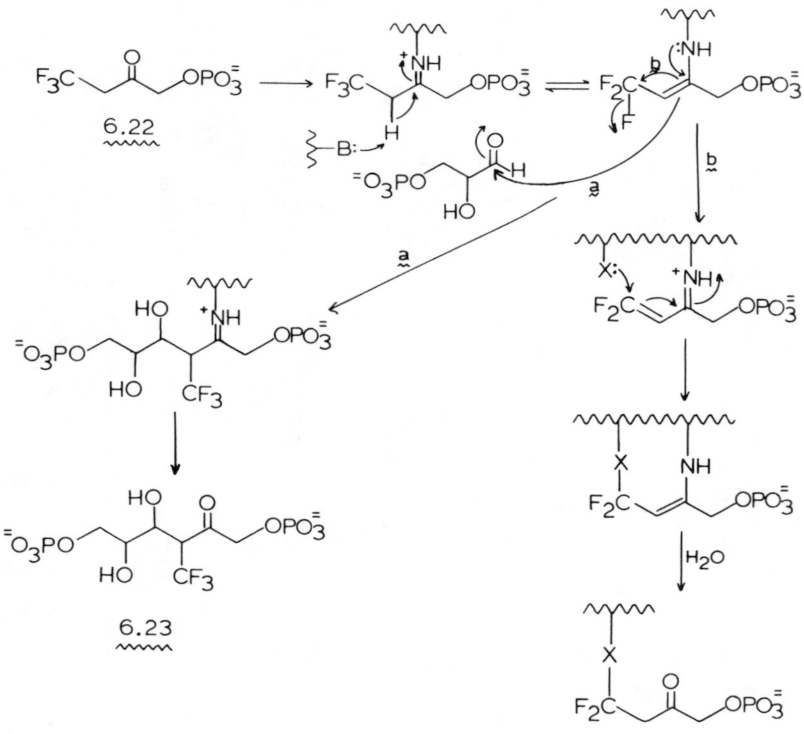

Scheme 29. Containing Structural Formula 6.23.

D. Monochlorinated Inactivators

1. β-Chloroalanine and Related

β-Chloro-L-alanine is both a substrate and inactivator for L-aspartate β-decarboxylase from *Alcaligenes faecalis* (strain N).[45] Time-dependent inactivation occurs with [U-^{14}C]- and [3-^{14}C]-β-chloro-L-alanine with concomitant incorporation of one mole of [^{14}C] per 60,000 g of enzyme. The radioactivity remains bound after dialysis, heat denaturation, or acid denaturation. *threo*-3-Chloro-L-α-aminobutyrate gives similar inactivation results, albeit at a substantially slower rate. No inactivation by either compound results with apoenzyme (followed by reconstitution with PLP). Treatment of the enzyme labeled with [1-^{14}C]β-chloroalanine with ceric sulfate leads to loss of 30% of the radioactivity as ^{14}CO$_2$; treatment with ninhydrin results in a 20% loss of ^{14}CO$_2$. Tate et al.[45] suggested that the enzyme adduct has multiple structures; 33% as an α-ketoacid and 20% as an α-amino acid (see **6.2a** and **6.2b** in Scheme 2). Further studies by Relyea et al.[46] on the structure of the enzyme adduct labeled with β-chloro-L-[^{14}C]alanine were carried out. When enzyme containing [^{32}P]PLP was used, dialysis of the inactivated enzyme resulted in release of about half of the cofactor without release of [^{14}C] from the enzyme. Cyanogen bromide digestion of the [^{14}C]-labeled enzyme results in a radioactive peptide. Treatment of this peptide with alkali or acid hydrolysis produces β-hydroxy[^{14}C]pyruvate. Ammonolysis releases the radioactivity and produces a new amino acid in the peptide, namely, glutamine. These results suggest that a glutamate residue is alkylated during inactivation and that ammonolysis converts the glutamate ester to glutamine. It was suggested that release of half of the cofactor indicates an aldimine-ketimine equilibrium in which release of the PMP from the ketimine form further stabilizes the aldimine form (Scheme 30). Treatment of β-chloro-L-[U-^{14}C]alanine-labeled enzyme with ceric sulfate resulted in release of 57% of the label as ^{14}CO$_2$ not 30%, as previously reported.[45] Treatment with ninhydrin did not give any ^{14}CO$_2$; previously 20% release was reported,[45] but this was shown to be due to free [^{14}C]chloroalanine. Therefore, half of the cofactor is released and half of the [^{14}C] adduct is an α-ketoacid. These results are consistent with an equal mixture of ketimine and aldimine forms.

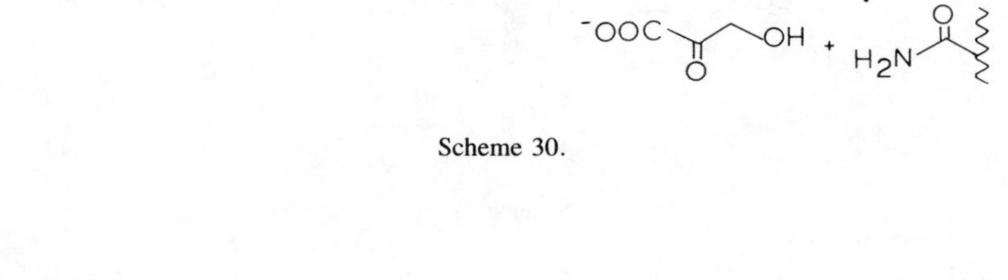

Scheme 30.

β-Chloro-L-alanine-induced inactivation of aspartate aminotransferase isozymes from pig heart was greatly accelerated by the presence of formate ions.[47,82] John and Tudball,[82] studying the cytoplasmic enzyme, proposed an anionic binding site for the β-carboxyl group of aspartate which is not occupied by β-chloroalanine. When formate is added, it binds to that site in the presence of β-chloroalanine. Both groups, Morino and Okamoto[47] and John and Tudball,[82] suggest that this binding results in a conformational change of the enzyme that aligns the appropriate active site base responsible for α-proton abstraction, thereby accelerating the rate of inactivation. Acetate and propionate are less effective at enhancing the β-chloroalanine inactivation rate with the soluble enzyme, and ineffective with the mitochondrial enzyme.[48] Formate does not affect binding of the inactivator, only the rate. α,β-Elimination of HCl and enzyme inactivation involve the same intermediate. Inactivation by β-chloro-L-[U-¹⁴C]alanine in the presence of 3 *M* formate, followed by borohydride reduction, results in incorporation of 1 mol of inactivator per monomer; the cofactor is covalently bound to the inactivator as well. The labeled enzyme complex, prior to borohydride treatment, is stable to dialysis, but decomposes upon denaturation, suggesting a labile covalent attachment. Inactivation of pig heart cytoplasmic aspartate aminotransferase by β-chloro[U-¹⁴C]-L-alanine results in the incorporation of one equivalent of radioactivity per enzyme molecule.[49] Trypsin digestion of the borohydride-reduced, inactivated enzyme produces a peptide containing the radioactivity and the phosphopyridoxyl group and having a positive circular dichroism; linkage of the ε-amino group of lysine to the radioactive fragment was demonstrated by Morino and Okamoto.[49] This modified lysine is identical to the one involved in pyridoxal phosphate binding to the apoenzyme. The mechanism proposed (Scheme 31) is an isomerization/substitution mechanism, but the mechanisms shown in Schemes 2 and 3 also are reasonable (see Section II.G.3.a. for evidence supporting the mechanism in Scheme 3). This same study also was carried out by Morino and Tanase[50] on the pig heart mitochondrial aspartate aminotransferase isozyme; very similar observations were made for both isozymes. The ε-amino group of Lys-250 is attached to the inactivator; in the case of the cytosolic enzyme, Lys-258 is attached. These are the lysine residues involved in Schiff base formation with PLP. The NaBH₄-reduced inactivated enzyme was tryptic digested; the radioactive peptide contains both the phosphopyridoxyl group and the 3-carbon moiety of the inactivator. This tryptic peptide and that obtained from the cytosolic enzyme have a considerable degree of homology.

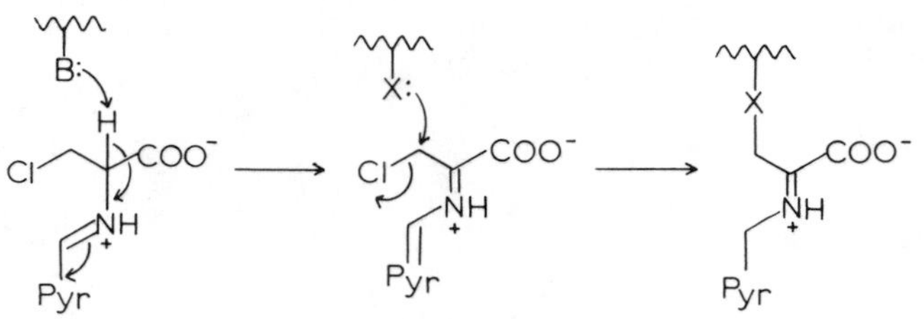

Scheme 31.

Pig heart alanine aminotransferase also is irreversibly inactivated by β-chloro-D,L-alanine[22] and by β-chloro-L-alanine.[51] Concomitant with inactivation there is a bleaching of the pyridoxal phosphate absorbance spectrum, which mostly is regenerated about 12 hr later **without** regeneration of enzyme activity.[51] The same observation was made by Morino et al.[48] for the inactivation of aspartate aminotransferase by chloroalanine. These observations are consistent with an inactivation mechanism similar to that proposed for the inactivation of aspartate aminotransferase by L-serine *O*-sulfate, namely, that shown in Scheme 3 (see Section II.G.3.a. for details). Morino et al.[52] also showed that β-chloro-L-alanine is a time-dependent inactivator of alanine aminotransferase from pig heart. The whole carbon fragment of β-chloro-L-[U-^{14}C]alanine becomes incorporated into the enzyme stoichiometrically. These results are similar to those found previously with aspartate aminotransferases.[48] A 325-nm absorption band results from addition of the inactivator; slowly, a 435-nm band appears at a rate which coincides with the rate of enzyme inactivation. The V_{max}/K_m values for β-chloro-L-alanine in the α,β-elimination reaction catalyzed by the cytosolic and mitochondrial aspartate aminotransferases and by alanine aminotransferase are 1.1 and 7 mM^{-1} min^{-1} for the cytosolic and mitochondrial aspartate aminotransferases,[48] respectively, and 5000 mM^{-1} min^{-1} for alanine aminotransferase. Whereas the presence of formate ion results in a major enhancement of both elimination and inactivation rates for the aspartate aminotransferase isozymes with β-chloro-L-alanine, only a 1.4-fold increase in α,β-elimination with a slight decrease in inactivation rate was observed with L-alanine aminotransferase. The 435-nm absorption band was proposed[52] to be derived from complex **6.24**. Upon standing, this band decreases and the 335-nm band increases, an observation similar to that made by Golichowski and Jenkins.[51] This new complex was suggested to be either **6.25** or **6.26** (see Scheme 32). However, the studies by Metzler and co-workers,[15,16] described in Section II.G.3.a., support the mechanism shown in Schemes 3 and 4 and also can account for these spectral changes.

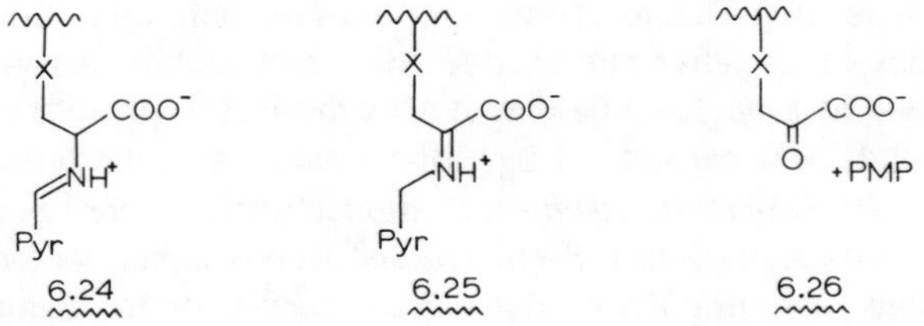

Scheme 32.   Structural Formulas 6.24 to 6.26.

D-Amino acid aminotransferase is important in the biosynthesis of bacterial D-amino acids and, therefore, is a target for antimicrobial agents. The enzyme from *B. sphaericus* catalyzes the elimination of HCl from β-chloro-D-alanine to give pyruvate, ammonia, and chloride ion.[53] One out of 1500 turnovers, however, the enzyme is inactivated without transamination. The L-isomer is neither a substrate nor an inactivator. The formation of products from β-chloro-D-alanine is not linear as a result of inactivation. Although reduced glutathione does not prevent inactivation of the enzyme by β-chloro-D-alanine, several other smaller thiols do prevent inactivation. However, in addition to preventing inactivation, they also prevent the production of pyruvate, even though β-chloro-D-alanine continues to be consumed. With β-mercaptoethylamine, the D-isomer of S-(β-aminoethyl)cysteine is produced stereospecifically, indicating that the reaction, probably a Michael addition to the aminoacrylate-PLP Schiff base, takes place inside the active site. The inactivation mechanism shown in Scheme 2 was suggested by Soper et al.[53] Experiments were carried out with β-bromo-D-alanine to differentiate elimination/addition vs. isomerization/substitution mechanisms (see Section II.F.2.). [^{14}C]-β-Chloro-D-alanine inactivates the enzyme with incorporation of two molecules of radioactivity per dimer;[25] spectral studies indicate that the inactive enzyme is in the ketimine form.[54]

β-Chloro-L-alanine also is a time-dependent inactivator of aminotransferase B from *Salmonella typhimurium*; enzyme activity is not restored by dialysis or gel filtration.[55]

β-Chloro-D-alanine and to a lesser extent, the L-isomer, were shown by Manning et al.[56] to be irreversible inactivators of D-glutamate-D-alanine aminotransferase and alanine racemase activities in a variety of bacteria. A kinetic analysis of the inhibition of alanine racemase from *B. subtilis* by D- and L-β-chloroalanine showed that the D-isomer inhibits at low concentration (K_i = 5 μM) competitively, but the L-isomer inhibits with a much higher K_i (1.71 mM) noncompetitively, suggesting an asymmetric active site.[57]

Both D- and L-β-chloroalanine inactivate the alanine racemase encoded by the *dadB* gene in *S. typhimurium*.[17] The D-isomer has a K_m value 200 times lower than that for the L-isomer, yet the partition ratios for these compounds and for the fluoro analogues are quite similar (all of the partition ratios vary from 730 to 790). These results indicate a common intermediate in the inactivation, namely, an aminoacrylate-PLP species, produced by H–X elimination after binding to PLP. With D- or L-β-chloro[^{14}C]alanine, a stoichiometric amount of radioactivity is incorporated that remains bound after gel filtration. However, heat denaturation results in release of an aldol adduct of pyruvate and PLP (**6.4**, Scheme 4), the same compound isolated by inactivation of L-aspartate aminotransferase[16] or glutamate decarboxylase[15] by β-substituted alanines, followed by base treatment (see Section II.G.3.a.). If inactivated enzyme is treated with borohydride prior to denaturation, however, the adduct becomes stabilized. Tryptic digestion and peptide purification gives a single labeled peptide that corresponds to residues 30 to 45 of the racemase sequence. The lysine at the 6 position of the peptide (Lys-35; the one that forms a Schiff base to the PLP) was shown to be involved in a ternary complex with C-3 of the inactivator linked to the active site lysine via the C-4' position of enzyme-bound PMP (**6.3**, Scheme 3). The mechanism shown in Scheme 3 is consistent with these results. This mechanism reverses the polarity of the inactivator; instead of nucleophilic attack of an active site residue on an electrophilic activated inactivator, the inactivator is released as a nucleophile that attacks the PLP-lysine Schiff base.

A very similar study was carried out by Roise et al.[18] with the broad specificity amino acid racemase from *Pseudomonas striata* and similar results were obtained. The partition ratios for the four isomers, D- and L-β-chloro- and fluoroalanine varied from 830 to 890. The same conclusions regarding the mechanism as were made for alanine racemase[17] were made in this case. Alanine racemase from *E. coli* B is irreversibly inactivated by the D- and L-isomers of β-chloroalanine.[21] Both compounds partition between α,β-elimination to give pyruvate, ammonia, and chloride ion and inactivation; no racemization occurs.

β-Chloroalanine was incorporated into dipeptides by Cheung et al.[58] in order to evaluate their prodrug antibacterial properties. Dipeptide combinations containing the inactivator with an amino acid were tested as antibacterial agents against a wide variety of bacteria. A similar approach for drug delivery was taken by Mobashery et al.[58a] in which β-chloro-L-alanyl-β-chloro-L-alanine replaces the acetoxy substituent at C-3' of cephalothin (Structural Formula 6.26a, Scheme 33). The dipeptide antibiotic is released by TEM β-lactamase-catalyzed hydrolysis of the cepham. Compound **6.26a** was shown by Mobashery and Johnston[58b] to act as a multi-enzyme-activated inactivator (see Chapter 1, Section III.D.) of *E. coli* alanine racemase. First, it is a substrate for *E. coli* TEM β-lactamase, which releases β-chloro-L-alanyl-β-chloro-L-alanine; then this dipeptide is hydrolyzed by leucine- or alanine aminopeptidase to give β-chloro-L-alanine,[58c] a mechanism-based inactivator of *E. coli* alanine racemase.

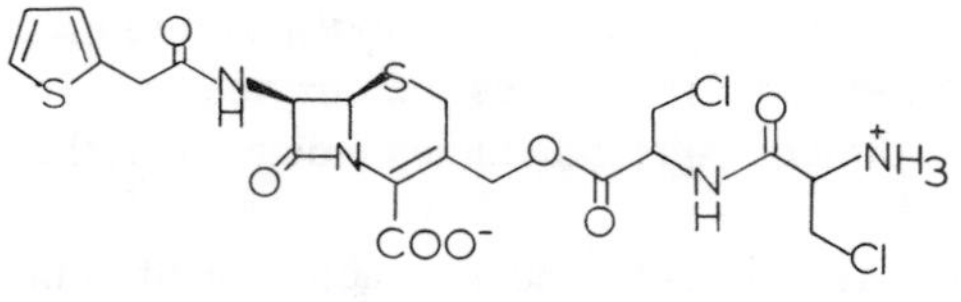

Scheme 33.   Structural Formula 6.26a.

Serine transhydroxymethylase from lamb or rabbit liver is inactivated by β-chloro-D-alanine, but at a slower rate than by the corresponding fluoro compound; the partition ratios, however, are comparable.[23]

β-Chloro-D-alanine also inactivates 1-aminocyclopropane-1-carboxylate deaminase from *Pseudomonas* sp.; the partition ratio is 68.[24] Because the partition ratios for the corresponding fluoro- and acetoxy compounds are different, an isomerization/substitution mechanism (see Scheme 5, for example) was suggested.

β-Substituted amino acids, including β-chloro-D-alanine, are both substrates and pseudo first-order time-dependent inactivators of kynureninase from *P. marginalis*; the partition ratio is 500.[59] Since there is no absorption in the 320- to 500-nm range after inactivation, Kishore[59] excluded the enamine mechanism.[15,16] Inactivation by [14C]β-chloro-L-alanine leads to the incorporation of two molecules of inactivator per dimer of kynureninase. Because of the lability of radioactivity at pH 9.2 and with hydroxylamine, a carboxylate nucleophile was suggested.

2. 3-Chloropropionyl CoA

3-Chloropropionyl CoA was shown by Miziorko and Behne[60] to be a pseudo first-order time-dependent inactivator of 3-hydroxy-3-methylglutaryl CoA synthase (HMG CoA synthase); dialysis does not restore activity. Inactivation with [1-14C]-3-chloropropionyl CoA leads to incorporation of radioactivity, which, after Pronase digestion, gives a single component by HPLC and TLC, comigrating with carboxyethylcysteine. The stoichiometry of labeling of HMG CoA synthase with [1-14C]-3-chloropropionyl CoA is 50% higher than when 3-chloropropionyl CoA is used, indicating that acylation does occur. Since HMG CoA synthase catalyzes the exchange of the α-proton of acetyl CoA, it is reasonable that elimination of HCl is involved in the inactivation mechanism (Scheme 34). Tryptic digestion of the enzyme inactivated with 3-chloro[14C]propionyl CoA gives one radioactively labeled peptide having the structure Glu–Ser–Gly–Asn–Thr–Asp–Val–Glu–Gly–Ile–Asp–Thr–(Thr)–Asn–Ala–S-[14C]carboxyethylcysteine–Tyr–Gly–Gln–Thr–(Ala).[61] The inactivator can be synthesized by the reaction of 3-chloropropionyl chloride with coenzyme A and potassium bicarbonate.[60]

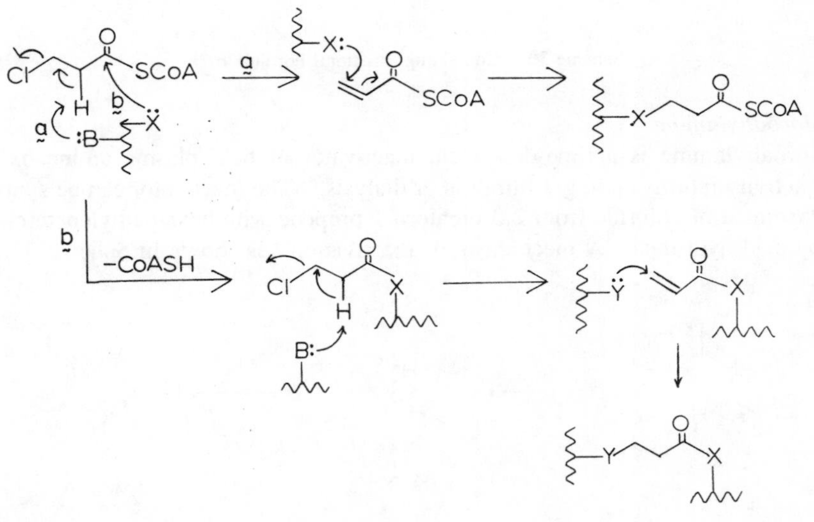

Scheme 34.

3. (3-Chloro-3-butenoyl)pantetheine

Fendrich and Abeles[62] synthesized (3-chloro-3-butenoyl)pantetheine (Structural Formula 6.27, Scheme 35) and found that it irreversibly inactivates butyryl CoA dehydrogenase from *Megasphaera elsdenii* without modification of the flavin cofactor; the partition ratio is 2 to 4. The green color of the enzyme is bleached upon inactivation, suggesting a description of a charge-transfer interaction between the flavin and a CoA persulfide at the active site. A carboxylate was proposed as the active site nucleophile involved in the inactivation because of the lability of the adduct toward hydroxylamine and because sodium borohydride treatment of the radioactively labeled enzyme produces 1,3-butanediol. The mechanism proposed is shown in Scheme 36 (SPant represents pantetheine). Structure **6.28** was preferred because there is no 260-nm absorbance for an α,β-unsaturated thioester.

$$CH_2{=}C(Cl){-}CH_2Cl \xrightarrow[\text{2. HCl, }\Delta]{\text{1. NaCN}} CH_2{=}C(Cl){-}CH_2CO_2H \xrightarrow[\text{pantetheine}]{\text{DCC}} CH_2{=}C(Cl){-}CH_2C(=O)SPant$$

6.27

Scheme 35. Containing Structural Formula 6.27.

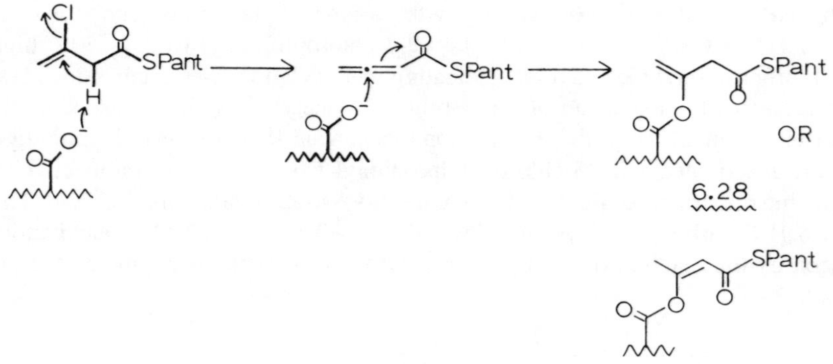

Scheme 36. Containing Structural Formula 6.28.

4. 2-Chloroallylamine

2-Chloroallylamine is a time-dependent inactivator of beef plasma amine oxidase; no enzyme activity returns upon gel filtration or dialysis.[63] The inactivator can be synthesized[64] by displacement of chloride from 2,3-dichloro-1-propene with hexamethylenetetramine followed by acid hydrolysis. A mechanism of inactivation[63] is shown in Scheme 37.

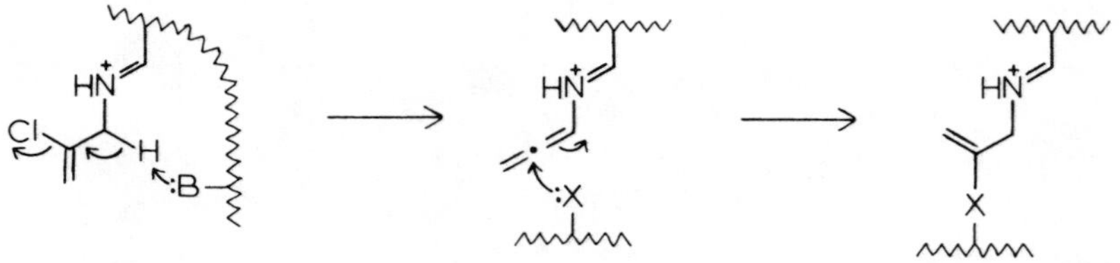

Scheme 37.

5. *(Z)- and (E)-4-Amino-3-halobut-2-enoic Acids*

A series of *Z* (**6.29**)- and *E* (**6.30**)4-amino-3-halobut-2-enoic acids was synthesized by Allan et al.[65] by the routes shown in Schemes 38 and 39. All of the halogenated amino acids except for (*Z*)-4-amino-3-chloro-2-butenoic acid are potent time-dependent inactivators of GABA aminotransferase from rat brain.

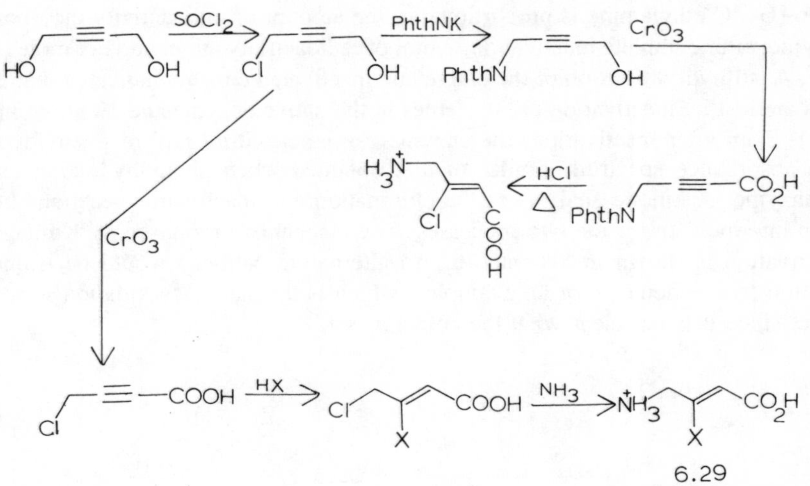

Scheme 38.   Containing Structural Formula 6.29.

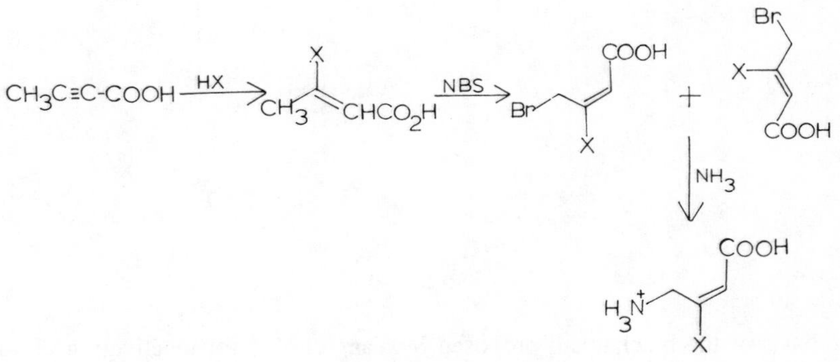

Scheme 39.   Containing Structural Formula 6.30.

6. *β-Chloroethylamine*

The amine oxidase from *Aspergillus niger* is irreversibly inactivated by β-chloroethylamine;[66] more details of the reaction are given in Section II.F.1.

β-Chloroethylamine also is a time-dependent inactivator of aortic lysyl oxidase;[67] more details of the inactivation are given in Section II.F.1.

E. Dichlorinated Inactivators

1. *β,β-Dichloro-D,L-alanine*

γ-Cystathionase (rat liver), tryptophanase (*E. coli*), and $\alpha_2\beta_2$ tryptophan synthase (*E. coli*) are irreversibly inactivated by β,β-dichloro-D,L-alanine; enzyme activity is not regenerated by gel filtration.[22]

F. Monobrominated Inactivators

1. 2-Bromoethylamine

The amine oxidase from *A. niger* is irreversibly inactivated by β-chloro- and β-bromo-ethylamine.[66] Inactivation requires the enzyme to be in its oxidized form, but does not require oxygen. β-Bromoethylamine also is turned over by the enzyme about 600 times with elimination of HBr to give acetaldehyde without inactivation. The amount of inactivation with β-bromo-[U-^{14}C]ethylamine is proportional to the amount of radioactivity incorporated into the enzyme; after complete inactivation, 2 mol of radioactivity are bound per mole of dimeric enzyme. A sulfhydryl titration of the native and inactivated enzyme indicated that 2 cysteine residues are lost by inactivation (12 cysteines in the native enzyme and 10 in the inactivated enzyme). Even after inactivation, the enzyme can react with 2 mol of phenylhydrazine to give an absorbance spectrum similar to that obtained when phenylhydrazine reacts with native enzyme. A kinetic analysis of the elimination and inactivation reactions indicates a common intermediate for the two processes. The mechanism proposed by Kumagai et al.[66] for inactivation is shown in Scheme 40. An alternative pathway would be isomerization/substitution (see Scheme 5 for an example), which is the same as oxidation/substitution in this case, since it is not clear what the cofactor is.

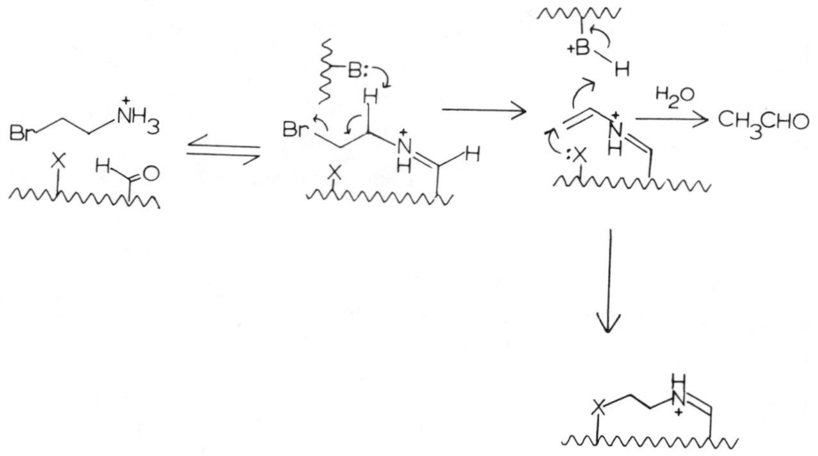

Scheme 40.

On the basis of the mechanism proposed by Tang et al.[68] for inactivation of aortic lysyl oxidase by β-aminopropionitrile (see Chapter 7 in Volume II, Section VII.C.2.), a series of β-substituted ethylamines was investigated by Tang et al.[67] as mechanism-based inactivators. The substituents chosen were bromo, chloro, nitro, hydroxyl, and β,β,β-trifluoro. The first three are time-dependent inactivators, which, like the β-cyano compound, exhibit biphasic kinetics. [1,2-^{14}C]-β-Bromoethylamine inactivates lysyl oxidase with the incorporation of only 0.17 equivalent of radioactivity. Low active enzyme was suggested as the cause for this. Acetaldehyde was identified as a turnover product after inactivation by the bromo compound. The mechanism proposed for all of the β-substituted ethylamines that inactivate lysyl oxidase is the same as shown in Scheme 40.

Another amine oxidase, beef plasma amine oxidase, also is inactivated by 2-bromoethylamine, but an oxidation/addition mechanism was proposed[69] (see Chapter 9 in Volume II, Section II.B.1.d. for details). It is now known that the coenzyme in lysyl oxidase[69a] and plasma amine oxidase[69b] is pyrroloquinoline quinone (methoxatin).

2. β-Bromoalanines

In order to differentiate the inactivation mechanism of *B. sphaericus* D-amino acid aminotransferase by β-chloro-D-alanine[53] via an aminoacrylate-PLP Schiff base intermediate (elimination/addition mechanism) or alkylation by β-chloropyruvate formed during the reaction (oxidation/substitution mechanism), experiments were carried out by Soper and Manning[70] with β-bromo-D-alanine and bromopyruvate in the presence of thiols. There is no kinetic difference between inactivation by β-chloro- and β-bromo-D-alanine, indicating loss of halide is not rate determining. This would be unexpected for an S_N2 reaction. Since bromopyruvate does not inactivate in the absence of a substrate, it is not responsible for inactivation of native enzyme; it apparently forms a Schiff base to the PMP form of the enzyme which then eliminates HBr (Scheme 41) to give the same aminoacrylate-PLP intermediate (**6.1**) shown in Scheme 2. Therefore, the elimination/addition mechanism for inactivation of D-amino acid aminotransferase by β-halo-D-alanines was favored. These results, however, also would be consistent with the enamine mechanism.[15,16]

β-Bromo-L-alanine was shown by Morino and Okamoto[71] to inactivate both isozymes of pig heart aspartate aminotransferase.

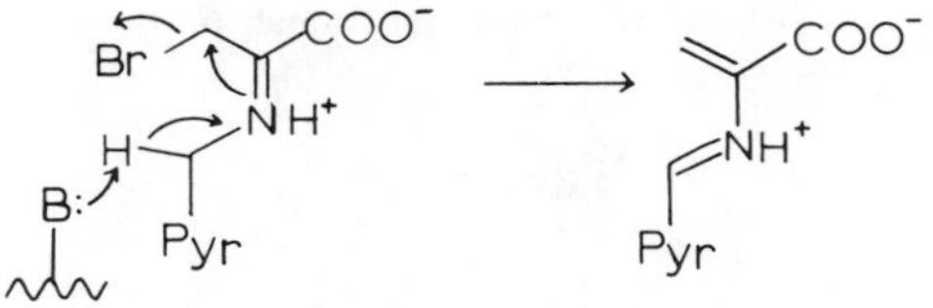

Scheme 41.

3. Bromopyruvate

In addition to the inactivation of D-amino acid aminotransferase (see Section II.F.2.), bromopyruvate slowly inactivates aspartate aminotransferase isozymes from pig heart; the inactivation rate is markedly enhanced by the addition of L-cysteine sulfinate.[71] Since the PMP form of the enzyme is rapidly converted to the PLP form by bromopyruvate, it was suggested by Morino and Okamoto[71] that the L-cysteine sulfinate functions to maintain the enzyme in its PMP form and that the actual inactivating species is β-bromo-L-alanine, generated from bromopyruvate and the PMP form of the enzyme. β-Bromo-L-alanine is an inactivator of the enzyme. These results indicate that bromopyruvate and β-bromo-L-alanine are mechanism-based inactivators of the pyridoxamine phosphate and pyridoxal phosphate forms of the enzyme, respectively. Differences in the mitochondrial and soluble isozymes also were noted. In a later report, Okomoto and Morino[72] note that neither the mitochondrial nor the soluble pig heart aspartate aminotransferase isozymes is inactivated by bromopyruvate alone; however, in the presence of L-cysteine sulfinate or L-aspartate, bromopyruvate rapidly inactivates both isozymes. β-Bromo-L-alanine was not detected as a product; rather, pyruvate and ammonia were isolated. With [2-^{14}C]bromopyruvate as the inactivator, inactivation followed by borohydride reduction and hydrolysis produces *S*-2-hydroxycarboxy[^{14}C] ethyl-cysteine (Structural Formula 6.31, Scheme 42). Inactivation was suggested to occur as shown in Scheme 43. If this is the case, though, it is affinity labeling of the enzyme. In analogy with the inactivation of GABA aminotransferase in its PMP form by 5-fluorolevulinate[33] (see Scheme 13 in Section II.A.2.) an alternative mechanism is elimination/addition (Scheme 44) or elimination followed by the enamine mechanism.[15,16] The bromopyruvate inactivation of pig heart cytoplasmic aspartate aminotransferase was reinvestigated by Birchmeier and Christen.[73] Again, it was found that the enzyme underoges rapid time-dependent inactivation by bromopyruvate in the presence of an amino acid; in the absence of an amino acid, inactivation is very slow. The presence of α-ketoglutarate at a concentration high enough to preclude binding of bromopyruvate to the active site does not protect the enzyme from inactivation. Furthermore, the inactivation rate increases linearly with the bromopyruvate concentration, even above saturation. This suggests that inactivation is not active site directed. After inactivation, there still is residual enzyme activity (about 1/3000 the specific activity with a 15-fold higher K_m for α-ketoglutarate). Inactivation results in alkylation of Cys-390. A syncatalytic modification of the enzyme is proposed in which nonidentical bromopyruvate molecules serve as substrate and as alkylating agents. As a substrate, it combines with an amino acid substrate to form covalent enzyme-substrate intermediates. In one of these intermediates, Cys-390, then acquires a markedly enhanced reactivity and is alkylated by a bromopyruvate molecules not attached to the active site. Since Cys-390 is not a functional residue, enzyme activity can persist after alkylation.

$$\text{Cys–SCH}_2^{14}\text{CHCOOH}$$
$$\mid$$
$$\text{OH}$$

Scheme 42. Structural Formula 6.31.

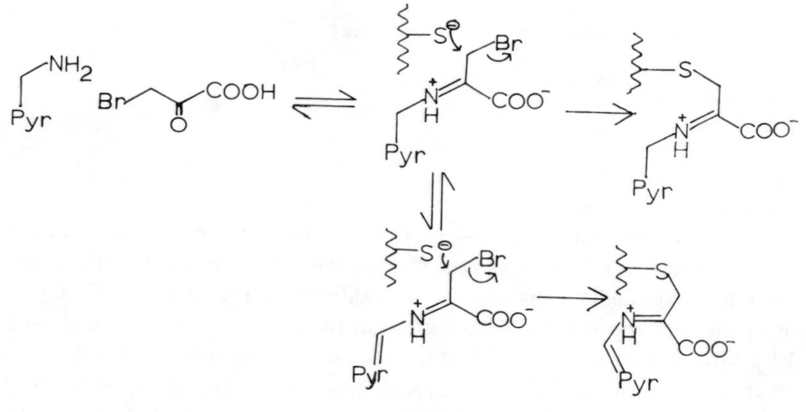

Scheme 43.

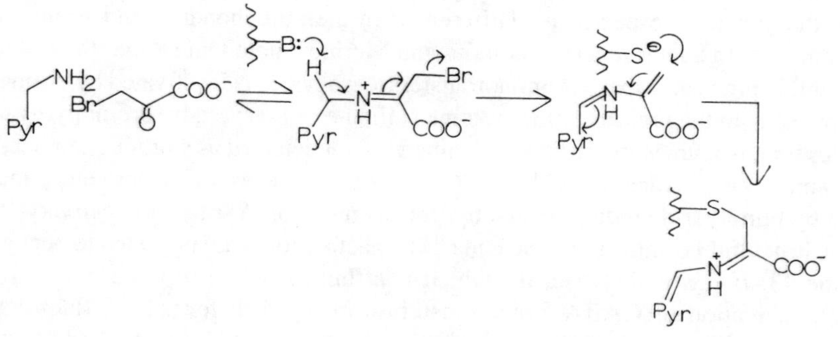

Scheme 44.

G. Inactivators with Other Leaving Groups

This section is arranged according to the atom that is eliminated form the inactivator, beginning with carbon, then nitrogen, oxygen, and sulfur.

1. Methylene as the Leaving Group
a. Methylenecyclopropylacetyl CoA

Methylenecyclopropylacetate is a time-dependent inactivator of rat liver butyryl CoA synthetase and ox liver butyryl CoA dehydrogenase in the presence of ATP, Mg(II), and CoA.[74,75] Hypoglycin, a toxin of unripe ackee fruit, produces severe hypoglycemia when ingested. A metabolite of hypoglycin, methylenecyclopropylacetyl CoA (Structural Formula 6.32, Scheme 45), produces time-dependent inactivation of pig kidney general acyl-CoA dehydrogenase, leading to irreversible modification of 80% of the FAD; the remainder of the inactivator most likely results in protein modification.[76] Racemic **6.32** inactivates general acyl CoA dehydrogenase with a partition ratio of 4 at the same rate as the natural (*R*) isomer; therefore, the *S* isomer is inactive.[76a] Several modified flavin derivatives, including a C4a-N5 dihydroflavin, were isolated as well as about 20% of unreacted FAD. A carbanion mechanism of inactivation (Scheme 45) was proposed by Wenz et al.[76] similar to that for 2-hydroxybutynoate inactivation of lactate dehydrogenase (see Chapter 9 in Volume II, Section II.C.1.a.(1)). Methylenecyclopropylacetyl CoA also inactivates butyryl CoA dehydrogenase. Methylenecyclopropaneacetic acid can be prepared from hypoglycin (Scheme 46),[77] or as shown in Scheme 46A.[76a]

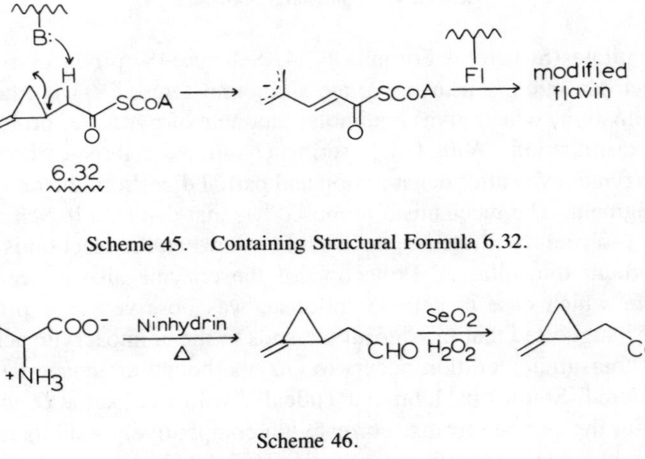

Scheme 45. Containing Structural Formula 6.32.

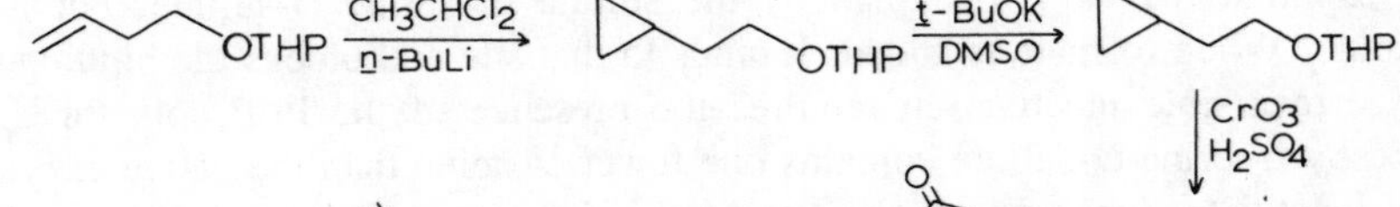

Scheme 46.

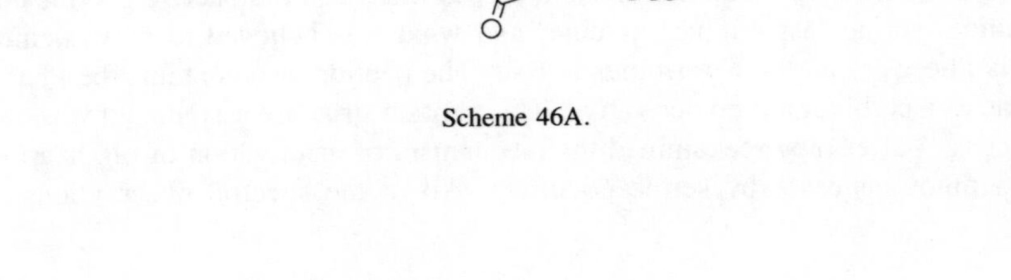

Scheme 46A.

2. Nitrite as the Leaving Group
a. 3-Nitro-D,L-alanine

3-Nitro-D,L-alanine (Structural Formula 6.33, Scheme 47) is a time-dependent inactivator of three aminotransferases, alanine-, aspartate-, and 4-aminobutyrate aminotransferase; the inactivation mechanism shown in Scheme 2 was suggested by Alston and Bright.[78] The inactivator was synthesized by the route shown in Scheme 47.[79]

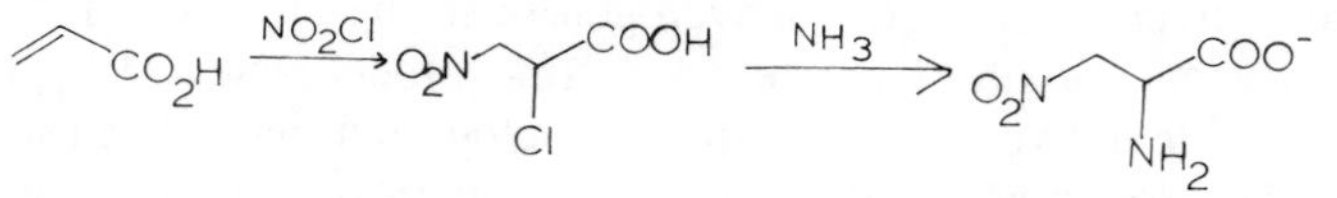

$$\text{CH}_2=\text{CH–CO}_2\text{H} \xrightarrow{\text{NO}_2\text{Cl}} \text{O}_2\text{N–CH}_2\text{–CHCl–COOH} \xrightarrow{\text{NH}_3} \text{O}_2\text{N–CH}_2\text{–CH(NH}_2)\text{–COO}^-$$

$$\underset{6.33}{}$$

Scheme 47.   Containing Structural Formula 6.33.

3. Sulfate as the Leaving Group
a. L-Serine O-Sulfate

$$\underset{\text{N H}_3\text{CHCH}_2\text{OSO}_3^-}{\overset{\text{COOH}}{+\ \ |}}$$

Scheme 48.   Structural Formula 6.34.

L-Serine O-sulfate (Structural Formula 6.34, Scheme 48) produces time-dependent inactivation of cytoplasmic pig heart aspartate aminotransferase.[80] Two other reactions occur as well, β-elimination, which gives equimolar amounts of pyruvate, protons, ammonia and sulfate, and transmination. With [^{14}C]L-serine O-sulfate, a three-carbon residue remains bound to the enzyme; even after denaturation and partial digestion, the radioactivity is bound to a peptide fragment. The mechanism proposed[80] is that depicted in Scheme 2. The rate of inactivation by L-serine O-sulfate is considerably less when the reaction is carried out in the presence of sodium thiosulfate.[81] Protection of the enzyme also is seen with L-cysteine sulfinic acid, in which case cysteine-S-sulfonate was observed as a product. Therefore, Cavallini et al.[81] suggested that the thiosulfate adds to the aminoacrylic acid generated. The possibility that thiosulfate addition occurs to enzyme-bound aminoacrylate prior to release also was considered. Studies by John and Tudball[82] with [^{14}C]serine O-sulfate showed that a 1:1 adduct with the enzyme forms. Formate ion competitively inhibits the inactivation by serine O-sulfate, but, as noted earlier (see Section II.D.1.), it increases the rate of inactivation by β-chloroalanine. It was proposed that there is an anionic binding site for the β-carboxyl group of aspartate which is occupied by the sulfate of serine O-sulfate, but not by β-chloroalanine. When formate is added, it binds to that site and blocks the binding of serine O-sulfate. Irreversible inactivation requires the presence of the PLP cofactor.[83] Enzyme inactivated by L-serine O-sulfate contains one fewer cysteine than the native enzyme. With the use of inactivator labeled with [^{14}C] and [^{35}S], John et al.[83] showed that 1 mol of [^{14}C] and no [^{35}S] is bound to the enzyme after inactivation, suggesting the loss of the sulfate group. Inactivation followed by thermolysin digestion gives a radioactive peptide containing phenylalanine, serine, asparagine, glycine, and what was believed to be cysteine; the N-terminus is Phe–Ser and the C-terminus is Asn. The peptide also contains the PLP, thought to be bound as a carbinol amino derivative. This peptide structure assignment was questioned by Ueno et al.[16] after they reexamined the mechanism of inactivation of pig heart cytosolic aspartate aminotransferase by serine O-sulfate. All of the spectral observations made by

John and Fasella[80] were reproduced by Ueno et al.;[16] initially there is a rapid production of a 336-nm absorption that slowly is converted to a 455-nm absorption. Ueno et al.,[16] however, found that if the pH of the inactivated and dialyzed enzyme solution is raised to 11, a bright yellow compound having structure **6.35** (see Scheme 49), identified by analytical comparison with a synthesized sample, is released. The mechanism for its formation was described previously (Schemes 3 and 4). Since it is known[84] that there is no cysteine-containing sequence in this enzyme that could give rise to a peptide with the structure indicated by John et al.,[83] Ueno et al.[16] suggested that the peptide actually contains Phe–Ser–-Lys*–Asn–Phe–Gly, where Lys* is the modified lysine (**6.36**, Scheme 49). This modified enzyme is believed to be the 336-nm absorption species and enzyme-bound **6.35** the 455-nm species. If inactivation of aspartate aminotransferase with β-chloroalanine produced the same adduct (**6.36**), which has a chiral center at the 4′-carbon, then that would explain the positive circular dichroism observed by Morino and Okamoto[49] and Morino and Tanase[50] for the peptides isolated by trypsin digestion of the inactivated enzyme. It is interesting to note that the circular dichroism of β-chloroalanine-inactivated alanine aminotransferase reported by Morino et al.[52] is negative; this may indicate the opposite chirality for **6.36** with this enzyme.

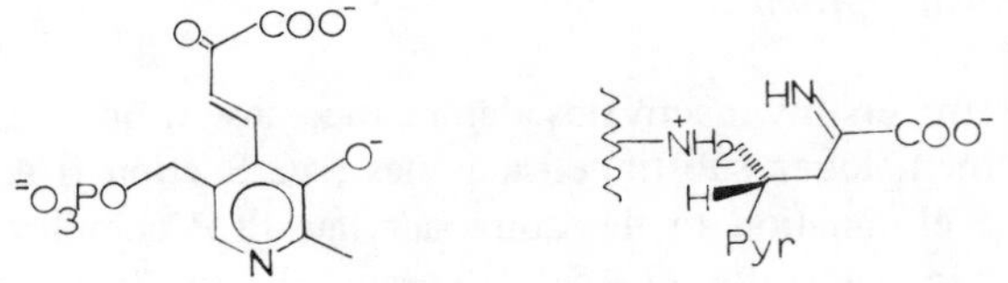

Scheme 49.   Structural Formulas 6.35 and 6.36.

Inactivation of glutamate decarboxylase from *E. coli* with L-serine *O*-sulfate also results in the appearance of a 336-nm absorption; dialysis at pH 4.6 has no effect on this spectrum.[15] Urea at pH 4.5 has a slight effect on the spectrum, but no coenzyme is lost. At pH 11, however, compound **6.35** is released from the enzyme. These results of Metzler and co-workers[15,16] indicate that **all** of the inactivators that have been proposed to involve an elimination-Michael addition mechanism should be reevaluated.

L-Serine *O*-sulfate, but not L-serine *O*-phosphate, is both a substrate and a pseudo first-order, time-dependent inactivator of kynureninase from *Pseudomonas marginalis*.[59] Because there is no absorption between 320 and 500 nm after inactivation, the enamine mechanism[15,16] was excluded by Kishore.[59]

b. Ethanolamine O-Sulfate

$$\overset{+}{N}H_3CH_2CH_2OSO_3^-$$

Scheme 50.   Structural Formula 6.37.

Ethanolamine *O*-sulfate (Structural Formula 6.37, Scheme 50) is a pseudo first-order, time-dependent inactivator of rat brain 4-aminobutyrate aminotransferase; dialysis does not reverse inactivation.[85] The compound does not inactivate pig heart aspartate aminotransferase or alanine aminotransaminase or rat brain glutamate decarboxylase. Fowler and John[86] also found that [2-¹⁴C]ethanolamine *O*-sulfate results in the incorporation of 1 mol of radioactivity per mole of rabbit brain 4-aminobutyrate aminotransferase active sites. When ethanolamine O-[^{35}S]sulfate was used, no radioactivity was incorporated. The amount of [^{35}S]sulfate eliminated is always an order of magnitude greater than the enzyme inactivated; the amount of ammonia produced is 11 times the amount of enzyme inactivated. The mechanism proposed is related to that shown in Scheme 2 (see Scheme 51). When the pro-R hydrogen of ethanolamine O-sulfate is replaced by a methyl- or methylthioethyl group, no inactivation of 4-aminobutyrate aminotransferase or ornithine aminotransferase was observed.[87]

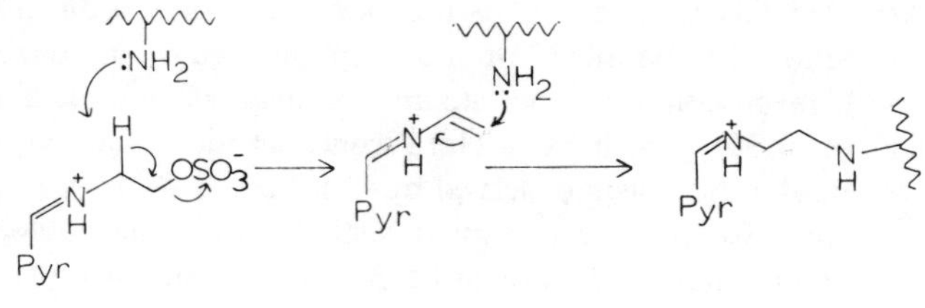

Scheme 51.

4. Acetate as the Leaving Group
a. O-Acetylserine

O-Acetyl-D-serine irreversibly inactivates alanine racemase from *E. coli* B with a partition ratio similar to that for halogen-substituted alanines (see Section II.A.1.); this supports a mechanism involving elimination to the aminoacrylate-PLP complex.[21] The L-isomer is neither a substrate nor an inactivator. O-Acetyl-D-serine also inactivates the alanine racemase encoded by the *dadB* gene in *Salmonella typhimurium*.[17] Both O-acetyl-D- and -L-serine (partition ratios 980 and 720, respectively) inactivate the broad specificity amino acid racemase from *Pseudomonas striata*.[18] With the use of the four possible chiral isomers of O-acetyl[3-^{3}H]serine, it was shown by Badet et al.[87a] that the *P. striata* enzyme in ^{2}H$_2$O catalyzes protonation of the aminoacrylate turnover product with a modest stereoselectivity (15 to 20% enantiomeric excess). O-Acetyl-L-serine inactivates kynureninase from *P. marginalis*.[59]

5. Carbamate as the Leaving Group
a. O-Carbamoyl-D-serine

O-Carbamoyl-D-serine inactivates *E. coli* B alanine racemase.[21]

6. Hydroxide as the Leaving Group
a. Serine

Serine is a time-dependent inactivator of *E. coli* threonine deaminase.[88] Threonine dehydratase from *Clostridium tetanomorphum* also is inactivated by L-serine in a time-dependent manner; it was postulated by Phillips[89] that a mechanism similar to that shown in Scheme 2 is responsible. L-Serine is a substrate and time-dependent inactivator of sheep liver L-threonine dehydratase;[90,91] gel filtration at 2°C does not result in enzyme reactivation.[91] However, upon incubation of inactivated enzyme with PLP[90] or upon standing at pH 8.9,[91] there is a time-dependent return of enzyme activity. Nishimura and Greenberg[90] and McLemore and Metzler[91] suggested that oxazolidine formation with the PLP is responsible for inactivation (Scheme 52). If this is the case, serine would not be a mechanism-based inactivator, because binding to PLP in this book is synonymous with enzyme binding.

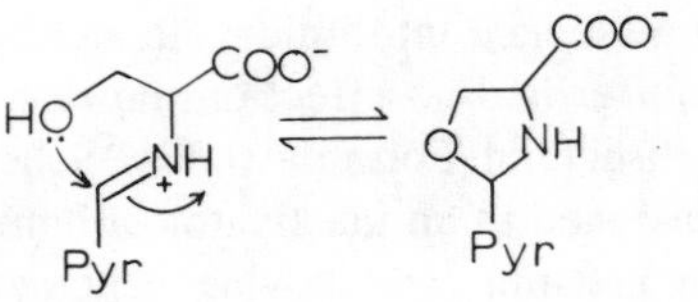

Scheme 52

7. Arylthiolate as the Leaving Group
a. S-(o-Nitrophenyl)-L-cysteine

S-(o-Nitrophenyl)-L-cysteine is both a substrate and pseudo first-order, time-dependent inactivator of kynureninase from *Pseudomonas marginalis;* S-(o-nitrophenyl)-L-homocysteine is not an inactivator.[59]

8. N-Methyldithiocarbamate as the Leaving Group
a. S-(N-Methylthiocarbamoyl)-L-cysteine

L-Methionine γ-lyase from *P. putida* is irreversibly inactivated by S-(N-methylthiocarbamoyl)-L-cysteine (Structural Formula 6.38, Scheme 53).[92] α,β-Elimination was demonstrated by Esaki et al.[92] by the isolation of pyruvate, ammonia and N-methyldithiocarbamate. The partition ratio is 800; however, in the presence of a high concentration of β-mercaptoethanol, the enzyme catalyzes a β-replacement reaction, producing S-(β-hydroxyethyl)-L-cysteine, without enzyme inactivation. This indicates that the β-mercaptoethanol, instead of the active site nucleophile, is trapping a reactive PLP-bound intermediate. Inactivation with [1,2,3-U-^{14}C]-inactivator leads to attachment of 4 mol of inactivator per tetrameric enzyme after gel filtration. However, half of the radioactivity is released, with concomitant reactivation of half of the enzyme, when the labeled enzyme is dialyzed against buffer containing PLP. This suggests that two different residues are modified, one of which forms an unstable adduct. The two mechanisms suggested are those depicted in Schemes 2 and 3. Compound **6.38** does not inactivate L-amino acid oxidase from *Crotalus adamanteus.*[93]

$$\underset{NH_2CHCH_2SCNHCH_3}{\overset{\overset{\displaystyle COOH\quad S}{\displaystyle |\qquad\ \ ||}}{}}$$

Scheme 53. Structural Formula 6.38.

P. putida amino acid racemase is inactivated by S-(N-methylthiocarbamoyl)-D- and L-cysteine.[94] Time-dependent, pseudo, first-order inactivation results, which is not affected by the presence of β-mercaptoethanol. Inactivation with [1,2,3-U-^{14}C]-inactivator, followed by dialysis results in the incorporation of 1 mol of [^{14}C] per mole of enzyme. No radioactivity becomes attached to enzyme that was pretreated with sodium borohydride. Treatment of inactivated enzyme at alkaline pH does not produce the yellow compound (based on spectral observation) that is produced when L-serine O-sulfate inactivated glutamate decarboxylase is treated with base[15] (see Section II.G.3.a.). Therefore, the mechanism proposed is the one related to that shown in Scheme 2.

III. ELIMINATION/ACYLATION

A. Imidazole and Related as the Leaving Group
1. Imidazole N-Carboxamides

Imidazole N-carboxamides were prepared by Groutas et al.[95] as mechanism-based inactivators of elastase; time-dependent loss of enzyme activity is pseudo first-order. Imidazole

N-carboxamides are known to dissociate into imidazole and isocyanates; however, addition of cysteine to the incubation buffer had no effect on the rate of inactivation. 1-(*N*-Amino-*n*-hexyl)carbamoylimidazole (Structural Formula **6.38′**, Scheme 53A), modeled after the work of Groutas et al.[95] was designed as an inactivator of human urokinase, pig kidney-cell plasminogen activator, human plasmin, and bovine pancreatic β-trypsin.[95a] All of these enzymes undergo time-dependent inactivation; bovine thrombin and Factor Xa are not inactivated by **6.38′**. Walker and Elmore[95a] favor a direct acylation mechanism over mechanism-based inactivation, in which case this would be an affinity labeling agent. This approach to the design of compounds having a latent isocyanate group was extended by Groutas et al.[96] to amino acid-derived azolides (*N*-(1*H*-imidazol-1-ylcarbonyl)amino acid methyl ester; **6.38a**), which were synthesized from amino esters (Scheme 54). Some of these compounds are time-dependent inactivators of porcine pancreatic elastase and human leukocyte elastase showing pseudo first-order, saturation kinetics; dialysis does not regenerate enzyme activity. Because of the known conversion of this class of compounds into isocyanates, an elimination/addition mechanism (Scheme 55) was suggested. However, since elastase is not meant to catalyze a deprotonation reaction, these compounds may be affinity labeling acylating agents.

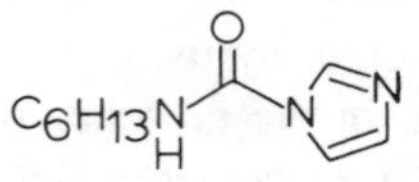

Scheme 53A.   Structural Formula 6.38′.

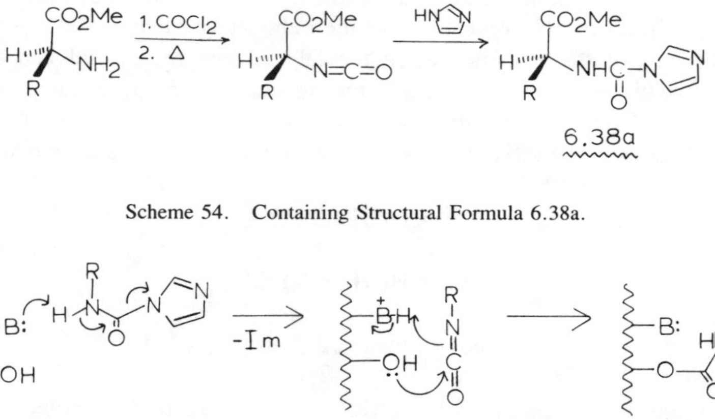

Scheme 54.   Containing Structural Formula 6.38a.

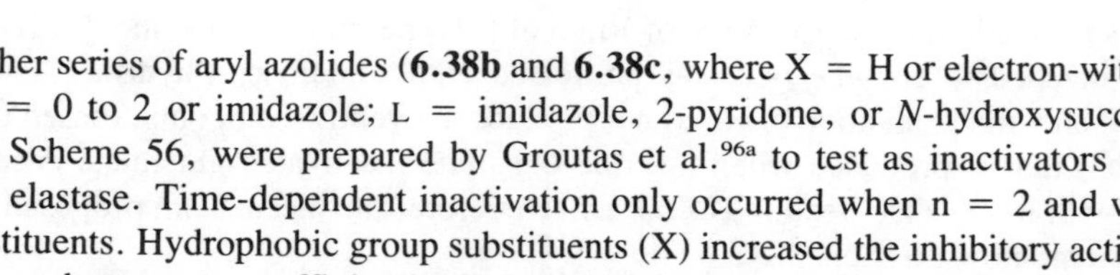

Scheme 55.

Two other series of aryl azolides (**6.38b** and **6.38c**, where X = H or electron-withdrawing group; n = 0 to 2 or imidazole; L = imidazole, 2-pyridone, or *N*-hydroxysuccinimide), shown in Scheme 56, were prepared by Groutas et al.[96a] to test as inactivators of human leukocyte elastase. Time-dependent inactivation only occurred when n = 2 and with all of the L substituents. Hydrophobic group substituents (X) increased the inhibitory activity since this enzyme shows greater affinity for hydrophobic substrates. The most potent inactivator had structure **6.38b** (n = 2, L = *N*-hydroxysuccinimide, X = *m*-CF₃).

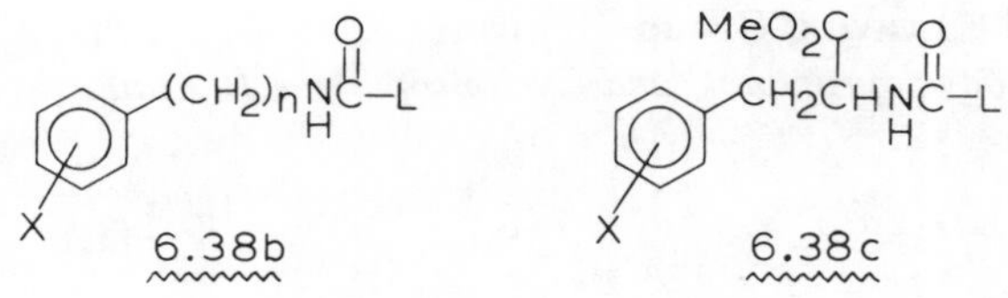

Scheme 56. Structural Formulas 6.38b and c.

B. Sulfite as the Leaving Group

1. Carbonyl Sulfonate Salts of Amino Acid Esters

A series of carbonyl sulfonate salts of amino acid esters (Structural Formula 6.38d, Scheme 57) were shown by Groutas et al.[96b] to be potent and selective time-dependent inactivators of human leukocyte elastase. The most active are the compounds derived from L-norleucine, L-norvaline, and L-valine; also active were analogues of L-leucine, L-methionine, and L-phenylalanine. The glycine, L-alanine, and D-norvaline analogues are not inactivators. The L-norvaline derivative has no effect on chymotrypsin and porcine pancreatic elastase, indicating that selectivity can be built into these inactivators by judicious choice of the side chain. The selectivity also suggests that a simple acylation is not occurring, and that the mechanism shown in Scheme 55 may be important.

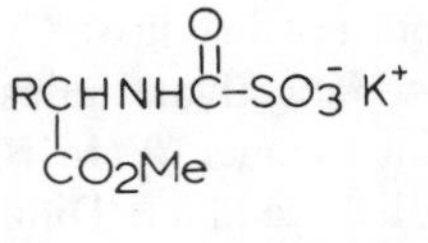

Scheme 57. Structural Formula 6.38d.

C. Aryloxide as the Leaving Group

1. Glycine Esters

On the basis of the ElcB mechanism for ester hydrolysis,[97] a series of glycine esters was tested by Maycock et al.[98] as mechanism-based inactivators of beef plasma amine oxidase. Only the esters with good leaving groups inactivate the enzyme. Of the two glycine esters that were inactivators (phenyl and *p*-nitrophenyl esters), the better the leaving group, the lower the partition ratio, indicating fewer turnovers prior to inactivation. Addition of a second aliquot of enzyme resulted in the same inactivation rate as the initial enzyme addition, suggesting that reactive species are not released during inactivation. [2-^{2}H] Phenyl glycinate inactivates plasma amine oxidase at the same rate as does the nondeuterated compound, indicating that proton removal is not rate determining. Inactivation with phenyl[1-^{14}C]glycinate results in incorporation of 2 mol of [^{14}C] per dimer after gel filtration. Inactivation with phenyl [2-^{3}H]glycinate releases one [^{3}H] per inactivation event. The mechanism proposed is shown in Scheme 58.

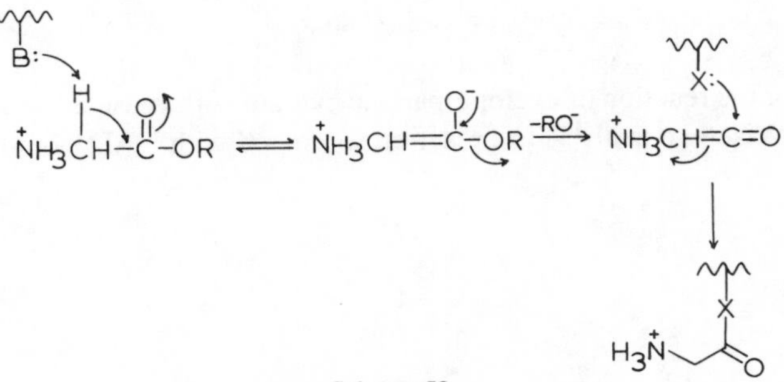

Scheme 58.

D. Hydroxide as the Leaving Group
1. Coprine, 1-Aminocyclopropanol, and Cyclopropanone Hydrate

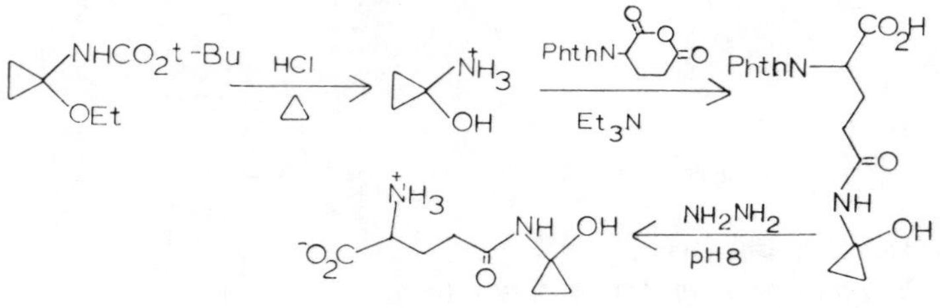

Scheme 59.   Containing Structural Formula 6.39.

Coprine (Structural Formula 6.39), N^5-(1-hydroxycyclopropyl)glutamine, is a disulfiram-like constituent of the inky cap mushroom *Coprinus atramentarius*,[99,100] which can be synthesized[100,100a] as shown in Scheme 59. Although coprine has no effect on the low K_m acetaldehyde dehydrogenase from rat liver, 1-aminocyclopropanol is a time-dependent inactivator.[101] Sephadex gel filtration does not restore enzyme activity, and 2-mercaptoethanol and dithiothreitol do not protect the enzyme from inactivation. Coprine does inactivate aldehyde dehydrogenase in vivo, but not in vitro. The hydrolysis product of coprine, cyclopropanone hydrate, however, does inactivate mouse liver and yeast aldehyde dehydrogenase in vitro in a time-dependent manner.[102] In both cases, 100% inhibition was not achieved, suggesting reversibility of the adduct. Dilution of inactivated enzyme results in pseudo first-order return of enzyme activity; the product released is cyclopropanone hydrate. With the use of [2-^{3}H]cyclopropanone hydrate, the stoichiometry of binding was found to be 0.82 mol per mole of active sites. Iodoacetamide inactivates aldehyde dehydrogenase and prevents binding of cyclopropanone hydrate. Therefore, it was suggested by Wiseman and Abeles[102] that the catalytically active sulfhydryl is involved in cyclopropanone hydrate binding, and all steps are reversible (Scheme 60). Cyclopropanone hydrate also was shown to inhibit many other enzymes that are sensitive to sulfhydryl reagents, e.g., alcohol dehydrogenase, glyceraldehyde 3-phosphate dehydrogenase, creatine kinase, pyruvate kinase, and papain.

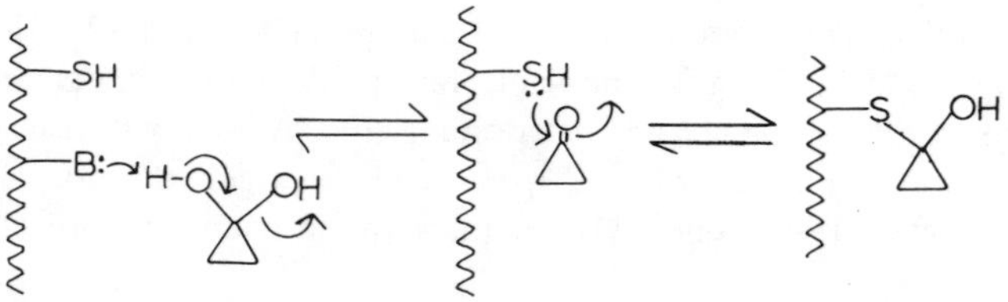

Scheme 60.

The kinetics for the reaction of cyclopropanone hydrate with yeast aldehyde dehydrogenase were studied by Wiseman et al.[103] The reaction is ordered with NAD$^+$ binding first (Scheme 61).

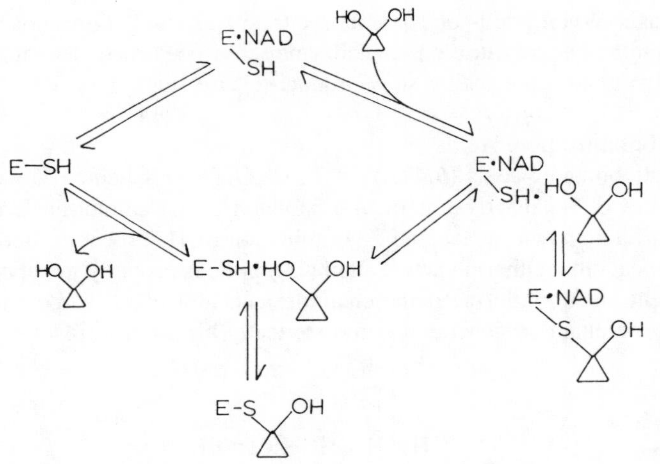

Scheme 61.

E. Coenzyme A as the Leaving Group

1. 2-Octynoyl CoA

Pig kidney general acyl-CoA dehydrogenase is irreversibly inactivated by 1 equivalent of 2-octynoyl CoA.[104] An 800-nm absorbance intermediate is produced which appears faster than inactivation of the enzyme; coenzyme A is released during decay of this intermediate. Two possibilities for this charge-transfer complex are **6.40** (Scheme 62) and **6.41**. Breakdown to the ketene with elimination of CoASH could follow. The pantetheine thioester does not produce this absorbance. Therefore, it is not clear if this is a mechanism-based inactivator (intermediate **6.40**) or an affinity labeling agent (intermediate **6.41**).

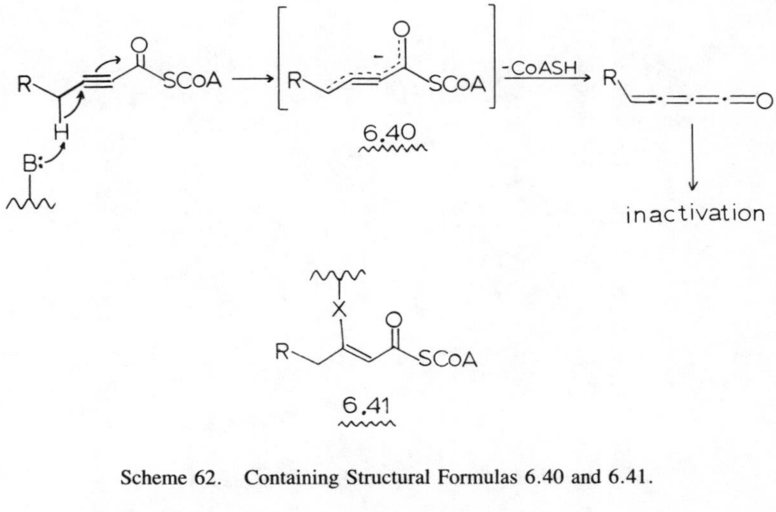

Scheme 62. Containing Structural Formulas 6.40 and 6.41.

IV. NOT MECHANISM-BASED INACTIVATION

A. (−)-*erythro*-Fluorocitrate

(−)-*erythro*-Fluorocitrate is a time-dependent inactivator of pig heart aconitase that exhibits no pseudo first-order kinetics.[105] Even with a 100-fold excess of inactivator, less than 100% inactivation is observed. During inactivation, 0.8 mol of fluoride ions is released per mole of aconitase. When [^{14}C]- or [^{3}H]-fluorocitrate was used, no radioactivity remained bound to the protein after gel filtration, and enzyme activity returned. Therefore, fluorocitrate appears to be just a substrate with a slow product release step.

B. 4-Amino-3-halobutanoic Acids

4-Amino-3-halobutanoic acids (**6.42**; Y = F, Cl, OH; see Scheme 63) were designed by Silverman and Levy[106] as inactivators of γ-aminobutyric acid aminotransferase via a mechanism similar to that shown in Scheme 2. Elimination of H−Y occurs, but no inactivation is evident. It appears that either no active site nucleophile is properly juxtaposed for reaction with the incipient electrophile or, if the enamine mechanism[15,16] is operative (see Section II.G.3.a.), that the enamine generated is too sterically hindered to react with the enzyme-bound PLP.

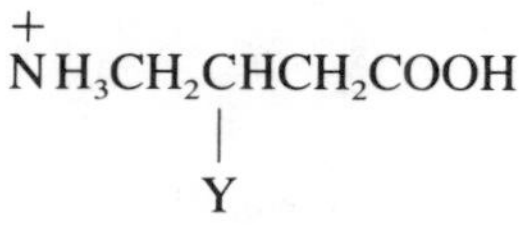

$$\overset{+}{N}H_3CH_2\underset{|}{C}HCH_2COOH$$
$$Y$$

Scheme 63.   Structural Formula 6.42.

C. 4-Amino-2-(Y-methyl)-2-butenoic Acids

A series of 4-amino-2-(Y methyl)-2-butenoic acids (**6.43**, Y = F, Cl, OH) were synthesized by Silverman et al.[107] as potential mechanism-based inactivators of GABA aminotransferase by the mechanism shown in Scheme 64. No time-dependent inactivation occurred, although elimination of HY was evident. The same reasons cited in Section IV.B. may be relevant to **6.43** as well.

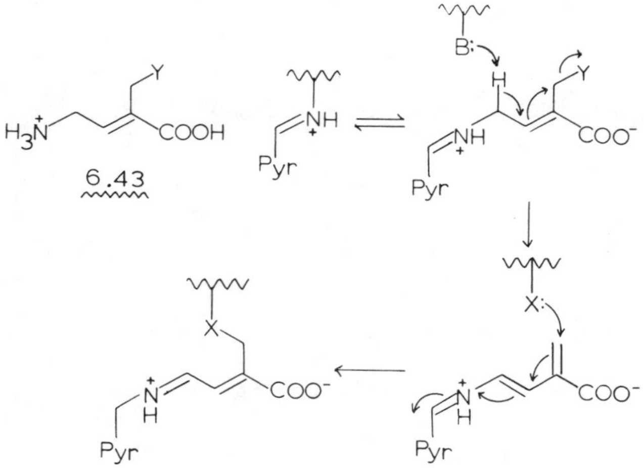

Scheme 64.   Containing Structural Formula 6.43.

D. Trifluoromethyl GABA and Related

The trifluoromethyl analogues of β-alanine (**6.44**, n = 1), γ-aminobutyrate (**6.44**, n = 2), and δ-aminopentanoate (**6.44**, n = 3) were prepared (note Structural Formula 6.44, Scheme 65) as potential inactivators of pig brain GABA aminotransferase; however, none of the compounds inactivates the enzyme.[30]

$$CF_3$$
$$|$$
$$NH_3CH(CH_2)_nCOOH$$

Scheme 65.   Structural Formula 6.44.

E. 3,3,3-Trifluoromethylamine

3,3,3-Trifluoromethylamine, designed as an inactivator of lysyl oxidase, does not inactivate the enzyme.[67]

F. β-Chloro-D-alanine

β-Chloro-D-alanine, an irreversible inactivator of alanine racemase[56] that does not inactivate D-alanine dehydrogenase, was used by Kaczorowski et al.[108,109] to show the coupling of amino acid racemization and oxidation to active transport of amino acids in bacteria. β-Chloro-D-alanine causes rapid time-dependent irreversible inactivation of D-alanine dehydrogenase-coupled transport and the phosphotransferase system. However, transport functions can be protected by dithiothreitol. β-Chloro-D-alanine does not inactivate transport because it is not racemized, but rather is converted to pyruvate following alanine racemase-catalyzed elimination of HCl. β-Chloro-D-alanine was shown to be oxidized by D-alanine dehydrogenase to chloropyruvate, which is responsible for the inactivation. Therefore, this is a metabolically activated inactivator, not a mechanism-based inactivator.

G. (1-Aminoethyl)phosphonic Acid

(1-Aminoethyl)phosphonic acid, a time-dependent inactivator of alanine racemase from *Streptococcus faecalis*, forms a very tight-binding enzyme complex.[110] The half-life for reactivation is 19 days, but the inhibitor is released intact. Furthermore, no tritium is released from [α-³H]-inactivator during inactivation. Therefore, this compound is a slow-tight-binding inhibitor, not a mechanism-based inactivator.

H. Fluorinated Aminophosphonic Acids

Fluorinated aminophosphonic acids (**6.45**; R = CH_2F, CHF_2, CF_3; see Scheme 66) were synthesized by Flynn et al.[111] as inactivators of alanine racemase that may have the enhanced binding affinity and specificity of aminophosphonic acid and retain the irreversibility component of fluorinated alanines. All three compounds are time-dependent inactivators of alanine racemase from *Pseudomonas*. However, given the time-dependent nature of 1-(aminoethyl)phosphonic acid, shown not to undergo deprotonation and to be a slow-tight-binding inhibitor of alanine racemase (see Section IV.G.), it is not clear that compounds **6.45** are mechanism-based inactivators.

$$RCHPO_3H^-$$
$$|$$
$$NH_3^+$$

Scheme 66.   Structural Formula 6.45.

I. (*E*)-3-(1-Aminocyclopropyl)-2-propenoic Acid

This compound was designed[32] as a potential inactivator of GABA aminotransferase, but it was only a noncompetitive reversible inhibitor (see Section II.A.2.).

J. 4-Fluoro-1-hydroxy-1-butanone Phosphate

4-Fluoro-1-hydroxy-1-butanone phosphate (Structural Formula 6.46, Scheme 67) was designed by Cook et al.[112] as a mechanism-based inactivator of spinach ribulose-1,5-bis-phosphate carboxylase/oxygenase, which was expected to undergo enzyme-catalyzed elimination of HF followed by active-site nucleophilic Michael addition. However, it was shown that nonenzymatic elimination of HF produced the Michael acceptor that inactivated the enzyme.

Scheme 67.   Structural Formula 6.46.

K. Inactivators That Utilize Abnormal Catalytic Mechanisms

As indicated in Chapter 1, a mechanism-based inactivator utilizes the normal catalytic mechanism of the enzyme. Several compounds described in this chapter are proposed to inactivate enzymes by mechanisms other than the normal catalytic mechanism, and, therefore, are not true mechanism-based enzyme inactivators. These include (but are not limited to) imidazole *N*-carboxamides (Schemes 53A and 55)[95,95a] and related (Scheme 56[96a] and 57[96b]) and 2-octynoyl CoA (Scheme 62).[104]

REFERENCES

1. **Parker, R. E.**, Mechanisms of fluorine displacement, *Adv. Fluorine Chem.*, 3, 63, 1963.
2. **Sakai, T. T. and Santi, D. V.**, Hydrolysis of hydroxybenzotrifluorides and fluorinated uracil derivatives. A general mechanism for carbon-fluorine bond labilization, *J. Med. Chem.*, 16, 1079, 1973.
3. **Kollonitsch, J., Marburg, S., and Perkins, L. M.**, Selective fluorination of hydroxy amines and hydroxy amino acids with sulfur tetrafluoride in liquid hydrogen fluoride, *J. Org. Chem.*, 40, 3808, 1975.
4. **Kollonitsch, J., Marburg, S., and Perkins, L. M.**, Fluorodehydroxylation, an novel method for synthesis of fluoroamines and fluoroamino acids, *J. Org. Chem.*, 44, 771, 1979.
5. **Kollonitsch, J., Barash, L., and Doldouras, G. A.**, Photofluorination with fluoroxytrifluoromethane, a general method for the synthesis of organic fluorine compounds. Direct fluorination of bioactive molecules, *J. Am. Chem. Soc.*, 92, 7494, 1970.
6. **Kollonitsch, J., Barash, L., Kahan, F., and Kropp, H.**, New antibacterial agent via photofluorination of a bacterial cell wall constituent, *Nature (London)*, 243, 346, 1973.
7. **Kollonitsch, J. and Barash, L.**, Organofluorine synthesis via photofluorination: 3-fluoro-D-alanine and 2-deuterio analogue, antibacterials related to the bacterial cell wall, *J. Am. Chem. Soc.*, 98, 5591, 1976.
8. **Kollonitsch, J., Marburg, S., and Perkins, L. M.**, Fluorodesulfurization. A new reaction for the formation of carbon-fluorine bonds, *J. Org. Chem.*, 44, 3107, 1976.
9. **Middleton, W. J.**, New fluorinating reagents. Dialkylaminosulfur fluorides, *J. Org. Chem.*, 40, 574, 1975.
9a. **Le Tourneau, M. E. and McCarthy, J. R.**, A novel synthesis of α-fluoroacetonitriles. Application to a convenient preparation of 2-fluoro-2-phenethylamines, *Tetrahedron Lett.*, 25, 5227, 1984.
10. **Markovskij, L. N., Pashinnik, V. E., and Kirsanov, A. V.**, Application of dialkylaminosulfur trifluorides in the synthesis of fluoroorganic compounds, *Synthesis*, p. 787, 1973.
11. **Bey, P. and Vevert, J. P.**, New approach to the synthesis of α-halogenomethyl-α-amino acids, *Tetrahedron Lett.*, p. 1215, 1978.
12. **Bey, P. and Schirlin, D.**, General approach to the synthesis of α-difluoromethyl amines as potential enzyme-activated irreversible inhibitors, *Tetrahedron Lett.*, p. 5225, 1978.
13. **Bey, P., Vevert, J.-P., Van Dorsselaer, V., and Kolb, M.**, Direct synthesis of α-halogenomethyl-α-amino acids from the parent α-amino acid, *J. Org. Chem.*, 44, 2732, 1979.

14. **Bey, P., Ducep, J. B., and Schirlin, D.,** Alkylation of malonates or Schiff base anions with dichloro-fluoromethane as a route to α-chlorofluoromethyl or α-fluoromethyl α-amino acids, *Tetrahedron Lett.*, 25, 5657, 1984.

15. **Likos, J. J., Ueno, H., Feldhaus, R. W., and Metzler, D. E.,** A novel reaction of the coenzyme of glutamate decarboxylase with L-serine *O*-sulfate, *Biochemistry*, 21, 4377, 1982.

16. **Ueno, H., Likos, J. J., and Metzler, D. E.,** Chemistry of the inactivation of cytosolic aspartate aminotransferase by serine *O*-sulfate, *Biochemistry*, 21, 4387, 1982.

17. **Badet, B., Roise, D., and Walsh, C. T.,** Inactivation of the *dadB Salmonella typhimurium* alanine racemase by D and L isomers of β-substituted alanines: kinetics, stoichiometry, active site peptide sequencing, and reaction mechanism, *Biochemistry*, 23, 5188, 1984.

18. **Roise, D., Soda, K., Yagi, T., and Walsh, C. T.,** Inactivation of the *Pseudomonas striata* broad specificity amino acid racemase by D and L isomers of β-substituted alanines: kinetics, stoichiometry, active site peptide, and mechanistic studies, *Biochemistry*, 23, 5195, 1984.

19. **Silverman, R. B. and Invergo, B. J.,** Mechanism of inactivation of γ-aminobutyrate aminotransferase by 4-amino-5-fluoropentanoic acid. First example of an enamine mechanism for a γ-amino acid with a partition ratio of 0, *Biochemistry*, 25, 6817, 1986.

20. **Kollonitsch, J., Perkins, L. M., Patchett, A. A., Doldouras, G. A., Marburg, S., Duggan, D. E., Maycock, A. L., and Aster, S. D.,** Selective inhibitors of biosynthesis of aminergic neurotransmitters, *Nature (London)*, 274, 906, 1978.

21. **Wang, E. and Walsh, C.,** Suicide substrates for the alanine racemase of *Escherichia coli* B, *Biochemistry*, 17, 1313, 1978.

21a. **Esaki, N. and Walsh, C. T.,** Biosynthetic alanine racemase of *Salmonella typhimurium:* purification and characterization of the enzyme encoded by the *alr* gene, *Biochemistry*, 25, 3261, 1986.

22. **Silverman, R. B. and Abeles, R. H.,** Inactivation of pyridoxal phosphate dependent enzymes by mono- and polyhaloalanines, *Biochemistry*, 15, 4718, 1976.

23. **Wang, E. A., Kallen, R., and Walsh, C.,** Mechanism-based inactivation of serine transhydroxymethylases by D-fluoroalanine and related amino acids, *J. Biol. Chem.*, 256, 6917, 1981.

24. **Walsh, C., Pascal, R. A., Jr., Johnston, M., Raines, R., Dikshit, D., Krantz, A., and Honma, M.,** Mechanistic studies on the pyridoxal phosphate enzyme 1-aminocyclopropane-1-carboxylate deaminase from *Pseudomonas* sp., *Biochemistry*, 20, 7509, 1981.

25. **Soper, T. S. and Manning, J. M.,** Different modes of action of inhibitors of bacterial D-amino acid transaminase, a target enzyme for the design of new antibacterial agents, *J. Biol. Chem.*, 256, 4263, 1981.

26. **Silverman, R. B. and Levy, M. A.,** Irreversible inactivation of pig brain γ-aminobutyric acid-α-ketoglutarate transaminase by 4-amino-5-halopentanoic acids, *Biochem. Biophys. Res. Commun.*, 95, 250, 1980.

27. **Silverman, R. B. and Levy, M. A.,** Mechanism of inactivation of γ-aminobutyric acid-α-ketoglutaric acid aminotransferase by 4-amino-5-halopentanoic acids, *Biochemistry*, 20, 1197, 1981.

28. **Silverman, R. B., Muztar, A. J., Levy, M. A., and Hirsch, J. D.,** *In vitro* and *in vivo* effects on brain GABA metabolism of (S)-4-amino-5-fluoropentanoic acid, a mechanism-based inactivator of γ-aminobutyric acid transaminase, *Life Sci.*, 32, 2717, 1983.

29. **Silverman, R. B. and Levy, M. A.,** Syntheses of (S)-5-substituted 4-aminopentanoic acids: a new class of γ-aminobutyric acid transaminase inactivators, *J. Org. Chem.*, 45, 815, 1980.

30. **Bey, P., Jung, M. J., Gerhart, F., Schirlin, D., Van Dorsselaer, V., and Casara, P.,** ω-Fluoromethyl analogues of ω-amino acids as irreversible inhibitors of 4-aminobutyrate: 2-oxoglutarate aminotransferase, *J. Neurochem.*, 37, 1341, 1981.

31. **Mathew, J., Invergo, B. J., and Silverman, R. B.,** An efficient synthesis of 3-amino-4-fluorobutanoic acid, an inactivator of GABA transaminase, *Syn. Commun.*, 15, 377, 1985.

32. **Silverman, R. B., Invergo, B. J., and Mathew, J.,** Inactivation of γ-aminobutyric acid aminotransferase by (S,E)-4-amino-5-fluoropent-2-enoic acid and effect on the enzyme of (E)-3-(1-aminocyclopropyl)-2-propenoic acid, *J. Med. Chem.*, 29, 1840, 1986.

32a. **Bey, P., Gerhart, F., and Jung, M.,** Synthesis of (E)-4-amino-2,5-hexadienoic acid and (E)-4-amino-5-fluoro-2-pentenoic acid. Irreversible inhibitors of 4-aminobutyrate-2-oxoglutarate aminotransferase, *J. Org. Chem.*, 51, 2835, 1986.

32b. **Schirlin, D., Baltzer, S., Heydt, J.-G., and Jung, M. J.,** Irreversible inhibition of GABA-T by halogenated analogues of β-alanine, *J. Enz. Inhib.*, 1, 243, 1987.

33. **Lippert, B., Metcalf, B. W., and Resvick, R. J.,** Enzyme-activated irreversible inhibition of rat and mouse brain 4-aminobutyric acid-α-ketoglutarate transaminase by 5-fluoro-4-oxopentanoic acid, *Biochem. Biophys. Res. Commun.*, 108, 146, 1982.

34. **Kuo, D. and Rando, R. R.,** Irreversible inhibition of glutamate decarboxylase by α-(fluoromethyl)glutamic acid, *Biochemistry*, 20, 506, 1981.

34a. **Maycock, A. L., Aster, S. D., and Patchett, A. S.,** Inactivation of 3-(3,4-dihydroxyphenyl)alanine decarboxylase by 2-(fluoromethyl)-3-(3,4-dihydroxyphenyl)alanine, *Biochemistry*, 19, 709, 1980.

35. **Kallio, A., McCann, P. P., and Bey, P.,** DL-α-Monofluoromethylputrescine is a potent irreversible inhibitor of *Escherichia coli* ornithine decarboxylase, *Biochem. J.,* 204, 771, 1982.

35a. **Bitonti, A. J., Bacchi, C. J., McCann, P. P., and Sjoerdsma, A.,** Catalytic irreversible inhibition of *Trypanosoma brucei brucei* ornithine decarboxylase by substrate and product analogs and their effects on murine trypanosomiasis, *Biochem. Pharmacol.,* 34, 1773, 1985.

36. **Danzin, C., Bey, P., Schirlin, D., and Claverie, N.,** α-Monofluoromethyl and α-difluoromethylputrescine as ornithine decarboxylase inhibitors: *in vitro* and *in vivo* biochemical properties, *Biochem. Pharmacol.,* 31, 3871, 1982.

37. **Bey, P., Casara, P., Vevert, J. P., and Metcalf, B.,** Synthesis of irreversible inhibitors of polyamine biosynthesis, *Methods Enzymol.,* 94, 199, 1983.

38. **Bey, P., Gerhart, F., Van Dorsselaer, V., and Danzin, C.,** α-(Fluoromethyl)dehydroornithine and α-(fluoromethyl)dehydroputrescine analogues as irreversible inhibitors or ornithine decarboxylase, *J. Med. Chem.,* 26, 1551, 1983.

38a. **Bitonti, A. J., Casara, P. J., McCann, P. P., and Bey, P.,** Catalytic irreversible inhibition of bacterial and plant arginine decarboxylase activities by novel substrate and product analogues, *Biochem. J.,* 242, 69, 1987.

39. **Wang, E. A., and Walsh, C.,** Characteristics of β,β-difluoroalanine and β,β,β-trifluoroalanine as suicide substrates for *Escherichia coli* B alanine racemase, *Biochemistry,* 20, 7539, 1981.

40. **Silverman, R. B. and Abeles, R. H.,** Mechanism of inactivation of γ-cystathionase by β,β,β-trifluoroalanine, *Biochemistry,* 16, 5515, 1977.

41. **Fearon, C. W., Rodkey, J. A., and Abeles, R. H.,** Identification of the active-site residue of γ-cystathionase labeled by the suicide inactivator β,β,β-trifluoroalanine, *Biochemistry,* 21, 3790, 1982.

42. **Abeles, R. H. and Walsh, C. T.,** Acetylenic enzyme inactivators. Inactivation of γ-cystathionase, in vitro and in vivo, by propargylglycine, *J. Am. Chem. Soc.,* 95, 6124, 1973.

43. **Metcalf, B. W., Holbert, G. W., and Lippert, B. J.,** α-Trifluoromethylhistamine: A mechanism-based inhibitor of mammalian histidine decarboxylase, *Bioorg. Chem.,* 12, 91, 1984.

44. **Magnien, A., LeClef, B., and Biellmann, J. -F.,** Suicide inactivation of fructose-1,6-bisphosphate aldolase, *Biochemistry,* 23, 6858, 1984.

45. **Tate, S. S., Relyea, N. M., and Meister, A.,** Interaction of L-asparatate β-decarboxylase with β-chloro-L-alanine. β-Elimination reaction and active-site labeling, *Biochemistry,* 8, 5016, 1969.

46. **Relyea, N. M., Tate, S. S., and Meister, A.,** Affinity labeling of the active center of L-asparate-β-decarboxylase with β-chloro-L-alanine, *J. Biol. Chem.,* 249, 1519, 1974.

47. **Morino, Y. and Okamoto, M.,** Evidence for the presence of a distinct subsite for binding the distal carboxyl group of dicarboxylate substrates and its role in the catalytic activity of aspartate aminotransferase, *Biochem. Biophys. Res. Commun.,* 47, 498, 1972.

48. **Morino, Y., Osman, A. M., and Okamoto, M.,** Formate-induced labeling of the active site of aspartate aminotransferase by β-chloro-L-alanine, *J. Biol. Chem.,* 249, 6684, 1974.

49. **Morino, Y. and Okamoto, M.,** Labeling of the active site of cytoplasmic aspartate aminotransferase by β-chloro-L-alanine, *Biochem. Biophys. Res. Commun.,* 50, 1061, 1973.

50. **Morino, Y. and Tanase, S.,** Chemical structure of the active site of pig heart mitochondrial aspartate aminotransferase labeled with β-chloro-L-alanine, *J. Biol. Chem.,* 253, 252, 1978.

51. **Golichowski, A. and Jenkins, W. T.,** Inactivation of pig heart alanine aminotransferase by β-chloroalanine, *Arch. Biochem. Biophys.,* 189, 109, 1978.

52. **Morino, Y., Kojima, H., and Tanase, S.,** Affinity labeling of alanine aminotransferase by 3-chloro-L-alanine, *J. Biol. Chem.,* 254, 279, 1979.

53. **Soper, T. S., Jones, W. M., Lerner, B., Trop, M., and Manning, J. M.,** Inactivation of bacterial D-amino acid transaminase by β-chloro-D-alanine, *J. Biol. Chem.,* 252, 3170, 1977.

54. **Manning, J. M. and Soper, T. S.,** Inactivation of D-amino acid transaminase and the antibacterial action of β-halo derivatives of D-alanine, in *Enzyme-Activated Irreversible Inhibitors,* Seiler, N., Jung, M. J., and Koch-Weser, J., Eds., Elsevier/North-Holland, Amsterdam, 1978, 163.

55. **Arfin, S. M. and Koziell, D. A.,** Inhibition of growth of *Salmonella typhimiurium* and of threonine deaminase and transaminase B by β-chloroalanine, *J. Bacteriol.,* 105, 519, 1971.

56. **Manning, J. M., Merrifield, N. E., Jones, W. M., and Gotschlich, E. C.,** Inhibition of bacterial growth by β-chloro-D-alanine, *Proc. Natl. Acad. Sci. U.S.A.,* 71, 417, 1974.

57. **Henderson, L. L. and Johnston, R. B.,** Inhibition studies of the enantiomers of β-chloroalanine on purified alanine racemase from *B. subtilis, Biochem. Biophys. Res. Commun.,* 68, 793, 1976.

58. **Cheung, K. -S., Wasserman, S. A., Dudek, E., Lerner, S. A., and Johnston, M.,** Chloroalanyl and propargylglycyl dipeptides. Suicide substrate containing antibacterials, *J. Med. Chem.,* 26, 1733, 1983.

58a. **Mobashery, S., Lerner, S. A., and Johnston, M.,** Conscripting β-lactamase for use in drug delivery. Synthesis and biological activity of a cephalosporin C_{10}-ester of an antibiotic dipeptide, *J. Am. Chem. Soc.,* 108, 1685, 1986.

58b. **Mobashery, S. and Johnston, M.**, Reactions of *Escherichia coli* TEM β-lactamase with cephalothin and with C_{10}-dipeptidyl cephalosporin esters, *J. Biol. Chem.*, 261, 7879, 1986.

58c. **Boisvert, W., Cheung, K. S., Lerner, S. A., and Johnston, M.**, Mechanisms of action of chloroalanyl antibacterial peptides. Identification of the intracellular enzymes inactivated on treatment of *Escherichia coli* JSR-O with the dipeptide βCl–LAla–βCl–LAla, *J. Biol. Chem.*, 261, 7871, 1986.

59. **Kishore, G. M.**, Mechanism-based inactivation of bacterial kynureninase by β-substituted amino acids, *J. Biol. Chem.*, 259, 10669, 1984.

60. **Miziorko, H. M. and Behnke, C. E.**, Active site directed inhibition of 3-hydroxy-3-methylglutaryl coenzyme A synthase by 3-chloropropionyl coenzyme A, *Biochemistry*, 24, 3174, 1985.

61. **Miziorko, H. M. and Behnke, C. E.**, Amino acid sequence of an active site peptide of avian liver mitochondrial 3-hydroxy-3-methylglutaryl-CoA synthase, *J. Biol. Chem.*, 260, 13513, 1985.

62. **Fendrick, G. and Abeles, R. H.**, Mechanism of action of butryl-CoA dehydrogenase: reactions with acetylenic, olefinic, and fluorinated substrate analogues, *Biochemistry*, 21, 6685, 1982.

63. **Hevey, R. C., Babson, J., Maycock, A. L., and Abeles, R. H.**, Highly specific enzyme inhibitors. Inhibition of plasma amine oxidase, *J. Am. Chem. Soc.*, 95, 6125, 1973.

64. **Speziale, A. J. and Hamm, P. C.**, Preparation of some new 2-chloroacetamides, *J. Am. Chem. Soc.*, 78, 2556, 1956.

65. **Allan, R. D., Johnston, G. A. R., and Twitchin, B.**, Synthesis of analogues of GABA. V. *trans* and *cis* isomers of some 4-amino-3-halogenobut-2-enoic acids, *Aust. J. Chem.*, 33, 1115, 1980.

66. **Kumagai, H., Uchida, H., and Yamada, H.**, Reaction of fungal amine oxidase with β-bromoethylamine, *J. Biol. Chem.*, 254, 10913, 1979.

67. **Tang, S. -S., Simpson, D. E., and Kagan, H. M.**, β-Substituted ethylamine derivatives as suicide inhibitors of lysyl oxidase, *J. Biol. Chem.*, 259, 975, 1984.

68. **Tang, S.-S., Trackman, P. C., and Kagan, H. M.**, Reaction of aortic lysyl oxidase with β-aminopropionitrile, *J. Biol. Chem.*, 258, 4331, 1983.

69. **Neumann, R., Hevey, R., and Abeles, R. H.**, The action of plasma amine oxidase on β-haloamines. Evidence for proton abstraction in the oxidative reaction, *J. Biol. Chem.*, 250, 6362, 1975.

69a. **Van der Meer, R. A. and Duine, J. A.**, Covalently bound pyrroloquinoline quinone is the organic prosthetic group in human placental amine oxidase, *Biochem. J.*, 239, 789, 1986.

69b. **Knowles, P. F., Pandeya, K. B., Ruis, F. X., Spencer, C. M., Moog, R. S., McGuirl, M. A., and Dooley, D. M.**, The organic cofactor in plasma amine oxidase: evidence for pyrroloquinoline quinone and against pyridoxal phosphate, *Biochem. J.*, 241, 603, 1987.

70. **Soper, T. S. and Manning, J. M.**, β-Elimination of β-halo substrates by D-amino acid transaminase associated with inactivation of the enzyme. Trapping of a key intermediate in the reaction, *Biochemistry*, 17, 3377, 1978.

71. **Morino, Y. and Okamoto, M.**, A comparative study on the affinity labelling of aspartate aminotransferase isozymes by β-bromopyruvate, *Biochem. Biophys. Res. Commun.*, 40, 600, 1970.

72. **Okamoto, M. and Morino, Y.**, Affinity labeling of aspartate aminotransferase isozymes by bromopyruvate, *J. Biol. Chem.*, 248, 82, 1973.

73. **Birchmeier, W. and Christen, P.**, The reaction of cytoplasmic aspartate aminotransferase with bromopyruvate. Syncatalytic modification simulates affinity labeling, *J. Biol. Chem.*, 249, 6311, 1974.

74. **Osmundsen, H. and Sherratt, H. S. A.**, The effect of pent-4-enoate and methylenecyclopropylacetate on some enzymes of β-oxidation in extracts of liver mitochondria, *Biochem. Soc. Trans.*, 3, 330, 1975.

75. **Osmondsen, H. and Sherratt, H. S. A.**, A novel mechanism for inhibition of β-oxidation by methylenecyclopropylacetyl-CoA, a metabolite of hypoglycin, *FEBS Lett.*, 55, 38, 1975.

76. **Wenz, A., Thorpe, C., and Ghisla, S.**, Inactivation of general acyl-CoA dehydrogenase from pig kidney by a metabolite of hypoglycin A, *J. Biol. Chem.*, 256, 9809, 1981.

76a. **Baldwin, J. E. and Parker, D. W.**, Stereospecific (methylenecyclopropyl)acetyl-CoA inactivation of general acyl-CoA dehydrogenase from pig kidney, *J. Org. Chem.*, 52, 1475, 1987.

77. **Van Holt, C.**, Methylenecyclopropaneacetic acid, a metabolite of hypoglycin, *Biochim. Biophys. Acta*, 125, 1, 1966.

78. **Alston, T. A. and Bright, H. J.**, Inactivation of pyridoxal 5'-phosphate-dependent enzymes by 5-nitro-L-norvaline, an analog of L-glutamate, *FEBS Lett.*, 126, 269, 1981.

79. **Porter, D. J. T. and Bright, H. J.**, 3-Carbanionic substrate analogues bind very tightly to fumarase and aspartase, *J. Biol. Chem.*, 255, 4772, 1980.

80. **John, R. A. and Fasella, P.**, The reaction of L-serine *O*-sulfate with aspartate aminotransferase, *Biochemistry*, 8, 4477, 1969.

81. **Cavallini, D., Federici, G., Bossa, F., and Granata, F.**, The protective effect of thiosulfate upon inactivation of aspartate aminotransferase by aminoacrylic-acid-producing substrates, *Eur. J. Biochem.*, 39, 301, 1973.

82. **John, R. A. and Tudball, N.**, Evidence for induced fit of a pseudo-substrate of aspartate aminotransferase, *Eur. J. Biochem.*, 31, 135, 1972.

83. **John, R., Bossa, F., Barra, D., and Fasella, P.,** Active-site labelling of cytoplasmic aspartate aminotransferase from pig heart by L-serine-O-sulphate, *Biochem. Soc. Trans.,* 1, 862, 1973.

84. **Ovchinnikov, Yu.A., Kiryushkin, A. A., Egorov, Ts. A., Abdulaev, N. G., Kiselev, A. P., Modyanov, N. N., Grishin, E. V., Sukhikh, A. P., Vinogradova, E. I., Feigina, M. Yu., Aldanova, N. A., and Lipkin, V. M.,** Cytoplasmic aspartate aminotransferase from pig heart muscle: partial sequence, *FEBS Lett.,* 17, 133, 1971.

85. **Fowler, L. J. and John, R. A.,** Active-site-directed irreversible inhibition of rat brain 4-aminobutyrate aminotransferase by ethanolamine O-sulphate *in vitro* and *in vivo, Biochem. J.,* 130, 569, 1972.

86. **Fowler, L. J. and John, R. A.,** The reaction of ethanolamine O-sulphate with 4-aminobutyrate aminotransferase, *Biochem. J.,* 197, 149, 1981.

87. **Williams, J. A., Hewlins, M. J., Fowler, L. J., and John, R. A.,** The reactions of aminobutyrate aminotransferase and ornithine aminotransferase with analogues of ethanolamine O-sulphate, *Biochem. Pharmacol.,* 32, 2350, 1983.

87a. **Badet, B., Lee, K., Floss, H. G., and Walsh, C. T.,** Stereochemical studies of processing of D- and L-isomers of suicide substrates by an amino acid racemase from *Pseudomonas striata, J. Chem. Soc. Chem. Commun.,* p. 838, 1984.

88. **Wood, W. A. and Gunsalus, I. C.,** Serine and threonine deaminases of *Escherichia coli:* activators for a cell-free enzyme, *J. Biol. Chem.,* 181, 171, 1949.

89. **Phillips, A.,** Mechanism of the inactivation of threonine dehydratase during dehydration of serine, *Biochim. Biophys. Acta,* 151, 523, 1968.

90. **Nishimura, J. S. and Greenburg, D. M.,** Purification and properties of L-threonine dehydratase of sheep liver, *J. Biol. Chem.,* 236, 2684, 1961.

91. **McLemore, W. O. and Metzler, D. E.,** The reversible inactivation of L-threonine dehydratase of sheep liver by L-serine, *J. Biol. Chem.,* 243, 441, 1968.

92. **Esaki, N., Kimura, T., Goto, J., Nakayama, T., Tanaka, H., and Soda, K.,** S-(N-Methylthiocarbamoyl)-L-cysteine, a suicide substrate of L-methionine γ-lyase, *Biochim. Biophys. Acta,* 785, 54, 1984.

93. **Kimura, T., Esaki, N., Tanaka, H., and Soda, K.,** Action of S-carbamoyl and S-thiocarbamoyl derivatives of L-cysteine on L-amino acid oxidase, *Agric. Biol. Chem.,* 48, 3157, 1984.

94. **Kimura, T., Esaki, N., Tanaka, H., and Soda, K.,** Inactivation of amino acid racemase by S-(N-methylthiocarbamoyl)-D,L-cysteine, *Agric. Biol. Chem.,* 48, 383, 1984.

95. **Groutas, W. C., Badger, R. C., Ocain, T. D., Felker, D., Frankson, J. and Theodorakis, M.,** Mechanism-based inhibitors of elastase, *Biochem. Biophys. Res. Commun.,* 95, 1890, 1980.

95a. **Walker, B. and Elmore, D. T.,** The irreversible inhibition of urokinase, kidney-cell plasminogen activator, plasmin and β-trypsin by 1-(N-6-amino-n-hexyl)-carbamoylimidazole, *Biochem. J.,* 221, 277, 1984.

96. **Groutas, W. C., Abrams, W. R., Theodorakis, M. C., Kasper, A. M., Rude, S. A., Badger, R. C., Ocain, T. D., Miller, K. E., Moi, M. K., Brubaker, M. J., Davis, K. S., and Zandler, M. E.,** Amino acid derived latent isocyanates: irreversible inactivation of porcine pancreatic elastase and human leukocyte elastase, *J. Med. Chem.,* 28, 204, 1985.

96a. **Groutas, W. C., Brubaker, M. J., Zandler, M. E., Mazo-Gray, V., Rude, S. A., Crowley, J. P., Castrisos, J. C., Dunshee, D. A., and Giri, P. K.,** Inactivation of leukocyte elastase by aryl azolide and sulfonate salts. Structure-activity relationship studies, *J. Med. Chem.,* 29, 1302, 1986.

96b. **Groutas, W. C., Brubaker, M. J., Zandler, M. E., Stanga, M. A., Huang, T. L., Castrisos, J. C., and Crowley, J. P.,** Sulfonate salts of amino acids: novel inhibitors of the serine proteinases, *Biochem. Biophys. Res. Commun.,* 128, 90, 1985.

97. **Pratt, R. F. and Bruice, T. C.,** The carbanion mechanism (ElcB) of ester hydrolysis. III. Some structure-reactivity studies and the ketene intermediate, *J. Am. Chem. Soc.,* 92, 5956, 1970.

98. **Maycock, A. L., Suva, R. H., and Abeles, R. H.,** Novel inactivators of plasma amine oxidase, *J. Am. Chem. Soc.,* 97, 5613, 1975.

99. **Hatfield, G. M., and Schaumberg, J. P.,** Isolation and structural studies of coprine, the disulfiram-like constituent of *Coprinus atramentarius, Lloydia,* 38, 489, 1975.

100. **Lindberg, P., Bergman, R., and Wickberg, G.,** Isolation and structrue of coprine, a novel physiologically active cyclopropanone derivative from *Coprinus atramentarius* and its synthesis via 1-aminocyclopropanol, *J. Chem. Soc. Chem. Commun.,* 946, 1975.

100a. **Lindberg, P., Bergman, R., and Wickberg, B.,** Isolation and structure of coprine, the *in vivo* aldehyde dehydrogenase inhibitor in *Coprinus atramentarius;* synthesis of coprine and related cyclopropanone derivatives, *J. Chem. Soc. Perkin Trans. I,* 684, 1977.

101. **Tottmar, O. and Lindberg, P.,** Effects on rat liver acetaldehyde dehydrogenases *in vitro* and *in vivo* by coprine, the disulfiram-like constituent of Coprinus atramentarius, *Acta Pharmacol. Toxicol.,* 40, 476, 1977.

102. **Wiseman, J. S. and Abeles, R. H.,** Mechanism of inhibition of aldehyde dehydrogenase by cyclopropanone hydrate and the mushroom toxin coprine, *Biochemistry,* 18, 427, 1979.

103. **Wiseman, J. S., Tayrien, G., and Abeles, R. H.,** Kinetics of the reaction of cyclopropanone hydrate with yeast aldehyde dehydrogenase: a model for enzyme-substrate interaction, *Biochemistry*, 19, 4222, 1980.

104. **Freund, K., Mizzer, J., Dick, W., and Thorpe, C.,** Inactivation of general acyl-CoA dehydrogenase from pig kidney by 2-alkynoyl coenzyme A derivatives: initial aspects, *Biochemistry*, 24, 5996, 1985.

105. **Villafranca, J. J. and Platus, E.,** Fluorocitrate inhibition of aconitase. Reversibility of the inactivation, *Biochem. Biophys. Res. Commun.*, 55, 1197, 1973.

106. **Silverman, R. B. and Levy, M. A.,** Substituted 4-aminobutanoic acids. Substrates for γ-aminobutyric acid α-ketoglutaric acid aminotransferase, *J. Biol. Chem.*, 256, 11565, 1981.

107. **Silverman, R. B., Durkee, S. C., and Invergo, B. J.,** 4-Amino-2-(substituted methyl)-2-butenoic acids: substrates and potent inhibitors of γ-aminobutyric acid aminotransferase, *J. Med. Chem.*, 29, 764, 1986.

108. **Kaczorowski, G., Shaw, L., Fuentes, M., and Walsh, C.,** Coupling of alanine racemase and D-alanine dehydrogenase to active transport of amino acids in *Escherichia coli* B membrane vesicles, *J. Biol. Chem.*, 250, 2855, 1975.

109. **Kaczorowski, G., Shaw, L., Laura, R., and Walsh, C.,** Active transport in *Escherichia coli* B membrane vesicles. Differential inactivating effects from the enzymatic oxidation of β-chloro-L-alanine and β-chloro-D-alanine, *J. Biol. Chem.*, 250, 8921, 1975.

110. **Badet, B. and Walsh, C.,** Purification of an alanine racemase from *Streptococcus faecalis* and analysis of its inactivation by (1-aminoethyl)phosphonic acid enantiomers, *Biochemistry*, 24, 1333, 1985.

111. **Flynn, G. A., Beight, D. W., Bohme, E. H. W., and Metcalf, B. W.,** The synthesis of fluorinated aminophosphonic acid inhibitors of alanine racemase, *Tetrahedron Lett.*, 26, 285, 1985.

112. **Cook, C. M., Tolbert, N. E., Smith, J. H., and Nelson, R. V.,** Inactivation of spinach ribulose-1,5-bisphosphate carboxylase/oxygenase by 1-hydroxy-3-buten-2-one phosphate, *Biochemistry*, 25, 4699, 1986.

INDEX

A

Pepsin, 37
Peptidase G, 140
Peptide fluoromethyl ketones, 73
Peptidylfluoromethanes, 74
Peptidyl transferase, 130
Phenelzine, 8
3-Phenyl-(5E)-(1-bromoethylidene)dihydro-2(3H)-
    furanone, 99
5-Phenyl-6-chloro-2-pyrone, 148
Phenyl glycinate, 191
Phenylhydrazine, 182
3-Phenyl-(5E)-(iodomethylidene)dihydro-2(3H)-
    furanone, 99
4-Phenyl-5(E)-(iodomethylidene)-dihydro-2(3H)-
    furanone, 148
4-Phenyl-5(E)-(iodomethylidene)-2-furanone, 148
4-Phenyl-6(E)-(iodomethylidene)-tetrahydro-2-
    pyranone, 148
5-Phenyl-6(E)-(iodomethylidene)-tetrahydro-2-
    pyranone, 148
Phenyl-substituted (halomethylidene)-2-furanones,
    148
Phenyl-substituted (halomethylidene)-tetrahydropyr-
    anones, 148
Phosphate esters, 49—50
Phosphinothricin (2-amino-4-
    (methylphosphinyl)butanoic acid), 57
Phosphoenol-3-bromopyruvate, 50
Phosphoenolpyruvate carboxylase, 50
1-(5-Phosphono-β-D-arabinofuranosyl)-5-fluoroura-
    cil, 70
5-Phosphoribosylpyrophosphate amidotransferase,
    32, 36
Phosphorylation reactions, 49—50, see also specific
    types
Phosphotransferase, 195
Pigmentation agents, 9, see also specific types
Pituitary suppressants, 9, see also specific types
Plasma amine oxidase, 180, 191
Plasmin, 190
Plasminogen activator, 190
PLP, see Pyridoxal phosphate
PMP, see Pyridoxamine phosphate
"Promechanism-based inactivators", 7
Proparglyglycine, 25, 172
α-Proparglyphenylalanine, 102
Propylthiouracil (thyroid peroxidase), 9
5(E)-(3-R'-2-Propynylidene)-3-R-tetrahydro-2-
    furanones, 143
5(Z)-(3-R'-2-Propynylidene)-3-R-tetrahydro-2-
    furanones, 143
6(E)-3-R'-2-Propynylidene-3-R-tetrahydro-2-
    pyrones, 143
Protease A, 106, 126
Protease V-8, 106
Protease I, 126
Protease II, 106, 126
DL-Prothionine R,S-sulfoximine, 57
DL-Prothionine R,S-sulfoximine S-n-propylhomocys-
    teine sulfoximine, 56
Protonation/acylation reactions, 41—42, see also

specific types
Protonation/substitution reactions, 31—41, see also
    specific types
    diazoketones with copper, 31
    diazoketones without copper, 31—38
PS-5, 142
Pyridoxal phosphate (PLP)-dependent aminotrans-
    ferases, 19
Pyridoxal phosphate (PLP)-dependent enzymes, 158,
    see also specific types
Pyridoxamine phosphate (PMP), 19
6-(2-Pyridyl)methylenepenicillanic acid sulfone, 121
Pyruvate decarboxylase, 87
Pyruvate dehydrogenase, 85, 86
Pyruvate kinase, 192

Q

Quinacillins, 115, 139, 140

R

Raman spectroscopy, 67
Rate constant terminology, 4—5
Resonance Raman spectroscopy, 67
Ribulose-1,5-bisphosphate carboxylase/oxygenase,
    196

S

Saligenin cyclic phosphorus esters, 49
Saturation, 10—11, 19—20
Secondary isotope effects, 61
Sedatives, 9, see also specific types
Serines, 3, 9, 160, 172, 176, 179, 186—188
Slow, tight-binding inhibitors, 5—6, see also specific
    types
Sparsomycin, 130—131, 150
Spectroscopy, see specific types
Spironolactone, 9
Steady-state kinetics, 16
Stoichiometry, 11, 21, 69
Streptococcal proteinase, 38
3-Substituted-6-chloro-2-pyrones, 107
5-Substituted-6-chloro-2-pyrones, 107
2-Substituted-4H-3,1-benzoxazin-4-ones, 147
3-Substituted-4-hydroxycoumarins, 41—42
2-Substituted-1,3-indanediones, 41—42
N-Substituted nitrosocarbamates, 129
Substrate-induced deactivation, 149
Substrate protection, 11, 20
Subtilisin, 123
Succinate dehydrogenase, 77, 79
"Suicide substrates", 3, 4
Sulbactam, see Penicillanic acid sulfone
Sulfamethazine acetyl CoA-dependent N-acetyltrans-
    ferase, 113
Sulfatase A, 149
Sulfate, 186—188
Sulfite, 191
6β-Sulfonamidopenicillanic acid sulfones, 117